Randomized Clinical Trials of Nonpharmacological Treatments

T0225646

Chapman & Hall/CRC Biostatistics Series

Chapman & Hall/CRC Biostatistics Series

Chapman & Hall/CRC Biostatistics Series

Randomized Clinical Trials of Nonpharmacological Treatments

Edited by

Isabelle Boutron
Philippe Ravaud
David Moher

CRC Press
Taylor & Francis Group
Boca Raton London New York

CRC Press is an imprint of the
Taylor & Francis Group, an **informa** business

A CHAPMAN & HALL BOOK

CRC Press
Taylor & Francis Group
6000 Broken Sound Parkway NW, Suite 300
Boca Raton, FL 33487-2742

First issued in paperback 2020

© 2012 by Taylor & Francis Group, LLC
CRC Press is an imprint of Taylor & Francis Group, an Informa business

No claim to original U.S. Government works

Version Date: 20111108

ISBN-13: 978-0-367-57680-6 (pbk)
ISBN-13: 978-1-4200-8801-4 (hbk)

Visit the Taylor & Francis Web site at
http://www.taylorandfrancis.com

and the CRC Press Web site at
http://www.crcpress.com

Contents

Part I   Assessing Nonpharmacological Treatments: Theoretical Framework

Part II   Assessing Nonpharmacological Treatments: Practical Examples

Introduction

Isabelle Boutron and Philippe Ravaud

Paris Descartes University
Public Assistance—Hospitals of Paris

Nonpharmacological treatments represent a wide range of treatments proposed to patients. They could be defined as all interventions involving not just the administration of pharmacological treatments. Nonpharmacological treatments concern technical interventions such as surgical procedures; technical interventions such as joint lavage and angioplasty; implantable devices such as stents and arthroplasty; nonimplantable devices such as orthoses, ultrasound treatments, and laser treatments; and participative interventions such as rehabilitation, education, behavioral interventions, and psychotherapy.

The number of published randomized controlled trials assessing nonpharmacological treatments is increasing with time. A cross-sectional assessment of randomized trials published in 2000 identified 25% of such trials assessing nonpharmacological treatments (10% surgery or procedures, 11% counseling or lifestyle interventions, and 4% equipment) [1]. A similar recent study showed that randomized trials assessing nonpharmacological treatments concerned 42% of the trials published in 2006 (21% surgery or procedures, 18% counseling or lifestyle interventions, and 3% equipment).

Assessing nonpharmacological treatments raises specific issues. An important issue is the funding source. Most assessments of nonpharmacological treatments, except perhaps implantable and nonimplantable devices, rely on public funding, or more restricted amounts of money [2,3]. Further, the regulatory requirements for nonpharmacological treatments are less stringent than for pharmacological treatments. In most cases, the drug approval process of the U.S. Food and Drug Administration requires demonstrated treatment effectiveness from at least two adequate and well-controlled clinical trials.

In contrast, most nonpharmacological treatments such as surgical procedures or participative interventions have no specific requirements for approval. Consequently, they can be widely proposed in clinical practice but may not have been adequately evaluated. This situation is an important barrier for the evaluation of the beneficial effects of these treatments and the conduct of randomized controlled trials.

Finally, assessing nonpharmacological treatments raises specific methodological issues [3]. First, blinding of patients, care providers, and outcome assessors is frequently not feasible, particularly because of a lack of placebo for most nonpharmacological treatments [4]. Second, nonpharmacological treatments are usually complex interventions made of several components that may all have an impact on the beneficial effect of the treatment [5]. These interventions are, consequently, difficult to describe, reproduce in the trial, and implement in clinical practice. Finally, care providers' expertise and centers' volume of care can have an important impact on the success of the interventions [6].

Nevertheless, it is essential to overcome these barriers and to adequately evaluate nonpharmacological treatments.

This book is divided in two parts. Part I is dedicated to specific issues when assessing nonpharmacological treatments. It highlights the difficulties of blinding and how these difficulties can be overcome. It discusses the placebos that can be used in such trials. It also addresses how the complexity of the intervention, the learning curve, and the clustering effect should be taken into account in trials. Issues of assessing harm and assessing the applicability of trials in this field are also raised. Different designs that are particularly useful in this context—cluster randomized controlled trials, expertise-based trials, pragmatic trials, and nonrandomized trials, as well as specific issues of systematic reviews in this field—are also presented.

Part II provides several examples of the planning, conduct, analyses, and reporting of trials in different fields. It is obviously impossible to cover all the different clinical areas, but these examples in the field of surgery, technical interventions, devices, rehabilitation, psychotherapy, behavioral interventions, etc., should be very useful for readers to learn and grasp some ideas from various domains.

References

1. Chan, A.W., Altman, D.G. Epidemiology and reporting of randomised trials published in PubMed journals. *Lancet*, 365(9465), 1159–1162, 2005.
2. Balasubramanian, S.P., Wiener, M., Alshameeri, Z., Tiruvoipati, R., Elbourne, D., Reed, M.W. Standards of reporting of randomized controlled trials in general surgery: can we do better? *Ann Surg*, 244(5), 663–667, 2006.
3. Boutron, I., Tubach, F., Giraudeau, B., Ravaud, P. Methodological differences in clinical trials evaluating nonpharmacological and pharmacological treatments of hip and knee osteoarthritis. *JAMA*, 290(8), 1062–1070, 2003.
4. Boutron, I., Tubach, F., Giraudeau, B., Ravaud, P. Blinding was judged more difficult to achieve and maintain in nonpharmacologic than pharmacologic trials. *J Clin Epidemiol*, 57(6), 543–550, 2004.
5. Glasziou, P., Meats, E., Heneghan, C., Shepperd, S. What is missing from descriptions of treatment in trials and reviews? *BMJ*, 336(7659), 1472–1474, 2008.
6. Halm, E.A., Lee, C., Chassin, M.R. Is volume related to outcome in health care? A systematic review and methodologic critique of the literature. *Ann Intern Med*, 137(6), 511–520, 2002.

Contributors

Charles Abraham
Peninsula College of Medicine and
 Dentistry
University of Exeter
Exeter, United Kingdom

Hidefumi Aoyama
Department of Radiology
Niigata University Graduate School
 of Medical and Dental Sciences
Niigata, Japan

Vincenzo Berghella
Department of Obstetrics and
 Gynecology
Thomas Jefferson University
Philadelphia, Pennsylvania

David Biau
Department of Biostatistics and
 Medical Informatics
Public Assistance—Hospitals
 of Paris
Paris, France

Isabelle Boutron
INSERM U738
Paris Descartes University
and
Center for Clinical Epidemiology
Public Assistance—Hospitals
 of Paris
Paris, France

Marion K. Campbell
Health Services Research Unit
University of Aberdeen
Aberdeen, United Kingdom

Jonathan A. Cook
Health Services Research Unit
University of Aberdeen
Aberdeen, United Kingdom

Nadine Elizabeth Foster
Arthritis Research UK Primary
 Care Centre
Keele University
Staffordshire, United Kingdom

Bruno Giraudeau
Center for Clinical Investigation
 (INSERM CIC 0202)
University Hospital of Tours
Tours, France

Paul Glasziou
Faculty of Health Sciences and
 Medicine
Bond University
Queensland, Australia

Asbjørn Hróbjartsson
The Nordic Cochrane Centre
Copenhagen, Denmark

John P.A. Ioannidis
Department of Hygiene and
    Epidemiology
University of Ioannina School
    of Medicine
and
Biomedical Research Institute
Foundation for Research and
    Technology–Hellas
Ioannina, Greece

and

Tufts University School
    of Medicine
Boston, Massachusetts

and

Stanford University School
    of Medicine
Stanford, California

Claire Jourdan
Cochin Hospital
Public Assistance—Hospitals
    of Paris
and
Paris Descartes University
and
Federative Research Institute
    on Disability
Paris, France

Arthur Kang'ombe
Department of Health Sciences
University of York
York, United Kingdom

Simon Lewin
Norwegian Knowledge Centre for
    the Health Services
Norway and Health Systems
    Research Unit
Medical Research Council of South
    Africa
Cape Town, South Africa

Isabelle Marc
Department of Pediatrics
Laval University
Quebec City, Quebec, Canada

Hugh MacPherson
Department of Health Sciences
University of York
York, United Kingdom

Franklin G. Miller
Department of Bioethics
National Institutes of Health
Bethesda, Maryland

David Moher
Clinical Epidemiology Program
Ottawa Hospital Research Institute
Ottawa, Ontario, Canada

Remy Nizard
Department of Orthopedic Surgery
    and Trauma
Lariboisière Hospital
Public Assistance—Hospitals
    of Paris
and
Faculty of Medicine
Paris Diderot University
Paris, France

Panagiotis N. Papanikolaou
Department of Hygiene and
    Epidemiology
University of Ioannina School
    of Medicine
Ioannina, Greece

Rafael Perera
Department of Primary Health Care
University of Oxford
Oxford, United Kingdom

Serge Poiraudeau
Cochin Hospital
Public Assistance—Hospitals
  of Paris
Paris Descartes University
and
Federative Research Institute
  on Disability
Paris, France

Craig R. Ramsay
Health Services Research Unit
University of Aberdeen
Aberdeen, United Kingdom

François Rannou
Cochin Hospital
Public Assistance—Hospitals
  of Paris
and
Paris Descartes University
and
Federative Research Institute
  on Disability
Paris, France

Keith G. Rasmussen
Department of Psychiatry and
  Psychology
Mayo Clinic
Rochester, Minnesota

Philippe Ravaud
INSERM U738
Paris Descartes University
and
Center for Clinical Epidemiology
Public Assistance—Hospitals
  of Paris
Paris, France

Barnaby C. Reeves
Bristol Heart Institute
University of Bristol
Bristol, United Kingdom

Katherine Sanchez
Cochin Hospital
Public Assistance—Hospitals
  of Paris
and
Paris Descartes University
and
Federative Research Institute
  on Disability
Paris, France

Paula P. Schnurr
National Center for Posttraumatic
  Stress Disorder
White River Junction, Vermont

and

Dartmouth Medical School
Hanover, New Hampshire

Larissa Shamseer
Clinical Epidemiology Program
Ottawa Hospital Research Institute
Ottawa, Ontario, Canada

Sasha Shepperd
Department of Public Health
University of Oxford
Oxford, United Kingdom

Michele Tansella
Department of Medicine and Public
  Health
University of Verona
Verona, Italy

Graham Thornicroft
Health Service and Population
  Research Department
King's College London
London, United Kingdom

Helen Tilbrook
Department of Health Sciences
University of York
York, United Kingdom

Jorge E. Tolosa
Department of Obstetrics and
    Gynecology
Oregon Health Science University
Portland, Oregon

David Torgerson
Department of Health Sciences
University of York
York, United Kingdom

Tom Treasure
Department of Mathematics
University College London
London, United Kingdom

Alexander Tsertsvadze
Clinical Epidemiology Program
Ottawa Hospital Research Institute
Ottawa, Ontario, Canada

Martin Utley
Department of Mathematics
University College London
London, United Kingdom

Patricia Yudkin
Department of Primary Health Care
University of Oxford
Oxford, United Kingdom

Merrick Zwarenstein
Sunnybrook Research Institute
and
Institute for Clinical Evaluative
    Sciences
University of Toronto
Toronto, Ontario, Canada

and

Division of Global Health (IHCAR)
Karolinska Institute
Stockholm, Sweden

Part I

Assessing Nonpharmacological Treatments: Theoretical Framework

1

Blinding in Nonpharmacological Randomized Controlled Trials

Isabelle Boutron and Philippe Ravaud

Paris Descartes University and
Public Assistance—Hospitals of Paris

CONTENTS

1.1 General Framework on Blinding

Blinding is a cornerstone of unbiased therapeutic evaluation [1,2]. Blinding refers to keeping key people, such as participants, healthcare providers, and outcome assessors, unaware of the treatment administered or of the true hypothesis of the trial [3,4].

Blinding of participants and healthcare providers in a trial prevents performance bias, which occurs when knowledge of the treatment assignment may affect the willingness of healthcare providers to prescribe and participants to take co-interventions, participants to be compliant with the assigned treatment, and participants to cross over or withdraw from the trial [5–7]. For example, in a randomized controlled trial comparing surgery for lumbar intervertebral disk herniation with usual care, blinding of patients, care providers, and outcome assessors was not feasible [8]. Lack of blinding was responsible for an important contamination between the two groups, with 50% of patients assigned to surgery receiving surgery within 3 months of enrolment, and 30% of those assigned to nonoperative treatment receiving surgery in the same period. Blinding of outcome assessors also minimizes the risk of detection bias (i.e., observer, ascertainment, assessment bias). This type of bias occurs if participant assignment influences the process of outcome assessment [5–7]. For example, nonblinded neurologists assessing the outcome of a trial demonstrated an apparent treatment benefit, whereas blinded neurologists did not [3]. Finally, blinding of data analysts can also prevent bias because knowledge of the intervention received may influence the choice of analytical strategies and methods [5].

There is some empirical evidence of bias when blinding is lacking. Schulz et al. [1] evaluated the association of estimates of treatment effect and lack of double-blinding. Trials that were not double-blinded yielded larger effect estimates, with odds ratios exaggerated by 17%. Moher et al. [9] performed a meta-epidemiological study to estimate the effect of different quality indicators such as adequate randomization generation, allocation concealment, and reporting of double-blinding. The authors showed an overestimation of treatment effect estimates for randomization, generation, and allocation concealment but not for double-blinding. Recently, Wood et al. [10] showed that the impact of blinding depended highly on the type of outcome evaluated; in trials with subjective outcomes that lacked blinding, treatment effect estimates were exaggerated by 25%. In contrast, trials with objective outcomes showed no evidence of bias. Nevertheless, these meta-epidemiological studies raise some issues. They are indirect evidence susceptible to a considerable risk of confounding. In fact, double-blind randomized trials can differ from other trials in other important aspects such as the treatment evaluated (pharmacological or nonpharmacological), the randomization procedure, the funding source, or other unknown factors. Further, meta-epidemiological studies do not take into account: who was blinded, whether blinding was efficient, or the possible risk of bias.

Direct evidence of bias demonstrating the influence of lack of blinding is sparse. A systematic review of >20 randomized controlled trials with blinded and nonblinded outcome assessment showed a substantial impact of the blinding of outcome assessors, especially in trials with more subjective outcomes [11].

The reporting of blinding in published reports of randomized controlled trials is frequently inadequate. Most publications use a common terminology of single-blind, double-blind, or triple-blind study. However, this terminology should be used with caution. In fact, the use of the terms is confusing because it means different things to different people [12,13]. For example, a single-blind randomized trial could imply that patients are blinded or that outcome assessors are blinded. Further, many authors neglect to report whether their trial was blinded, who was blinded, and how blinding was achieved [14–16]. Haahr and Hrobjartsson evaluated how blinding was reported in 200 blinded randomized clinical trials with articles published in 2001; 78% of the articles described the trial as "double-blind," with only 2% explicitly reporting the blinding status for each trial participant (patients, care providers, outcome assessor). After contacting the authors of the studies, Haahr and Hrobjartsson [15] showed that about one-fifth of the "double-blind" trials did not blind patients, care providers, or outcome assessors. Hróbjartsson et al. [17] showed that the reporting of data related to blinding was better in protocols of studies than the published results, but a large proportion of protocols still report blinding unclearly.

To improve the quality of reporting of blinding, an international group, the CONSORT group, developed reporting guidelines, or statements, first published in 1996 [18] and updated in 2001 [19,20] and 2010 [21,22]. These guidelines are now endorsed and cited in the recommendations to authors of most peer-review journals. The guidelines clearly indicate that the authors should report "If [blinding was] done, who was blinded after assignment to interventions (e.g., participants, care providers, those assessing outcomes) and how" and provide "If relevant, a description of the similarity of interventions."

1.2 Blinding and Nonpharmacological Trials

Blinding is essential to limit the risk of bias; however, blinding is not always feasible. A study of a sample of randomized controlled trials assessing pharmacological and nonpharmacological treatments in the field of osteoarthritis showed that blinding was almost always feasible for patients, care providers, and outcome assessors in trials assessing pharmacological treatments. However, in trials assessing nonpharmacological treatments, blinding was considered feasible in only 42% of the trials for patients, 12% for care providers, and 34% for outcome assessors. When blinding was judged feasible, the perceived risk of nonblinding was more often considered moderate or important in trials assessing nonpharmacological treatments. When blinding was judged feasible, it was reported less often in nonpharmacological reports. These differences are linked to the difficulties of finding a placebo for nonpharmacological treatments. In fact, the procedures of blinding mainly rely on the use of a placebo defined as a control intervention with similar appearance as the experimental treatment but devoid of the components in the experimental intervention whose effects the trial is designed to evaluate. A placebo is usually feasible for pharmacological treatments but raises important issues for nonpharmacological treatments. For example, what is an appropriate placebo for a surgical procedure? for psychotherapy? These issues will be discussed in Chapter 2. Frequently, providing a completely similar placebo intervention in the control group is not possible, and the use of partial blinding could be proposed. Such a procedure should not be considered a panacea because complete blinding is not achieved. However, these procedures aim to limit the risk of bias. Patients could be blinded to the study hypothesis, that is, patients are aware that they will have a 50% chance of receiving one of the two interventions being evaluated, that they do not know which intervention is the most effective, and that for scientific reasons, they cannot be informed of all the hypotheses of the trial. In other situations, patients could be aware that one of the interventions is a placebo but will not be informed of the nature of the placebo.

To overcome the difficulties of blinding patients and care providers, a prospective randomized open, blinded end-point (PROBE) study could be proposed [23]. Such a study limits the risk of detection bias because of blinded assessment of the outcome. This method has been proposed for different types of outcomes. It mainly relies on a centralized and blinded assessment of the outcome. For physician-driven data, the study could be a centralized assessment of clinical examinations through the use of photography, video, or audio of an interview. For example, in a trial evaluating the efficacy of multi-wavelength light therapy to treat pressure ulcers in subjects with disorders of the spinal cord [24], photographs of the ulcers were taken at the beginning and end of treatment and at 14 days after the last session. All evaluations were performed by a blinded outcome assessor. To assess the effects of a treatment for verbal communication in aphasia after stroke [25], patient responses were tape-recorded and scored by two independent blinded observers. When the outcome is a complementary test, a centralized assessment of the test will avoid bias. In a trial evaluating off-pump versus conventional coronary artery bypass grafting—early and 1 year graft patency—three cardiologists who were blinded to group assignment simultaneously reviewed angiograms [26].

For clinical events such as occurrence of myocardial infarction, a blinded adjudication committee is useful. However, this situation still entails a risk of bias, particularly if the adjudication committee evaluates and adjudicates only the events identified and transmitted by nonblinded investigators. Therefore, what was evaluated by the blinded adjudication committee must be considered: Did the adjudication committee evaluate all the patients included in the study (which is difficult to achieve because of time and cost)? Were patients systematically screened by a routine check of biochemical markers and electrocardiographic analyses by core laboratories? Were specific computer algorithms used to identify events? Mahaffey et al. [27] used a computer algorithm to show that 270 cases of myocardial infarction (5.2% of all patients enrolled) were not identified by site investigators and 134 cases (2.6%) identified by site investigators were not confirmed by the adjudication committee. Similarly, an independent review of case-report forms from the Rosiglitazone Evaluated for Cardiac Outcomes and Regulation of Glycemia in Diabetes (RECORD) trial [28] by the U.S. Food and Drug Administration (FDA) provided evidence of bias in that investigators were aware of the treatment allocated. In the RECORD study, the method for selecting cases to be assessed by the adjudication committee relied on whether the nonblinded investigators identified and reported the case. Only 12.5% of the case-report forms were reviewed. Errors such as a patient with an event not referred for adjudication were systematically bias with more errors in the experimental group and 81% of errors favoring the experimental group [27].

A frequent and difficult situation in assessing nonpharmacological treatments is the comparison of the treatment with usual care. In this situation, blinding of patients and care providers is not feasible. The risk of bias may be particularly important because of the deception of patients who will not receive any treatments, which is particularly problematic when the primary

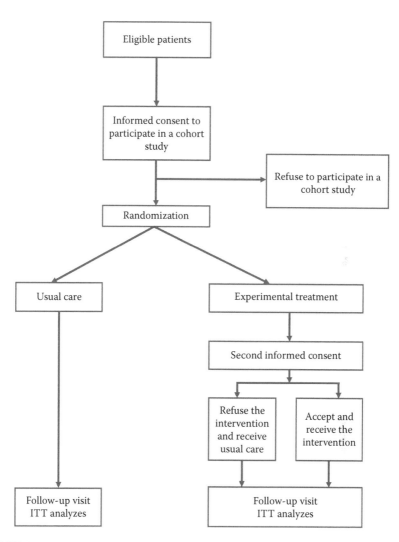

FIGURE 1.1
Modified Zelen design.

outcome is a patient-reported outcome (e.g., pain, quality of life). Some specific designs may be proposed in these situations. A modified Zelen design has been proposed, although it has been criticized for ethical issues [29,30] (see Figure 1.1). In a first step, patients are invited to participate in a cohort study. They are informed of the different follow-up visits and sign a consent form. In a second step, patients are randomized. Patients randomized to receive the experimental treatments are informed and sign a second consent form. Such a design avoids deceiving patients. However, this design raises some issues. An ethical issue is that some ethics committee will not agree to approve studies in which patients will be blinded to study hypotheses.

However, this issue has to be balanced with the ethical issue of conducting a trial knowing that there will be a high risk of bias. Other issues are related to logistics in terms of ensuring that patients from both groups do not meet. Finally, this design is not adequate in studies with a high risk that patients will refuse the experimental intervention because the analysis will have to be an intention-to-treat analysis and the high rate of refusal will decrease the power of the trial. Recently, Relton and colleagues proposed the cohort multiple randomized controlled trial design, which could also help address these methodological issues [31].

When blinding is not performed, the risk of bias must be evaluated. In fact, this approach is necessary for the critical appraisal of published results of randomized controlled trials but also when assessing the risk of bias in trial results included in systematic reviews and meta-analysis. The Cochrane collaboration has developed a specific tool to evaluate the risk of bias in randomized controlled trials, the Risk of Bias Tool (the RoB tool) [5]. The RoB tool recommends evaluating the risk of performance bias and the risk of detection bias. The risk of performance bias is high if patients and care providers are not blinded and if the outcome is likely to be influenced by the lack of blinding (e.g., crossover trials, differential co-interventions, and differential attrition). The risk of detection bias will be high if the outcome assessor is not blinded and the outcome measurement is likely to be influenced by lack of blinding (e.g., subjective outcomes).

1.3 Conclusion

Blinding is essential to limit the risk of bias. However, blinding is more difficult to achieve and maintain in trials assessing nonpharmacological treatments. There is a need to use creative methods of blinding, and in some situations to accept that the only way to limit the risk of bias is to attempt partial blinding. Nevertheless, frequently, blinding is not feasible, and evaluating the risk of bias in such studies is necessary.

References

1. Schulz, K.F., Chalmers, I., Hayes, R.J., Altman, D.G. Empirical evidence of bias. Dimensions of methodological quality associated with estimates of treatment effects in controlled trials. *JAMA*, 273(5), 408–412, 1995.
2. Schulz, K.F., Altman, D.G., Moher, D. CONSORT 2010 statement: Updated guidelines for reporting parallel group randomized trials. *Ann Intern Med*, 152(11), 726–732, 2010.

3. Noseworthy, J.H., Ebers, G.C., Vandervoort, M.K., Farquhar, R.E., Yetisir, E., Roberts, R. The impact of blinding on the results of a randomized, placebo-controlled multiple sclerosis clinical trial. *Neurology*, 44(1), 16–20, 1994.
4. Day, S.J., Altman, D.G. Statistics notes: Blinding in clinical trials and other studies. *BMJ*, 321(7259), 504, 2000.
5. Julian, H., Sally, G. *Cochrane Handbook for Systematic Reviews of Interventions.* Wiley-Blackwell, Hoboken, NJ, 2008.
6. Schulz, K.F., Grimes, D.A. Blinding in randomised trials: Hiding who got what. *Lancet*, 359(9307), 696–700, 2002.
7. Schulz, K.F., Grimes, D.A., Altman, D.G., Hayes, R.J. Blinding and exclusions after allocation in randomised controlled trials: Survey of published parallel group trials in obstetrics and gynaecology. *BMJ*, 312(7033), 742–744, 1996.
8. Weinstein, J.N., Tosteson, T.D., Lurie, J.D., et al. Surgical vs nonoperative treatment for lumbar disk herniation: The Spine Patient Outcomes Research Trial (SPORT): A randomized trial 10.1001/jama.296.20.2441. *JAMA*, 296(20), 2441–2450, 2006.
9. Moher, D., Pham, B., Jones, A., et al. Does quality of reports of randomised trials affect estimates of intervention efficacy reported in meta-analyses? *Lancet*, 352(9128), 609–613, 1998.
10. Wood, L., Egger, M., Gluud, L.L., et al. Empirical evidence of bias in treatment effect estimates in controlled trials with different interventions and outcomes: Meta-epidemiological study. *BMJ*, 336(7644), 601–605, 2008.
11. Hróbjartsson, A., Thomsen, A., Emanuelsson, F., et al. Empirical evidence of observer bias in randomised trials: A meta-analysis, 2011 (submitted).
12. Devereaux, P.J., Manns, B.J., Ghali, W.A., et al. Physician interpretations and textbook definitions of blinding terminology in randomized controlled trials. *JAMA*, 285(15), 2000–2003, 2001.
13. Devereaux, P.J., Bhandari, M., Montori, V.M., Manns, B.J., Ghali, W.A., Guyatt, G.H. Double blind, you are the weakest link—Good-bye! *ACP J Club*, 136(1), A11, 2002.
14. Montori, V., Bhandari, M., Devereaux, P., Manns, B., Ghali, W., Guyatt, G. In the dark: The reporting of blinding status in randomized controlled trials. *J Clin Epidemiol*, 55(8), 787, 2002.
15. Haahr, M.T., Hrobjartsson, A. Who is blinded in randomized clinical trials? A study of 200 trials and a survey of authors. *Clin Trials*, 3(4), 360–365, 2006.
16. Chan, A.W., Altman, D.G. Epidemiology and reporting of randomised trials published in PubMed journals. *Lancet*, 365(9465), 1159–1162, 2005.
17. Hróbjartsson, A., Pildal, J., Chan, A.W., Haahr, M.T., Altman, D.G., Gotzsche, P.C. Reporting on blinding in trial protocols and corresponding publications was often inadequate but rarely contradictory. *J Clin Epidemiol*, 62(9), 967–973, 2009.
18. Altman, D.G. Better reporting of randomised controlled trials: The CONSORT statement. *BMJ*, 313, 570–571, 1996.
19. Altman, D.G., Schulz, K.F., Moher, D., et al. The revised CONSORT statement for reporting randomized trials: Explanation and elaboration. *Ann Intern Med*, 134(8), 663–694, 2001.
20. Moher, D., Schulz, K.F., Altman, D.G. The CONSORT statement: Revised recommendations for improving the quality of reports of parallel-group randomised trials. *Lancet*, 357(9263), 1191–1194, 2001.

21. Schulz, K.F., Altman, D.G., Moher, D. CONSORT 2010 statement: Updated guidelines for reporting parallel group randomised trials. *PLoS Med*, 7(3), e1000251, 2010.
22. Schulz, K.F., Altman, D.G., Moher, D., Fergusson, D. CONSORT 2010 changes and testing blindness in RCTs. *Lancet*, 375(9721), 1144–1146, 2010.
23. Hansson, L., Hedner, T., Dahlof, B. Prospective randomized open blinded endpoint (PROBE) study. A novel design for intervention trials. Prospective randomized open blinded end-point. *Blood Press*, 1(2), 113–119, 1992.
24. Taly, A.B., Sivaraman Nair, K.P., Murali, T., John, A. Efficacy of multiwavelength light therapy in the treatment of pressure ulcers in subjects with disorders of the spinal cord: A randomized double-blind controlled trial. *Arch Phys Med Rehabil*, 85(10), 1657–1661, 2004.
25. Doesborgh, S.J., van de Sandt-Koenderman, M.W., Dippel, D.W., van Harskamp, F., Koudstaal, P.J., Visch-Brink, E.G. Effects of semantic treatment on verbal communication and linguistic processing in aphasia after stroke: A randomized controlled trial. *Stroke*, 35(1), 141–146, 2004.
26. Puskas, J.D., Williams, W.H., Mahoney, E.M., et al. Off-pump vs conventional coronary artery bypass grafting: Early and 1-year graft patency, cost, and quality-of-life outcomes: A randomized trial. *JAMA*, 291(15), 1841–1849, 2004.
27. Mahaffey, K.W., Roe, M.T., Dyke, C.K., et al. Misreporting of myocardial infarction end points: Results of adjudication by a central clinical events committee in the PARAGON-B trial. Second platelet IIb/IIIa antagonist for the reduction of acute coronary syndrome events in a global organization network trial. *Am Heart J*, 143(2), 242–248, 2002.
28. Home, P.D., Pocock, S.J., Beck-Nielsen, H., et al. Rosiglitazone evaluated for cardiovascular outcomes in oral agent combination therapy for type 2 diabetes (RECORD): A multicentre, randomised, open-label trial. *Lancet*, 373(9681), 2125–2135, 2009.
29. Quilty, B., Tucker, M., Campbell, R., Dieppe, P. Physiotherapy, including quadriceps exercises and patellar taping, for knee osteoarthritis with predominant patello-femoral joint involvement: Randomized controlled trial. *J Rheumatol*, 30(6), 1311–1317, 2003.
30. Bravata, D.M., Smith-Spangler, C., Sundaram, V., et al. Using pedometers to increase physical activity and improve health: A systematic review. *JAMA*, 298(19), 2296–2304, 2007.
31. Relton, C., Torgerson, D., O'Cathain, A., Nicholl, J. Rethinking pragmatic randomised controlled trials: Introducing the "cohort multiple randomised controlled trial" design. *BMJ*, 340, c1066, 2010.

2

Placebo in Nonpharmacological Randomized Trials*

Asbjørn Hróbjartsson
The Nordic Cochrane Centre

Franklin G. Miller
National Institutes of Health

CONTENTS

2.1 Introduction

One of the first clinical trials to explicitly use a placebo control group was published in 1938 by Diel et al. [1]. The trial compared patients treated with capsules containing vaccine for common cold with patients treated with placebo capsules containing lactose. In contrast to previous trials, Diel et al. found no effect of oral vaccine.

Since then, a vast multitude of placebo-controlled trials have been conducted. A search for "placebo*" in The Cochrane Central Register of Controlled Trials in April 2009 provided 111,592 references. The majority of hits are to pharmacological trials comparing a drug with a placebo, in some cases as an addition to a standard care regime, or as part of a "doubly dummy" design.

Placebo interventions are difficult to define unambiguously [2,3]. However, within a clinical trial, a placebo can be characterized as a control intervention

---

* The opinions expressed are the views of the author and do not necessarily reflect the policy of the National Institutes of Health, the Public Health Service, or the U.S. Department of Health and Human Services.

with similar appearance as the experimental treatment, but void of the components in the experimental intervention whose effects the trial is designed to evaluate.

Roughly 24% of contemporary randomized trials indexed in PubMed are nonpharmacological [4]. Nonpharmacological trials involve diverse types of interventions, and equally diverse types of placebo control groups. For the sake of clarity in the following, we will primarily address three types of nonpharmacological placebo control interventions: devices, surgical interventions, and psychological interventions. These three types of interventions exemplify the most typical challenges posed to nonpharmacological trials in general.

In this chapter, we will describe and discuss nonpharmacological placebo interventions as used in clinical trials. In general, device placebos and surgical placebos are conceptually similar to pharmacological placebos, though they pose some distinctive methodological and ethical issues. In both types of trials, there is a noticeable risk of unblinding, and surgical placebo trials carry risks to subjects from the placebo intervention itself. Psychological placebos are generally of a different type conceptually, as they are often designed to control for specific factors, such as expectancy, and not as a tool for blinding. When designing a nonpharmacological trial, or when interpreting the results of such a trial, the exact nature of the placebo intervention deserves considerable attention.

2.2 Placebo-Controlled Trials

The main aim of a placebo-controlled trial is to establish whether the components of the experimental intervention, hypothesized to be effective, in fact can produce clinically significant benefit in patients with a given medical condition. Accordingly, the experimental intervention is compared with a placebo control that appears indistinguishable and lacks the components of the intervention hypothesized to be responsible for its therapeutic efficacy.

There are two fundamental differences between a placebo control group and a no-treatment control group. First, a trial using a no-treatment control group tests whether an intervention as a whole has an effect. The design can say nothing clear about which component within a treatment package is the main causal factor. Second, the design is unreliable, especially when outcomes are subjective, because it does not permit masking of the study intervention and comparator.

The notion of a placebo control is historically linked with the idea that placebo interventions cause large effects. Especially after Beecher in 1955 published a review of the improvements reported in the placebo groups of 14 trials, it became a standard notion that placebo interventions had large effects on many patients on both objective and subjective outcomes [5]. His assessment of "the powerful placebo" was based on a comparison between baseline and posttreatment in placebo control groups, and did not control for natural fluctuations in the patient's condition, including spontaneous remission, and regression to the mean. Nonetheless, this article was an important factor in persuading clinicians that randomized trials were necessary and ethical.

A recent update of a systematic review of 202 randomized trials with both placebo groups and no-treatment control groups found small to moderate differences between no treatment and placebo overall, but effects were more pronounced in trials with patient-reported outcomes [6]. For pain the mean effect corresponded to roughly 6 mm on a 100 mm visual analogue scale; however, in certain settings the effect was larger. Four well-performed German acupuncture trials reported an effect of placebo acupuncture of roughly 17 mm on a 100 mm visual analogue scale. It is noteworthy that the patients involved in the trials were falsely informed that the study involved a comparison between two types of acupuncture, and not between acupuncture and placebo acupuncture [7]. The effect of placebo was considerably smaller in other acupuncture trials [8]. When all trials with continuous outcomes were examined in a regression analysis, there was a clear tendency for larger effects in device placebos and psychological placebos as compared with pharmacological placebos. The general pattern of results from the review, when disregarding the obvious risk of reporting bias, is that placebo interventions can affect subjective outcomes, but that this effect is quite variable, and dependent on underlying causal factors, for example, patient information and type of placebo.

Besides controlling for placebo effects, there are additional compelling reasons for implementing masked placebo control groups. Reporting bias occurs when patients report their symptoms more favorably than they otherwise would have, for example, because of courtesy to the doctor who offers them treatment. In the case of placebo-controlled trials of invasive treatments, patients may be disposed to perceive or report benefit as a result of having undergone a burdensome or seemingly powerful intervention. This type of bias is much more likely to occur when an intervention is compared with a no-treatment control group instead of a placebo group. Similarly, attrition bias occurs when patients stop the trial, or do not adhere to the treatment, because they wanted to be in the other treatment group. This type of behavior is also likely to be more pronounced in no-treatment group as compared with a placebo group.

It is a common misunderstanding that placebo control interventions have to be "inert" [9]. Strictly speaking, classic placebos such as sugar pills and saline infusions are not inert, since they contain biologically active ingredients. They are "inert" only in the relative sense that there is no scientific reason to think that the sugar or salt in the placebo intervention will have an effect on the outcomes of interest in a clinical trial. Similarly, in sham surgery trials, comparing a real to a fake surgical procedure, the invasive placebo control obviously is not inert. But as long as the placebo surgical intervention does not include the surgical manipulation hypothesized to be responsible for the outcomes under investigation, it counts as a valid control. Likewise, a sham acupuncture intervention (whether superficial needling at non-acupuncture points or a retractable device) constitutes a valid control for detecting whether the needling techniques characteristic of traditional acupuncture are responsible for clinical effects, regardless of the possibility that the physical stimulus provided by the sham acupuncture intervention might itself have an effect.

The ethics of placebo-controlled trial has been debated intensely [10]. There is no doubt that placebo-controlled trials are used in many situations where there is an established treatment. For example, it is routine to use of placebo controls in many psychiatric conditions and conditions in which pain is the outcome, despite proven effective treatment [11]. The fifth revision (1996) of the Helsinki Declaration stated that: "The benefits, risks, burdens and effectiveness of a new method should be tested against those of the best current prophylactic, diagnostic, and therapeutic methods. This does not exclude the use of placebo, or of no-treatment, in studies where no proven prophylactic, diagnostic or therapeutic method exist." [12]. Taken at face value, this means that a large number of placebo-controlled trials would be in violation with the fifth revision of the declaration.

However, the sixth revision (2008) is importantly different: "The benefits, risks, burdens and effectiveness of a new intervention must be tested against those of the best current proven intervention, except in the following circumstances: The use of placebo, or no treatment, is acceptable in studies where no current proven intervention exists; or where for compelling and scientifically sound methodological reasons the use of placebo is necessary to determine the efficacy or safety of an intervention and the patients who receive placebo or no treatment will not be subject to any risk of serious or irreversible harm. Extreme care must be taken to avoid abuse of this option." [13].

This shift toward endorsing placebo-controlled trials finds support from three strange bedfellows. First, the pharmaceutical industry is generally very interested in establishing effect beyond placebo, and much less interested in a trial that risk showing a drug to be less effective than a standard therapy. Second, the U.S. Food and Drug Administration (FDA) are concerned with the so-called "assay sensitivity," by which they mean the ability of trials to detect an effect compared with an ineffective therapy, and normally require new drugs to show superiority to placebo in at least two trials, before they approve new drugs [14]. Third, researchers writing from an evidence-based medicine perspective have pointed out that "proven" effective may quite often be illusive [15]. What clinicians have thought to be an effective therapy has later often been shown to have no effect, or that the harmful effects outweighed the beneficial effects. If an intervention falsely is regarded effective, and new interventions are compared with this false positive yardstick, we risk introducing a number of equally ineffective interventions.

2.3 Nonpharmacological Placebo Interventions: Device Placebos

To illustrate the nature of contemporary nonpharmacological placebo interventions, a search on PubMed from November 2008 to February 2009 for publications containing the terms placebo* or sham* and indexed as "randomized controlled trial" provided 21 references (Table 2.1). Twelve trials used various forms of placebo devices: two trials using ineffective ultrasound machines, two trials using ineffective lasers, three trials using ineffective magnetic or electric stimulation, and five trials using other forms of sham devices, for example, paper filters in an air cleaner. There were six trials with placebo acupuncture or acupressure procedures. One trial used a manual placebo procedure, and there were two psychological trials.

The typical nonpharmacological placebo intervention thus seems to be a device. An illustrative example is the trial by Sulser et al., investigating the effect of high-efficiency particulate arresting (HEPA) air cleaner filters on the symptoms of asthma in children and adolescents sensitized to cat and dog allergens. The machines containing the filters were identical, and the only difference between the "active" and the "placebo" machine was that the active air cleaners contained HEPA filters, and that the placebo machines contained paper filters.

The trial is very similar to the standard pharmacological placebo-controlled trial. It is fairly easy to construct two machines that appear to be identical, one with a true HEPA filter and one with a paper filter. The active component in the trial is clearly defined and delineated. There are similarly no conceptual challenges in constructing placebo devices for ultrasound, or magnetic/electronic devices, that appear identical to the real devices, but without their magnetic or electronic property.

The difference between a pharmacological and a device placebo is most often of a practical kind, often concerning the risk of unblinding. Patients may try to check whether their intervention is placebo or not and this may be easier when treated with a device placebo than a pharmacological placebo. For example, in Chen et al.'s trial of magnetic knee wrappers (Table 2.1), it would be easy for patients to test whether their wrappers attracted metal sometimes during the 12-week period of the trial.

Another difference appears when it is difficult to construct a device placebo void of the active component tested in the experimental group. For example, in Chermont et al.'s trial of the effect of continuous positive airway pressure (CPAP) for chronic heart failure, a CPAP placebo was used with a low air pressure (0–1 mm H_2O) as compared with the higher pressure in the real CPAP group (3 mm). This is different from the classic pharmacological placebo trial, in that placebos differ from the active treatment in dose, not in nature. The procedure is meaningful only as long as the assumption of subtherapeutic dose is correct.

TABLE 2.1

Nonpharmacological Randomized Trials with Placebo Groups[a]

Trial	Clinical Problem	Experimental Procedure	Placebo Procedure
Stowell et al. [1]	Phlebotomy pain	Ultrasound	Unclear
Özgönenel et al. [2]	Knee osteoarthritis	Ultrasound	The applicator was disconnected to the ultrasound machine
Deng et al. [3]	Pain after thoracotomy	Acupuncture	Sham studs not penetrating the skin, and placed at sites not … true acupuncture sites
Elden et al. [4]	Pelvic girdle pain in pregnancy	Acupuncture	Nonpenetrating needles and no attempt to evoke "Qi"
Gaudet et al. [5]	Labor initiation	Acupuncture	Needling in sites not known to have an effect on the initiation of labor
Nordio and Romanelli [6]	Insomnia	Acupressure	Application of wrist pressure at a site different from the true HT 7 Shenmen acupuncture point
Sima and Wang [7]	Cisplatin-induced nausea	Acupuncture	Needling at points not regarded effective for nausea and vomiting
Desantana et al. [8]	Postoperative pain	Transcutaneous electrical nerve stimulation	No electronic stimulation, but the machine displayed an active indicator light
Schutter et al. [9]	Mood in healthy subjects	Transcranial magnetic stimulation	The device mimics the sound click … but the brain is shielded from actual stimulation [with an aluminum plate]
Lisanby et al. [10]	Major depression	Transcranial magnetic stimulation	A magnetic shield "limited the magnetic energy reaching the cortex to 10% …." Active and sham coils had "similar appearance, placement, and acoustic properties"
Koenigs et al. [11]	Emotional function in healthy subjects	Transcranial direct current stimulation	Stimulation for only 30 s
Chen et al. [12]	Knee osteoarthritis	Magnetic knee wrap	No magnetic activity
Kuhn et al. [13]	Oral mucositis	Low-level infrared laser therapy	The laser was turned off, but patients were blindfolded

TABLE 2.1 (continued)

Nonpharmacological Randomized Trials with Placebo Groups[a]

Trial	Clinical Problem	Experimental Procedure	Placebo Procedure
Teggi et al. [14]	Chronic tinnitus	Low-level laser	The [laser] device was pointed into the ear canal but the laser remained inactive
Azarpazhooh et al. [15]	Dentin hypersensitivity	Ozone machine	Machine delivered air
Sulser et al. [16]	Asthmatic children	Air cleaners with HEPA filters	Air cleaners with paper filters
Chermont et al. [17]	Chronic heart failure	Continuous positive airway pressure (3 mm H_2O)	Continuous positive airway pressure with low pressure (0–1 mm H_2O)
Lettieri and Eliasson [18]	Restless legs syndrome	Pneumatic compression (40 mm H_2O)	Pneumatic compression with low air pressure (3–4 mm H_2O)
Perry and Green [19]	Nervous activity in the lower limbs	Lumbar oscillatory mobilization	Same hand positioning but without any oscillatory movements
Edinger et al. [20]	Insomnia	Cognitive behavioral therapy and relaxation	Quasi-desensitization procedure
Walkup et al. [21]	Childhood anxiety	Cognitive behavioral therapy and sertraline	Placebo pill

References

1. Stowell, C.P., Trieu, M.Q., Chuang, H., Katz, N., Quarrington, C. Ultrasound-enabled topical anesthesia for pain reduction of phlebotomy for whole blood donation. *Transfusion*, 49(1), 146–153, 2009.
2. Ozgönenel, L., Aytekin, E., Durmuşoglu, G. A double-blind trial of clinical effects of therapeutic ultrasound in knee osteoarthritis. *Ultrasound Med Biol*, 35(1), 44–49, 2009.
3. Deng, G., Rusch, V., Vickers, A., et al. Randomized controlled trial of a special acupuncture technique for pain after thoracotomy. *J Thorac Cardiovasc Surg*, 136(6), 1464–1469, 2008.
4. Elden, H., Fagevik-Olsen, M., Ostgaard, H.C., Stener-Victorin, E., Hagberg, H. Acupuncture as an adjunct to standard treatment for pelvic girdle pain in pregnant women: Randomised double-blinded controlled trial comparing acupuncture with non-penetrating sham acupuncture. *BJOG*, 115(13), 1655–1668, 2008.
5. Gaudet, L.M., Dyzak, R., Aung, S.K., Smith, G.N. Effectiveness of acupuncture for the initiation of labour at term: A pilot randomized controlled trial. *J Obstet Gynaecol Can*, 30(12), 1118–1123, 2008.
6. Nordio, M., Romanelli, F. Efficacy of wrists overnight compression (HT 7 point) on insomniacs: Possible role of melatonin? *Minerva Med*, 99(6), 539–547, 2008.

(continued)

TABLE 2.1 (continued)

Nonpharmacological Randomized Trials with Placebo Groups[a]

7. Sima, L., Wang, X. Therapeutic effect of acupuncture on cisplatin-induced nausea and vomiting. *Zhongguo Zhen Jiu*, 29(1), 3–6, 2009.

8. Desantana, J.M., Sluka, K.A., Lauretti, G.R. High and low frequency TENS reduce postoperative pain intensity after laparoscopic tubal ligation: A randomized controlled trial. *Clin J Pain* 25(1), 12–19, 2009.

9. Schutter, D.J., Enter, D., Hoppenbrouwers, S.S. High-frequency repetitive transcranial magnetic stimulation to the cerebellum and implicit processing of happy facial expressions. *J Psychiatry Neurosci*, 34(1), 60–65, 2009.

10. Lisanby, S.H., Husain, M.M., Rosenquist, P.B., et al. Daily left prefrontal repetitive transcranial magnetic stimulation in the acute treatment of major depression: Clinical predictors of outcome in a multisite, randomized controlled clinical trial. *Neuropsychopharmacology*, 34(2), 522–534, 2009.

11. Koenigs, M., Ukueberuwa, D., Campion, P., Grafman, J., Wassermann, E. Bilateral frontal transcranial direct current stimulation: Failure to replicate classic findings in healthy subjects. *Clin Neurophysiol*, 120(1), 80–84, 2009.

12. Chen, C.Y., Chen, C.L., Hsu, S.C., Chou, S.W., Wang, K.C. Effect of magnetic knee wrap on quadriceps strength in patients with symptomatic knee osteoarthritis. *Arch Phys Med Rehabil*, 89(12), 2258–2264, 2008.

13. Kuhn, A., Porto, F.A., Miraglia, P., Brunetto, A.L. Low-level infrared laser therapy in chemotherapy-induced oral mucositis: A randomized placebo-controlled trial in children. *J Pediatr Hematol Oncol*, 31(1), 33–37, 2009.

14. Teggi, R., Bellini, C., Piccioni, L.O., Palonta, F., Bussi, M. Transmeatal low-level laser therapy for chronic tinnitus with cochlear dysfunction. *Audiol Neurotol*, 14(2), 115–120, 2009.

15. Azarpazhooh, A., Limeback, H., Lawrence, H.P., Fillery, E.D. Evaluating the effect of an ozone delivery system on the reversal of dentin hypersensitivity: A randomized, double-blinded clinical trial. *J Endod*, 35(1), 1–9, 2009.

16. Sulser, C., Schulz, G., Wagner, P., et al. Can the use of HEPA cleaners in homes of asthmatic children and adolescents sensitized to cat and dog allergens decrease bronchial hyperresponsiveness and allergen contents in solid dust? *Int Arch Allergy Immunol*, 148(1), 23–30, 2009.

17. Chermont, S., Quintão, M.M., Mesquita, E.T., Rocha, N.N., Nóbrega, A.C. Noninvasive ventilation with continuous positive airway pressure acutely improves 6-minute walk distance in chronic heart failure. *J Cardiopulm Rehabil Prev*, 29(1), 44–48, 2009.

18. Lettieri, C.J., Eliasson, A.H. Pneumatic compression devices are an effective therapy for restless legs syndrome: A prospective, randomized, double-blinded, sham-controlled trial. *Chest*, 135(1), 74–80, 2009.

19. Perry, J., Green, A. An investigation into the effects of a unilaterally applied lumbar mobilisation technique on peripheral sympathetic nervous system activity in the lower limbs. *Man Ther*, 13(6), 492–499, 2008.

20. Edinger, J.D., Carney, C.E., Wohlgemuth, W.K. Pretherapy cognitive dispositions and treatment outcome in cognitive behavior therapy for insomnia. *Behav Ther*, 39(4), 406–416, 2008.

21. Walkup, J.T., Albano, A.M., Piacentini, J., et al. Cognitive behavioral therapy, sertraline, or a combination in childhood anxiety. *N Engl J Med*, 359, 2753–2766, 2008.

[a] Based on a PubMed search for placebo* or sham*, restricted to "randomized clinical trials" and publicized from November 2008 to February 2009.

A somewhat special type of device trials is sham-acupuncture studies. There were 5 placebo acupuncture or acupressure trials among the 12 device trials listed in Table 2.1, reflecting that acupuncture placebo trials are common. The type of placebo control (usually called sham control within acupuncture research) varied. In one trial, the placebo intervention was penetrative ("Needling at points not regarded effective for nausea and vomiting"), whereas that was not the case for another trial ("Sham studs not penetrating the skin, and placed at sites not … true acupuncture sites"). Placebo acupuncture procedures involve manual manipulation, and typically an intense doctor–patient interaction and they therefore differ from most other device placebos. The risk of unblinding the patient through subtle cues in the acupuncturist's behavior is considerable, and the effects of needling in non-acupuncture points cannot be ruled out [8,16,17].

Placebo-controlled trials of device interventions tend to be conceptually similar to trials of pharmacological trials, though in general the practical construction of the placebo devices may be more challenging, and the risk of unblinding is higher. Acupuncture trials involve additional challenges, for example, to deal with intense patient–provider interaction.

2.4 Nonpharmacological Placebo Interventions: Surgical Placebos

In the 1950s, ligation of the internal mammary arteries for angina pectoris became popular in the United States. However, two small trials comparing the effect of ligation of the internal mammary arteries with the effect of placebo operation only (skin incision only) concluded that the procedure had no effect, and the operation became unfashionable. In 1961, Beecher energetically described the classic story emphasizing the need for rigorous randomized trials of surgical procedures [18].

His call for surgical trials is equally relevant today. The number of randomized trials in surgery is still very low compared with medicine in general. Only 10% of trials indexed in PubMed in December 2000 were Surgical/procedure trials, whereas 76% of were pharmacological trials [4]. Though the scarcity of surgical trials may be explained for reasons of practicality, history, lack of regulatory oversight requiring clinical trials before new surgical procedures are introduced into practice, and sparse funding sources [19], it remains a public health scandal that numerous surgical procedures are not reliably evaluated.

In the comparatively few surgical trials conducted, placebo-controlled trials are rare, though they tend to be highly publicized [20]. An illustrative example of a surgical placebo trial is Moseley et al.'s trial of arthroscopic lavage vs. arthroscopic debridement vs. placebo surgery in patients with osteoarthritis of the knee [21]. The placebo surgery consisted in three 1 cm incisions in the skin after having received a short-acting intravenous tranquillizer and an opioid, and spontaneously breathed oxygen-enriched air. Patients were unaware of which treatment they had received. The main outcome was pain after 2 years, and the trial found no difference between the three groups.

In the trial by Moseley et al., both active interventions involved instrumentation within the knee joint, not performed on the patients in the placebo group. It is meaningful and comparatively easy to clearly define and delineate the active surgical components. From a conceptual point of view, many surgical placebos are similar to the trial by Moseley et al. It seems meaningful to construct a placebo surgery procedure as long as the active part of the surgical procedure is conducted on an anatomical entity that is either covered by skin, for example, a bone, or confined within an anatomical space, for example, a ligament within a joint, or procedures within the abdominal or thoracic cavities. However, some surgical procedures involve procedures that are impossible to mimic with a placebo surgery control group, for example, amputation of the lower limb.

The main problem with surgical placebo interventions is that, unlike pharmacological placebos, they can harm patients directly. The patients in the placebo surgical group may have to undergo a skin incision, have pain medication or anesthesia, with its potential harmful effects, and risk postoperative infection. In that sense, surgical placebo trials involve a more direct ethical challenge than pharmacological or device trials, where patients may be harmed, but more indirectly only if their participation precludes them from access to alternative effective treatment.

The result of the trial by Moseley et al. is much more reliable due to its placebo procedure than it would have been with a control group receiving usual medical therapy to treat pain or a nonblinded no-treatment control group. Still, Moseley et al. must have struggled somewhat with the unavoidable dilemma of when the added risk to the included patients was outweighed by the benefit to future patients. The authors implemented a quite strict informed consent procedure, stating that "placebo surgery will not benefit my knee arthritis" (44% of screened patients declined). Furthermore, the trialists made an effort to minimize the risk to placebo patients by not giving standard general anesthesia.

Surgical placebo trials are similar to the device placebo trials in that there often are potential problems of nonblinding. For example, it seems possible for some of the patients in the placebo knee surgery group to realize that their anesthesia procedure was different from standard procedures, and from there deduce that they had received placebo.

Another high-profile surgical placebo-controlled trial evaluated the effect of surgical implantation of fetal tissue to patients with Parkinson's disease [22]. The placebo procedure included "the placement of a stereotactic frame, target localization on magnetic resonance imaging, the administration of general anesthesia with a laryngeal-mask airway, and a skin incision with a partial burr hole that does not penetrate the inner cortex of the skull." Whether the added risk to the included patients was outweighed by the benefit to future patients in this trial has been discussed stormily [23,24].

The additional risk involved in surgical trials is not necessarily as dramatic as a scull burr hole. The risk involved in some surgical trials is similar to that of other generally accepted procedures, for example, "muscle biopsy, bronchoscopy, and phase 1 testing of experimental drugs in healthy volunteers, which do not offer participants a prospect of direct benefit" [25]. Providing that the trial authors minimize the risk of the included patients, justify the remaining risk necessary to produce scientifically valid results, and carefully think through the informed consent procedure, a surgical placebo control group is not unethical per se.

Surgery is always harmful and sometimes is also beneficial. Placebo-controlled surgery trials are needed to reliably assess this balance, especially when outcomes are subjective. The ethical considerations involved in any placebo surgery control group needs to be delicately and cautiously analyzed, and the conclusion will probably differ according to the clinical scenario and the type of placebo surgery involved. However, in general, a surgical placebo control group seems attractive when outcomes are subjective, and the risks of the surgical placebo procedure are minor.

2.5 Nonpharmacological Placebo Interventions: Psychological Placebos

Effects of placebo interventions and effects of psychological interventions are both psychologically mediated. The discussion of placebo control groups in psychological trials is closely linked to the discussion of what exactly distinguishes psychological placebo interventions from psychological "verum" interventions. Both issues have been hotly debated in psychology for years, from Frank's classic characterization of placebo as a form of psychotherapy [26] to a more recent theme issue in *Journal of Clinical Psychology* [27].

One of the psychological trials listed in Table 2.1 is illustrative: Edinger et al. compared cognitive behavioral therapy with a placebo intervention for insomnia (there was also a third arm of relaxation training). The cognitive behavior intervention consisted of sessions providing practical information on sleep and stimulus control, and instructions to establish standard wake-up times, to get out of bed during extended awakenings, to avoid sleep-incompatible behavior in the bedroom, and to eliminate daytime napping. Furthermore, patients were given an initial time in bed prescription, which was modified during the sessions.

The placebo intervention ("quasi-desensitization") was presented to patients as a method to overcome "conditioned arousal." Therapists helped each patient to develop a hierarchy of common activities he/she did on awakening at night (e.g., opening eyes, clock watching). Therapists also helped them develop scenes of themselves engaged in neutral activities (e.g., reading the newspaper). In each session, patients were taught to pair neutral scenes with items on the hierarchy. The exercise was tape-recorded and the patient was given this tape locked in a player. The patients were told to practice their exercises at home once each day, but to avoid using the tape or exercise during sleep periods.

It is clear that the cognitive behavioral therapy intervention and the placebo intervention were not identical in appearance. In fact, the two treatments though vaguely similar were quite diverse, and the trial is more like a trial of two different, and differently appearing, interventions.

Psychological placebo control groups differ, but will rarely appear similar to the experimental intervention. Thus, though they are called "placebo groups," and have some similarities with standard placebo control groups, they are dissimilar, and in this title we use the term "placebo analogue control group."

In its most pragmatic, and primitive form, a placebo analogue control group consists of a pill placebo. For example, Walkup et al. compared the effect of cognitive behavioral therapy, sertraline, and combined therapy, with pill placebo (Table 2.1). Other trials compare psychological interventions with attention placebos, a kind of basic psychotherapy, often described as "neutral nondirective" void of any "specific" content. For example, in one trial of the effect of cognitive behavioral therapy for depression after stroke, the "attention placebo" was described as "a conversation that focused on day-to-day occurrences and discussions regarding the physical effects of stroke and life changes" [28].

Finally, a third group of placebo control groups have been laboriously developed as to be as similar as possible to an experimental psychological treatment. Such "authoritative" placebos, after having been successful as comparators to a highly specialized psychological intervention, are then sometimes used as controls for other, quite diverse, interventions. Examples are "quasi-desensitization" described earlier, and "pseudo-meditation" [29].

The basic idea behind such psychological placebo analogues is that patients are presented to a treatment with equal credibility as the experimental treatment, and often also equal patient–provider time, and thus positively control for these two important factors. Placebo analogue control groups are fundamentally different from the classic placebo control groups in many pharmacological or device trials, because they aim to control for specifically defined factors (typically patient–provider time and treatment credibility, or sometimes what is called "common factors"), whereas classic placebos aim to control for all known and unknown factors.

The major challenge is that placebo analogues cannot control for unknown factors. However, as long as patients can perceive the differences between the compared treatments, and have been informed that one is considered a placebo, there is a perceivable risk that patients may not regard the two treatments with equal credibility, or that the dynamics of the patient–provider interaction may be different in a placebo session as compared with a session of cognitive behavioral therapy. In many cases intervention credibility is assessed, but the sensitivity and reliability of such tests are often difficult to establish.

Those trying to design an improved psychological placebo control group more like device placebos face two problems. First, effects of placebo and effects of psychological interventions are both psychologically mediated, thus making a clear theoretical distinction between placebo and verum interventions very important. Second, to a larger extent than medicine and biology, psychology consists of overlapping, competing, and partly contradicting theories, thus making a clear theoretical distinction between placebo and verum controversial, and very dependent on a theoretical point of view. It is therefore often difficult to define a meaningful or uncontroversial psychological placebo intervention. However, in special situations, this is possible. One example is a trial of biofeedback for headache. Because biofeedback involves a device that functions as a psychological intervention, placebo and real biofeedback can be made to appear alike in almost every aspect of the treatment, and to differ only in the actual type of feedback [30].

The controversy between those in favor and those against using placebo-controlled trials in psychology seems futile. In many cases, placebo-controlled trials unfortunately cannot meaningfully be applied. There is no doubt that the risk of bias is considerable in psychological trials using other types of control groups, like pill placebos or attention placebos, especially as the vast majority use subjective outcomes. The risk of bias may be somewhat reduced as compared with no-treatment control groups, but that is a small comfort. Rifkin puts it aptly: "psychotherapy research is not for the faint of heart" [31].

2.6 Conclusion

Some nonpharmacological placebo control interventions are similar conceptually to pharmacological placebos, for example, many device placebos and many surgical placebo procedures. Methodologically surgical and device placebo trials will often involve a high risk of unblinding, especially acupuncture trials. Surgical placebo procedures are rare, and will often pose ethical problems, but are not unethical per se, and can probably be implemented much more commonly. Psychological placebo control interventions are generally of a different kind conceptually, as they often are designed to control for specific factors, like expectancy, and not as a tool for blinding. When designing a nonpharmacological trial, or when interpreting the result of such trials, the exact nature of the placebo intervention deserves considerable attention.

References

1. Diel, H.S., Baker, A.B., Cowan, D.V. Cold vaccines. An evaluation based on a controlled study. *JAMA*, 111, 1168–1173, 1938.
2. Grünbaum, A. The placebo concept in medicine and psychiatry. *Psychol Med*, 16, 19–38, 1986.
3. Hróbjartsson, A. What are the main methodological problems in the estimation of placebo effects? *J Clin Epidemiol*, 55, 430–435, 2002.
4. Chan, A.-W., Altman, D.G. Epidemiology and reporting of randomised trials published in PubMed journals. *Lancet*, 365, 1159–1162, 2005.
5. Beecher, H.K. The powerful placebo. *JAMA*, 159, 1602–1606, 1955.
6. Hróbjartsson, A., Gøtzsche, P.C. Placebo interventions for all clinical conditions. *Cochrane Database Syst Rev*, (1), CD003974, 2010.
7. Miller, F.G., Kaptchuk, T.J. Acupuncture trials and informed consent. *J Med Ethics*, 33, 43–44, 2007.
8. Madsen, M.V., Gøtzsche, P.C., Hróbjartsson, A. Acupuncture treatment for pain. Systematic review of randomised clinical trials with acupuncture, placebo acupuncture and no-acupuncture groups. *BMJ*, 338, a3115, 2009.
9. Lund, I., Lundeberg, T. Are minimal, superficial or sham acupuncture procedures acceptable as inert placebo controls? *Acupunct Med*, 24, 13–15, 2006.
10. Michels, K.B., Rothman, K.J. Update on unethical use of placebos in randomised trials. *Bioethics*, 17, 188–204, 2003.
11. Temple, R., Ellenberg, S.E. Placebo-controlled trials and active-control trials in the evaluation of new treatments: Part 1: Ethical and scientific issues. *Ann Intern Med*, 133, 455–463, 2000.
12. Carlson, R.V., Boyd, K.M., Webb, D.J. The revision of the Declaration of Helsinki: Past, present and future. *Br J Clin Pharmacol*, 57, 695–713, 2004.
13. World Medical Association Declaration of Helsinki. Ethical Principles for Medical Research Involving Subjects (8th Revision). Available at: www.wma. net/en/30publications/10policies/b3/17c.pdf (accessed August 15, 2011).
14. Laughren, T.P. The scientific and ethical basis for placebo-controlled trials in depression and schizophrenia: An FDA perspective. *Eur Psychiatry*, 16, 418–423, 2001.
15. Stang, A., Hense, H.W., Jöckel, K.H., Turner, E.H., Tramèr, M.R. Is it always unethical to use a placebo in a clinical trial? *PLoS Med*, 2, e72, 2005.
16. Dincer, F., Linde, K. Sham intervention in randomized clinical trials of acupuncture—A review. *Complement Ther Med*, 11, 235–242, 2003.
17. Streitberger, K., Kleinhenz, J. Introducing a placebo needle into acupuncture research. *Lancet*, 352, 364–365, 1998.
18. Beecher, H.K. Surgery as placebo. A quantitative study of bias. *JAMA*, 176, 1102–1107, 1961.
19. McCulloch, P., Taylor, I., Sasako, M., Lovett, B., Griffin, D. Randomised trials in surgery: Problems and possible solutions. *BMJ*, 324(7351), 1448–1451, 2002.
20. Hervé, C., Moutel, G., Meningaud, J.P., Wolf, M., Lopes, M. Controversies over the surgical placebo: Legal issues and the ethical debate. *Presse Med*, 29, 1180–1183, 2000.

21. Moseley, J.B., O'Malley, N.J., Petersen, N.J., et al. A controlled trial of arthroscopic surgery for osteoarthritis of the knee. *N Engl J Med*, 347, 81–88, 2002.
22. Freed, C.R., Greene, P.E., Breeze, R.E., et al. Transplantation of embryonic dopamine neurons for severe Parkinson's disease. *N Engl J Med*, 344(10), 710–719, 2001.
23. Freed, C.R., Vawter, D.E., Leaverton, P.E. Use of placebo surgery in controlled trials of a cellular based therapy for Parkinson's disease. *N Engl J Med*, 344, 988–991, 2001.
24. Macklin, R. The ethical problems with sham surgery in clinical research. *N Engl J Med*, 341, 992–996, 1999.
25. Horng, S., Miller, F.G. Is placebo surgery ethical? *N Engl J Med*, 347, 137–149, 2002.
26. Frank, J.D. The placebo is psychotherapy. *Behav Brain Sci*, 6, 291–292, 1983.
27. Herbert, J.D., Gaudiano, B.A. Introduction to the special issue on the placebo concept in psychotherapy. *J Clin Psychol*, (61), 787–790, 2005.
28. Lincoln, N.B., Flannaghan, T. Cognitive behavioral psychotherapy for depression following stroke: A randomized controlled trial. *Stroke*, 34, 111–115, 2003.
29. Blanchard, E.B., Schwarz, S.P., Suls, J.M., et al. Two controlled evaluations of multicomponent psychological treatment of irritable bowel syndrome. *Behav Res Ther*, 30, 175–189, 1992.
30. Rains, J.C., Penzien, D.B. Behavioral research and the double-blind placebo-controlled methodology: Challenges in applying the biomedical standard to behavioral headache research. *Headache*, 45(5), 479–486, 2005.
31. Rifkin, A. Randomized controlled trials and psychotherapy research. *Am J Psychiatry*, 164, 7–8, 2007.

3

Complexity of the Intervention

Patricia Yudkin and Rafael Perera

University of Oxford

Paul Glasziou

Bond University

CONTENTS

3.1 Introduction

3.1.1 What Is a Complex Intervention?

While we intuitively sense that psychological or social care interventions are more complex than prescribing a drug, there is no clear dividing line between complex and noncomplex interventions. The term is to some extent subjective, and there may be disagreement as to whether a particular intervention is complex or not, or whether it is very complex or mildly so.

We can generalize by saying that a complex intervention must include several components that act both independently and in interaction with others. The complexity of an intervention increases with the number of components, the number and difficulty of personal actions and interactions involved in them, and the degree of flexibility allowed in their delivery. Additional aspects adding to complexity are the number of groups targeted by the intervention, and the number and variability of outcomes [1,2]. All these features make the intervention difficult to design, to describe, to reproduce, and to evaluate [3]. Types of complex intervention range widely, and include not only those interventions in which the targeted individuals are active participants but also surgical and technical interventions [4,5].

Interventions that have a complex construction also tend to work in a complex way. Any effect may be mediated through a complicated pathway, which varies with the characteristics of the recipients, the skills of those delivering the intervention, and the setting and circumstances of its delivery. A complex intervention that works in one setting and at one time may not work at another; a complex intervention that is effective for one group of patients may not work with another [6]. While these features may also be true of some pharmacological treatments (e.g., multidrug cancer chemotherapy that repeatedly adjusts dose and cycle length based on individual response), they are more prominent in complex nonpharmacological treatments. One reason for this is that external factors (such as a change in the delivery of usual care or economic pressures) can have a stronger influence on the outcome than the intervention being tested. These issues are especially evident with participative interventions that aim to change beliefs or behaviors.

3.1.2 Developing a Complex Intervention

There are several stages in the development of a complex intervention, comprehensively explained in a publication from the U.K. Medical Research Council [2]. The process of development often requires a wide range of expertise, and a suitable team of investigators should be gathered at the outset. This will include experts in the relevant field of medicine (who may be clinicians, lay people, or other professionals), statisticians, administrators, and often economists, social scientists, and psychologists. Development of a complex intervention may take several years [7].

An intervention that is intended to affect beliefs or behavior should be consistent with relevant theory and with findings from qualitative research. A new qualitative study may be needed to develop the contents of the intervention or to test the effectiveness of theory-based techniques [8,9]. It is important to think through the way in which the intervention is intended to work and to examine whether the assumptions underlying the expected effect size are realistic [10]. If some parts of the intervention are new and untried, it is essential to pilot them before the trial. These novel aspects might include the process of recruitment and consent, the content of educational sessions, interview techniques, and booklets or questionnaires for patients. Finally, the trial should aim to standardize the delivery by clear and detailed specification in the protocol and appropriate pre-trial training of those delivering the intervention.

Two contrasting examples of trials of complex interventions are now described to illustrate the development and evaluation process.

3.1.3 Surgical Intervention: The MARS Trial

The mesothelioma and radical surgery (MARS) trial was a pilot randomized trial of radical surgery (extrapleural pneumonectomy [EPP]) in patients with malignant pleural mesothelioma [4]. A systematic review of EPP had found claims of longer than expected survival in surgical follow-up series, but there were no randomized trials [11]. Nevertheless, EPP was being offered to patients in hope of cure [12].

MARS compared chemotherapy plus EPP and radiotherapy against chemotherapy and usual oncological management. Usual management might include diagnostic and palliative surgery if appropriate, but not EPP. EPP is very major surgery, and was only claimed to be effective in highly selected patients and in the context of multimodality therapy. The aim of this pilot trial was mainly to assess the feasibility of the trial process. This included the recruitment of patients, the acceptability of randomization between major surgery and a palliative treatment known to have very little influence on survival, the acceptability of the trimodal treatment, and in view of such differences between the trial arms, compliance with randomized allocation.

To avoid pointless extensive surgery potential EPP patients needed a diagnostic operation (mediastinoscopy), which otherwise would not have been performed. Hence, the MARS trial had a two-stage consent process. At first consent, a potentially eligible patient was registered for the trial, but with the commitment to randomization, on the part of both patient and clinicians, remaining open. Staging of disease was then done by mediastinoscopy and biopsy, with optional staging also by positron emission tomography (PET). Patients eligible after staging received at least three cycles of platinum-based chemotherapy after which they were re-staged with computerized tomography (CT) and any uncertainties resolved by further investigation. Finally, a multidisciplinary team, including the referring clinician, met by teleconference to review the suitability of the patient for surgery and subsequent radiotherapy, using a standard set of criteria. Only then were eligible patients invited to consent to randomization. Thus, patients in both arms of the trial were balanced in terms of staging of disease and fitness to undergo radical surgery, and all had undergone chemotherapy.

All surgeons undertaking the EPP surgery had to satisfy the trial management group that they had both sufficient experience of this operation and satisfactory technical results. With 25 patients expected to be allocated to EPP, it was decided that not more than four or five surgeons should operate within the trial. A detailed protocol set out both the operative procedure and the procedure for postoperative radical radiotherapy. Some flexibility was allowed in purely "craft" considerations such as the choice of incision and details of suture technique and materials for reconstruction of the diaphragm and pericardium. Primary outcome was survival at 1, 2, and 5 years.

3.1.4 Participative Intervention: The DiGEM Trial

A quite different complex intervention—self-monitoring—was evaluated in the Diabetes Glycaemic Education and Monitoring (DiGEM) trial that took place in non-insulin treated patients with type 2 diabetes in primary care. This trial aimed to assess whether blood glucose self-monitoring, either alone, or with training in interpreting and using measurements to guide behavior, was more effective than usual care in improving glycemic control [13,14]. At the time, self-monitoring in these patients was increasing despite inconclusive evidence of its effectiveness, and the costs of the consumable test strips used in self-monitoring had risen to significant levels [15,16].

Previous qualitative studies suggested that in some patients awareness of fluctuations in blood glucose levels may promote adherence to medication, diet, and exercise regimens and thus improve glycemic control [17]. Furthermore, patients who understood the consequences of diabetes and the effectiveness of treatment were more likely to adhere to their recommended lifestyle [18,19]. This was in line with psychological theory (the common sense model) that postulates that the way in which people perceive threats to their health is a key factor in determining the strength of their efforts to minimize these threats [20]. These qualitative findings informed the content of the intervention.

The intervention was delivered by trained research nurses. Before randomization, all participants attended a 45 min assessment visit at which beliefs about diabetes were elicited and patients helped to understand how diabetes might be a threat to their health. The roles of diet, physical activity, and drugs were discussed according to the common sense model, and blood taken for HbA1c measurement to assess eligibility for the trial.

The intervention was initiated 2 weeks after the assessment visit, with follow-up visits 1, 3, 6, and 9 months later. For patients in all three groups, the initiation visit with the nurse took about 45 min and follow-up visits about 20 min. Throughout, patients in the usual care group were encouraged to set behavioral goals and to monitor and review physical activity, eating patterns, and medication. Their HbA1c was measured trimonthly by their doctor, and the results reviewed as showing the impact of these self-care activities.

The less-intensive self-monitoring group, in addition to goal setting and review, were shown how to carry out blood glucose self-monitoring according to a specified protocol. They were advised to consider contacting their doctor if readings were persistently high or low, but were not told how to interpret their blood glucose readings. The more intensive self-monitoring group, in addition to goal setting and review and blood glucose self-monitoring, were given training and support in understanding their blood glucose results and using them to enhance motivation and maintain adherence to diet, physical activity, and drug regimens. The primary outcome was HbA1c at 1 year.

3.2 Describing a Complex Nonpharmacological Intervention

3.2.1 General Principles

Trial interventions should be reproducible. If they cannot be reliably delivered to each trial participant, the source of any observed benefit (or harm) due to "the intervention" will be unknown. It will also be impossible to reliably transfer the intervention outside the trial setting. At the same time, the effectiveness of a complex intervention often depends on its adaptability to the differing needs and circumstances of patients and care providers. The protocol for a complex intervention must therefore balance standardization and uniformity with flexibility.

General guidance for describing nonpharmacological interventions is provided by the extended CONSORT statement for trials of nonpharmacological treatment [21]. This emphasizes the need to give precise details of both the experimental and control interventions. The components should be described, making clear how they were standardized, and if and how they were tailored to individual participants. In practice, this is difficult to achieve and there is ample evidence that reporting of complex interventions is often inadequate [9,22–25].

3.2.2 Pre-Trial Training

Many nonpharmacological interventions are delivered by people who need specific training for the purposes of the trial. Although this training is not literally a part of the intervention being evaluated, it is nevertheless an essential precursor to that evaluation, and must be described with as much care as the intervention itself.

A full description of any pre-trial training should be given, noting who was trained, who delivered the training, and what the training consisted of. Training programs should be based on manuals or other media that can be described and referenced [26,27].

3.2.3 Delivering the Intervention

The delivery of a complex intervention can be seen as a set of different activities carried out by those delivering the intervention, and often involving the participation of patients. These activities range widely in type and in their degree of complexity. They may include carrying out a delicate technical procedure—selective head cooling—on a sick newborn baby [5], giving a course of acupuncture to treat back pain [28], training patients to manage their own oral anticoagulant therapy [29], conducting a home-based rehabilitation program for patients with chronic fatigue syndrome [27], or teaching an exercise class [30].

The trial protocol must give the following information about each of these activities:

- Its "contents" (including materials and equipment used, and if relevant, the intensity of interaction between deliverer and recipient)
- The degree of flexibility allowed (including options and criteria for their use)
- Who performs it (including their training, expertise, and support received during the trial)
- Where it happens and at what time point(s) during the trial
- How long it takes

The contents of a relatively simple activity can be simply described. An exercise class with a structured program was part of the intervention evaluated in the trial of Mutrie et al. [30]. The contents of each class were: warm-up of 5–10 min, 20 min of exercise (e.g., walking, cycling, low-level aerobics, muscle-strengthening exercises, or circuits of specifically tailored exercises), and a cool down and relaxation period. Each class lasted for 45 min and women were monitored throughout the class to ensure that they were exercising at a moderate level (50%–75% of age adjusted maximum heart rate).

Many participative activities, such as educational sessions, discussion groups, face-to-face consultations, or telephone contacts, are more interactive and dynamic than a structured exercise class [7,13,27,31]. Consequently, they require more comprehensive and detailed description. The aim of any one contact session may be manifold, including imparting information, helping patients to set goals, giving support, motivating patients to stick to agreed goals, and so on. To ensure that all topics are covered and that each session is conducted in a uniform way for all participants, it must be based on a detailed protocol.

Where individual interviews or telephone calls are involved, scripts may be provided for facilitators, with reference notes to remind them of the intentions of each encounter. These aids help to ensure that sessions are standardized and reproducible.

The intensity and nature of interaction between facilitator and participants should also be described, together with the degree of flexibility allowed in the sessions. In describing flexibility, the available options and criteria for invoking them should be stated [32].

The requirements for certain treatments such as psychotherapy or acupuncture are somewhat different. These treatments are individually tailored to patients and their integrity rests on basic assumptions and techniques rather than on standard content. Practitioners may come from different traditions, have different training, and follow different principles, so the particular type of treatment being assessed must be clearly specified [28,33].

Delivery of an intervention will often require the use of certain equipment and this must be described in detail. The equipment may be central to the delivery of a treatment, such as acupuncture needles, sutures, or stents, or it may be a medical device for the patient's own use, such as a blood glucose monitor. Interventions that aim to change beliefs and behavior generally include material for patients, such as manuals, booklets, or computer programs. Educational sessions may include case vignettes, role play scenarios, and test exercises. In practice, there will rarely be enough space in a published work to describe all components of an intervention in detail, but supplementary material can be placed on the web (see, e.g., www.hta. ac.uk/1330 for material used in the DiGEM trial; [29] appendix).

3.2.4 Graphical Representation of a Complex Intervention

A complex intervention may be inadequately described on two levels: its basic design and structure, or the precise details of its components. Graphical representation enables the basic structure of an intervention to be clear and obvious.

Perera et al. [34] found that reports of complex intervention trials often failed to identify the components of experimental and comparator interventions. Also, the timing, duration, and frequency of delivery of different components were often not stated or were unclear. Thus, the points of difference between the interventions in the trial arms were obscured, making it difficult to interpret the trial results.

The components and timing of interventions within a trial may be clarified by depicting these graphically using a PaT (Parts and Timing) plot [34]. A tool for drawing the plot is available at http://www.cebm.net/index. aspx?o=4200. The interventions assessed in the MARS trial are shown in Figure 3.1 and those evaluated in the DiGEM trial shown in Figure 3.2. Each intervention is represented within its own column, with the intervention components shown as icons in that column. Pre-randomization components are shown as common to all participants. Components are conceptualized either as objects or activities, and represented by squares (objects) or circles (activities). Different objects are labeled with different letters and activities with numbers. A brief description of each component is given in the key following the diagram. The time scale of the trial runs from top to bottom, with the time of randomization and outcome measurement clearly shown. Components delivered concurrently are shown side by side, whereas those delivered consecutively are shown one beneath the other.

Graphical depiction of an intervention allows its structure to be quickly understood. Aspects that may be missed in a long verbal description are clearly shown, and with the experimental and control interventions placed side by side on the diagram, differences between them become obvious.

Timeline	Intervention—EPP	Control—No EPP
Registration, 1st consent		
Investigations, chemotherapy, and assessment of eligibility, approx. 4 months	(1) (2) (3) (4) (5)	
Randomization, 2nd consent		
Baseline, soon after randomization	(6)	
Approx. 2 months from baseline	(7)	(9)
Subsequently	(8)	
1, 2, 5 years from baseline	Primary outcome (survival)	

(1)	PET scan (recommended)
(2)	Mediastinoscopy
(3)	Three cycles of platinum-based chemotherapy
(4)	Repeat staging by CT
(5)	Multidisciplinary team meeting for final assessment of eligibility
(6)	Extrapleural pneumonectomy (EPP)
(7)	Radical radiotherapy
(8)	Any further oncological treatment considered appropriate
(9)	Oncological management including surgery if appropriate, but not EPP

FIGURE 3.1
MARS-1 trial: PaT plot.

The diagram can also be useful in the design and monitoring stages of a trial. It should help to ensure that the control intervention has been adequately considered and that the difference between the experimental and control arms is appropriate for measuring the effect of the intervention. It is a useful adjunct to the very detailed descriptions of interventions that are an essential part of the trial design, delivery, and reporting.

Timeline	Control	Meter: less intensive self-monitoring	Meter: more intensive self-monitoring
Pre-randomization nurse training	①		
Pre-randomization assessment visit	②		
Randomization			
Baseline, 2 weeks after assessment	③ a	④ a b c	⑤ b d
2 weeks after baseline			⑥
2 weeks before 3, 6, and 9 month consultation	⑦		
1, 3, 6, and 9 months from baseline	⑧	⑨	⑩
12 months	Measurement of outcomes		

①	Nurse training: 6 days of case-based training over 5 weeks.
②	Assessment visit, 45 min.
③	Nurse consultation, 45 min. Setting of goals and review of diet, physical activity and medication. Patients asked not to use a meter.
a	Action planning diary to record self care goals.
④	Nurse consultation, 45 min. Goal setting and review. Patients taught to use meter according to a specified protocol. Not told how to interpret blood glucose readings.
b	Blood glucose meter.
c	Blood glucose monitoring diary, to record blood glucose results.
⑤	Nurse consultation, 45 min. Goal setting and review. Patients taught to use meter and to interpret results. Encouraged to explore effect of specific activities on blood glucose levels.
d	Diary combining action planning and blood glucose monitoring. Each page sets goals, achievements, barriers, etc., against concurrent blood glucose results.
⑥	Phone call from nurse to reinforce use of monitor and discuss any problems.
⑦	Patient attends blood test at surgery to measure HbA1c.
⑧	Nurse consultation, 20 min. Goal setting and review. HbA1c results reviewed as showing impact of self-care activities on glycemic control.
⑨	Nurse consultation, 20 min. Blood glucose values reviewed as showing impact of self-care activities on glycemic control.
⑩	Nurse consultation, 20 min. Blood glucose values reviewed and diary used to encourage recognition of relationship between behavior and blood glucose results.

FIGURE 3.2
DIGEM trial: PaT plot.

3.3 Monitoring and Assessment of Protocol Fidelity

Even when an intervention has been specified to the correct level of detail, it cannot be assumed that its delivery will go according to plan. Any discrepancies between the treatment planned and that actually administered must be recorded; without this knowledge, it is impossible to interpret the trial results or to understand how the intervention could be improved in future.

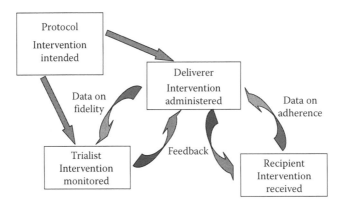

FIGURE 3.3
Monitoring of intervention fidelity and treatment adherence during trial.

The trialist must set up procedures for monitoring the delivery and reception of the intervention during the trial, and for correcting incipient lapses from protocol (Figure 3.3). This process evaluation may include both quantitative and qualitative methods [7,35,36].

In monitoring the intervention, the trial management committee and the Data Monitoring and Ethics Committee perform essential roles. Meeting at intervals throughout the trial, these committees review the trial's progress and recommend action if aspects of the protocol are in danger of slipping. They will ask, for example, whether the method of recruitment and consent is working well, or whether it needs adjusting in order to meet the projected sample size. They will question whether any of the protocol demands are proving untenable (such as weekly telephone calls to patients), and if so, whether they should be dropped. They will investigate whether the monitoring procedures to ensure that the intervention is being delivered according to plan are actually being applied, and will seek assurance that the interventions being compared are maintaining their distinctive features.

Participative interventions that involve contacts between patients and professionals require particularly careful overview. The contacts should (with patient consent) be recorded and independently assessed for protocol fidelity [7,37]. The difficulty of maintaining protocol fidelity in such interventions is illustrated by findings from the trial of a home-based program to increase physical activity among people at high risk of diabetes [38]. Trained facilitators delivered the intervention in a series of home visits and telephone calls to participants. The facilitators followed detailed protocols for all contacts, were supervised by a lead facilitator throughout the year-long program, and met fortnightly for team discussions. They also taped some contacts for quality assurance. In spite of these quality control measures, a study based on transcripts of a number of contacts found that fidelity to the protocol was only moderate, and lower than that reported by the facilitators themselves [39]. Furthermore, the level of fidelity fell during the course of the intervention.

The study highlighted the stringent demands of keeping to the protocol in a behavioral intervention of this sort. It identified 14 different techniques that facilitators were supposed to use, including building support, strengthening motivation, setting goals, planning action, self-monitoring, using rewards, reviewing goals, using prompts, and preventing relapse.

Where there is a relatively subtle distinction between two interventions within a trial, the danger of blurring or overlap should be recognized. Ward et al. [33] assessed the effectiveness of nondirective counseling and cognitive behavior therapy for treating depression. Samples of the therapeutic sessions throughout the intervention were taped, and rated by an independent auditor on a cognitive therapy rating scale, which assessed whether therapists made substantial use of cognitive techniques. In this trial, all the cognitive behavior sessions, but none of the counseling sessions, were assessed by the auditor as adequate cognitive therapy.

3.4 Monitoring and Assessment of Treatment Adherence

Interpretation of the trial results demands knowledge of both protocol fidelity on the part of those delivering the intervention and treatment adherence on the part of the patient (Figure 3.3). These two aspects interlock where the protocol instructions for the deliverers of the intervention concern treatment adherence. For example, one aspect of treatment adherence is attendance by patients at interviews or therapeutic sessions. The protocol may specify what action the deliverer of the intervention should take if a patient fails to attend (e.g., contact the patient again and make up to two further appointments). If the intervention requires the patient to use equipment (such as a coagulometer), the protocol will expect the deliverer of the intervention to ensure that the patient uses the equipment correctly. Where patients are asked to complete daily diaries, the protocol may include instruction to the deliverer to stress the importance of this activity and to encourage compliance. In all these examples, it is the deliverers of the intervention rather than the trialists who can identify signs of poor treatment adherence, and who are able to respond to them, following procedures set out in the protocol.

Whatever the level of treatment adherence that is achieved, it must be recorded at the end of the trial. As mentioned earlier, a simple measure of treatment adherence is attendance at group or individual sessions. However, patients will also be expected to involve themselves in the behaviors that the intervention promotes, and there is a need to record their level of involvement.

In a trial of self-management of oral coagulant therapy, international normalized ratio (INR) readings were supposed to be taken weekly. All readings were saved electronically in the portable coagulometers used by patients, providing a record both of the actual INR values and of the frequency of measurement [29]. In the DiGEM trial, self-monitoring patients were asked to measure three values of blood glucose daily on 2 days in the week. Their diary record of blood values provided an inbuilt measure of adherence to this protocol.

For many behaviors, however, there is no inbuilt mechanism for measuring the level of patient adherence, although devices such as pedometers and electronic pill bottle caps may prove useful. Validated questionnaires or diaries are often used to determine to what extent patients follow advice on medication, diet, and exercise [37,38,40]. The summary information provided by these instruments is, however, limited and a deeper understanding of a patient's involvement may be gained from qualitative research [35]. This approach was used in the Diabetes Manual trial, in patients with type 2 diabetes. The intervention aimed to develop patients' confidence in managing their own condition and reducing diabetes-related distress [7]. It was delivered by trained nurses, and was based on a manual that patients were expected to work through independently. Three support telephone calls were made by nurses during the trial, and the content of these calls was analyzed to examine the degree of engagement of patients with the program. The analysis identified three types of patient. "Embracers" reported adherence to the program and described changes in attitudes and behavior; "dippers" reported dipping into components of the program, and attempting behavior change that was not sustained; and "non-embracers" reported deriving no benefit from the program, whether or not they had read the manual.

3.5 Future Directions

Complex interventions as a specific group have been acknowledged only recently. Thanks to the MRC's seminal document in 2000 [41] and to their follow-up document published in 2008 [2], methods for developing and evaluating these interventions are improving rapidly and developing in scope.

Interventions that aim to change beliefs and behavior perhaps present the greatest challenge; in spite of advances in understanding how these interventions should be developed and assessed, they still fail to perform as well as is hoped [7,38]. Much work needs to be done to improve the way in which these interventions are designed, delivered, and reported. Descriptions of behavior change techniques need to be made more precise and standardized, so that interventions can be replicated faithfully, and the contribution of specific techniques to the effectiveness of an intervention can be examined [22]. We need to understand better how to tailor interventions effectively for different individuals. As electronically based interventions become more common and more individualized, they present an opportunity for exploring this question in a controlled context [32]. What is designed is not always what is delivered. As we have noted, process evaluation is essential for monitoring the delivery and reception of an intervention and should be a standard adjunct to a complex intervention trial [8,35].

Where theory suggests that behavior change is mediated by particular psychological changes, does it actually work that way in practice? Mediation analysis was used to explore this question in the DiGEM trial [37], but such analyses are rarely performed. Measures of theory-based antecedents of behavior change should be collected during a trial so that mediation analysis can be used to explore change mechanisms [9].

While some generic approaches have been developed, each discipline area may need to adapt and refine methods to improve the effectiveness, reporting, and reproducibility of their own complex interventions.

References

1. Craig, P., Dieppe, P., Macintyre, S., Michie, S., Nazareth, I., Petticrew, M. Developing and evaluating complex interventions: The new Medical Research Council guidance. *BMJ*, 337, a1655, 2008.
2. Medical Research Council. Developing and evaluating complex interventions: New guidance. www.mrc.ac.uk/complexinterventionsguidance, 2008 (accessed August 12, 2011).
3. Delaney, A., Angus, D.C., Bellomo, R., et al. Bench-to-bedside review: The evaluation of complex interventions in critical care. *Crit Care*, 12(2), 210–218, 2008.
4. Treasure, T., Waller, D., Tan, C., et al. The Mesothelioma and Radical Surgery randomized controlled trial: The MARS feasibility study. *J Thorac Oncol*, 4(10), 1254–1258, 2009.
5. Gluckman, P.D., Wyatt, J.S., Azzopardi, D., et al. Selective head cooling with mild systemic hypothermia after neonatal encephalopathy: Multicentre randomised trial. *Lancet*, 365(9460), 663–670, 2005.
6. Bonell, C., Oakley, A., Hargreaves, J., Strange, V., Rees, R. Assessment of generalisability in trials of health interventions: Suggested framework and systematic review. *BMJ*, 333, 346–349, 2006.
7. Sturt, J.A., Whitlock, S., Fox, C., et al. Effects of the Diabetes Manual 1:1 structured education in primary care. *Diabet Med*, 25(6), 722–731, 2008.
8. Lewin, S., Glenton, C., Oxman, A.D. Use of qualitative methods alongside randomised controlled trials of complex healthcare interventions: Methodological study. *BMJ*, 339, b3496, 2009.
9. Michie, S., Abraham, C. Interventions to change health behaviours: Evidence-based or evidence-inspired? *Psychol Health*, 19(1), 29–49, 2004.
10. Eldridge, S., Spencer, A., Cryer, C., Pearsons, S., Underwood, M., Feder, G. Why modelling a complex intervention is an important precursor to trial design: Lessons from studying an intervention to reduce falls-related injuries in elderly people. *J Health Serv Res Policy*, 10(3), 133–142, 2005.

11. Treasure, T., Sedrakyan, A. Pleural mesothelioma: Little evidence, still time to do trials. *Lancet*, 364(9440), 1183–1185, 2004.
12. Sugarbaker, D.J., Flores, R.M., Jaklitsch, M.T., et al. Resection margins, extrapleural nodal status, and cell type determine postoperative long-term survival in trimodality therapy of malignant pleural mesothelioma: Results in 183 patients. *J Thorac Cardiovasc Surg*, 117(1), 54–63, 1999.
13. Farmer, A., Wade, A., Goyder, E., et al. Impact of self monitoring of blood glucose in the management of patients with non-insulin treated diabetes: Open parallel group randomised trial. *BMJ*, 335(7611), 132–139, 2007.
14. Farmer, A., Wade, A., French, D., et al. on behalf of the DiGEM Trial Group. Blood glucose self-monitoring in Type 2 diabetes: A randomised controlled trial. *Health Technol Assess*, 13(15), 1–72, 2009.
15. Farmer, A.J., Neil, A. Variations in glucose self-monitoring during oral hypoglycaemic therapy in primary care (letter). *Diabet Med*, 22(4), 511–512, 2005.
16. Davidson, M.B. Counterpoint: Self-monitoring of blood glucose in type 2 diabetic patients not receiving insulin: A waste of money. *Diabetes Care*, 28(6), 1531–1533, 2005.
17. Peel, E., Parry, O., Douglas, M., Lawton, J. Blood glucose self-monitoring in non-insulin-treated type 2 diabetics: A qualitative study of patients' perspectives. *Br J Gen Pract*, 54(500), 183–188, 2004.
18. Hampson, S., Glasgow, R., Foster, L. Personal models of diabetes among older adults: Relationship to self-management and other variables. *Diabetes Educ*, 21(4), 300–307, 1995.
19. Hampson, S.E., Glasgow, R.E., Strycker, L.A. Beliefs versus feelings: A comparison of personal models and depression for predicting multiple outcomes in diabetes. *Br J Health Psychol*, 5(1), 27–40, 2000.
20. Leventhal, H., Nerenz, D.R., Steele, D.J. Illness representations and coping with health threats, in Baum, A., Taylor, S.E., Singer, J.E. (eds.), *Handbook of Psychology and Health*. Hillsdale, NJ: Erlbaum, pp. 219–252, 1984.
21. Boutron, I., Moher, D., Altman, D.G., Schulz, K.F., Ravaud, P. Extending the CONSORT statement to randomized trials of nonpharmacologic treatment: Explanation and elaboration. *Ann Intern Med*, 148(4), 295–309, 2008.
22. Abraham, C., Michie, S. A taxonomy of behavior change techniques used in interventions. *Health Psychol*, 27, 379–387, 2008.
23. Jacquier, I., Boutron, I., Moher, D., Roy, C., Ravaud, P. The reporting of randomized clinical trials using a surgical intervention is in need of immediate improvement: A systematic review. *Ann Surg*, 244(5), 677–683, 2006.
24. Glasziou, P., Meats, E., Heneghan, C., Sheppard, S. What is missing from descriptions of treatment in trials and reviews? *BMJ*, 336(7659), 1472–1474, 2008.
25. Ward, A.M., Heneghan, C., Peresa, R., et al. What are the basic self-monitoring components for cardiovascular risk management? *BMC Med Res Methodol.* 10, 105, 2010. Published online 2010 November 12. doi:10.1186/1471-2288-10-105.
26. Sturt, J., Hearnshaw, H., Farmer, A., Dale, J., Eldridge, S. The diabetes manual trial protocol—A cluster randomized controlled trial of a self-management intervention for type 2 diabetes. *BMC Fam Pract*, 7, 45, 2006.
27. Wearden, A.J., Riste, L., Dowrick, C., et al. Fatigue intervention by nurses evaluation—The FINE trial. A randomised controlled trial of nurse led self-help treatment for patients in primary care with chronic fatigue syndrome: Study protocol. *BMC Med*, 4, 9, 2006.

28. Leibing, E., Leonhardt, U., Köster, G., et al. Acupuncture treatment of chronic low-back pain—A randomized, blinded, placebo-controlled trial with 9-month follow-up. *Pain*, 96(1–2), 189–196, 2002.
29. Menéndez-Jándula, B., Souto, J.C., Oliver, A., et al. Comparing self-management of oral anticoagulant therapy with clinical management. *Ann Intern Med*, 142(1), 1–10, 2005.
30. Mutrie, N., Campbell, A.M., Whyte, F., et al. Benefits of supervised group exercise programme for women being treated for early stage breast cancer: Pragmatic randomised controlled trial. *BMJ*, 334(7592), 517–523, 2007.
31. Patterson, J., Barlow, J., Mockford, C., Klimes, I., Pyper, C., Stewart-Brown, S. Improving mental health through parenting programmes: Block randomised controlled trial. *Arch Dis Child*, 87, 472–477, 2002.
32. Lustria, M.L., Cortese, J., Noar, S.M., Glueckauf, R.L. Computer tailored health interventions delivered over the Web: Review and analysis of key components. *Patient Educ Counsel*, 74, 156–173, 2009.
33. Ward, E., King, M., Lloyd, M., et al. 2000. Randomised controlled trial of non-directive counselling, cognitive-behaviour therapy, and usual general practitioner care for patients with depression. I: Clinical effectiveness. *BMJ*, 321(7273), 1383–1388.
34. Perera, R., Heneghan, C., Yudkin, P. A graphical method for depicting randomised trials of complex interventions. *BMJ*, 334(7585), 127–129, 2007.
35. Oakley, A., Strange, V., Bonell, C., Allen, E., Stephenson, J. Process evaluation in randomised controlled trials of complex interventions. *BMJ*, 332(7538), 413–416, 2006.
36. Glasziou, P., Chalmers, I., Altman, D.G., et al. Taking healthcare interventions from trial to practice. *BMJ*, 341, c3852, 2010.
37. French, D.P., Wade, A.N., Yudkin, P., Neil, H.A.W., Kinmonth, A.L., Farmer, A.J. Self-monitoring of blood glucose changed non-insulin-treated type 2 diabetes patients' beliefs about diabetes and self-monitoring in a randomized trial. *Diabet Med*, 25(10), 1218–1228, 2008.
38. Kinmonth, A.L., Wareham, N.J., Hardeman, W., et al. Efficacy of a theory-based behavioural intervention to increase physical activity in an at-risk group in primary care (ProActive UK): A randomised trial. *Lancet*, 371(9606), 41–48, 2008.
39. Hardeman, W., Michie, S., Fanshawe, T., Prevost, A.T., McLoughlin, K., Kinmonth, A.L. Fidelity of delivery of a physical activity intervention: Predictors and consequences. *Psychol Health*, 23(1), 11–24, 2008.
40. Lautenschlager, N.T., Cox, K.L., Flicker, L., et al. Effect of physical activity on cognitive function in older adults at risk for Alzheimer disease. *JAMA*, 300(9), 1027–1037, 2008.
41. Medical Research Council. *A Framework for Development and Evaluation of RCTs for Complex Interventions to Improve Health*. London, U.K.: MRC, 2000.

4

Learning Curves

Jonathan A. Cook and Craig R. Ramsay

University of Aberdeen

CONTENTS

4.1 What Is a Learning Curve?

A learning curve can be defined simply as an improvement in performance over time. This improvement tends to be most rapid at first and then tails off over time. Learning curves have been observed for many health technologies [1] from relatively simple procedures such as fiber-optic intubation [2] to complex surgical procedures (for example, minimal access interventions [3]) and also the assessment of diagnostic technologies [1]. In surgery, this issue has been described as the most intractable of the obstacles to conducting randomized trials [4,5].

4.2 How Do Learning Curves Impede the Conduct of Randomized Controlled Trials?

Randomized controlled trials (RCTs) of skill-dependent interventions have been impeded by concerns that improvements in the technical performance over time (a "learning curve") may distort comparisons. When RCTs are successfully mounted, generalizability is questioned, especially when a new technically difficult intervention is compared with a standard approach. It has been argued that the trial results are not a "fair" reflection on the new interventions and what can be expected in clinical practice. Additionally, subsequent technological advancements may make the relevance of previous studies to clinical practice uncertain. Some of these concerns are illustrated in the trial example below.

4.3 Trial Example: Dutch Gastric Surgical Randomized Trial

The Dutch gastric surgical randomized trial compared the conventional surgery for D1 gastric cancer against the newer D2 surgery [6]. The trial showed a higher rate of complications for D2 surgery and concluded that without further evidence the newer D2 intervention should not be the standard surgery treatment for western countries. The trial provoked substantial correspondence with the most common and sustained criticism of the trial being related to the expertise the participating surgeons had in the D2 surgery. It was suggested that they were still learning the intervention during the trial and this accounted for at least some of the excess morbidity observed [7–10]. It was therefore argued that the comparison was flawed and the "true" performance of D2 surgery had not been assessed. The authors remained confident that their conclusion concerning the newer D2 surgery was correct and found no evidence of a learning effect in a subsequent analysis [6]. Shortly afterward, a single surgeon series of 38 D2 gastrectomies was published. As surgical experience was gained, the observed rate of complications reduced [11].

4.4 What Are the Possible Impacts of Learning on Randomized Trials?

Figure 4.1 illustrates three potential scenarios for how the existence of a learning process can impact upon an RCT assessment. Following the above trial example, a lower level of the performance indicator corresponds to a

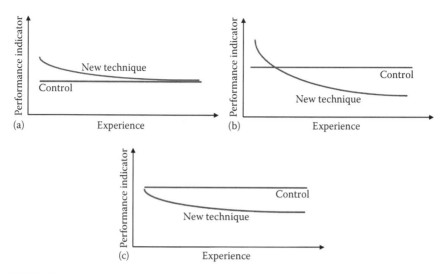

FIGURE 4.1
RCT learning curve scenarios. (From Cook, J.A. et al., *Clin Trials*, 1, 421–427, 2004.)

higher level of performance (e.g., complications). Under scenario (a), the new intervention (technique) has a lower level of performance compared with the control intervention. However, once sufficient expertise is acquired, it is as effective as the control. Under scenario (b), though the new intervention initially appears to be inferior to the control, once experience is gained, it has the superior performance. Potentially, this is the most concerning of the three scenarios where ignoring the presence of learning could potentially lead to a conclusion of no difference between the intervention. Under scenario (c), at first, the new intervention has a similar performance to the control but then improves beyond it as experience is gained; here a conclusion of benefit in favor of the new intervention may only be identified once adequate experience is obtained.

4.5 How Can We Measure Learning?

Measuring learning requires two items: a performance measure by which the impact of learning can be quantified and an indicator of the experience.

4.5.1 Choosing a Measure of Performance

Choosing a performance measure for learning is particularly difficult. The absence, for example, of reliable measures of surgical skill has been highlighted [12]. In general, there are two types of outcomes used: patient outcome and measures of intervention process [3]. Common learning measures used for surgery are complications and survival (patient outcomes), the amount of blood loss during an operation, and the length of operation (surgical process). For psychotherapy, it might be a self-report instrument that evaluated symptoms such as anxiety [13] and assessment of the relationship between the therapist and the client, respectively. As they are more readily collected and more directly related to learning, process outcomes are more convenient to collect and more often used to assess the impact of learning. However, process measures are only indirectly related to patient outcome [14] and therefore a suboptimal measure in terms of the impact of learning upon overall performance. Measures of learning can be confounded with patient health and the context within which the intervention is delivered. So, although assessment of the impact on patient outcome measures is clearly more desirable, it is more difficult to undertake. An area that has received little attention but is perhaps a key element in the learning process is that of nontechnical skills including decision making [15]. More research into identifying suitable measures is required.

4.5.2 Choosing an Indicator of Experience

The number of cases performed prior to the RCT is commonly used to quantify experience. There are a number of difficulties with such an approach, such as the contribution of cases where the delivery was supervised or only partly conducted by the interventionalist, the contribution of undertaking other related intervention and ascertaining a complete record of cases performed. The latter being surprisingly difficult to collect as it is often not formally recorded (e.g., in surgery). Nevertheless, restricting measured "experience" to the specific experience of the intervention under assessment is a practical and useful way to proceed. It should be acknowledged that measurement is conditional on previous training and experience and may not necessarily apply to another, otherwise similar, interventionalist. We consider the factors which influence individual learning using surgery though noting the relevance to other skill-dependent interventions.

4.6 Which Factors Influence Individual Learning?

Many factors contribute to an individual's learning process in addition to the number of cases performed [16,17]. A hierarchy of influences on individual performance can be hypothesized (see Figure 4.2). At the lowest level, the clinical community develops the intervention and informs the guidelines and protocols for the procedure. Above this, the institution is which an interventionalist belongs will impact upon the individual learning process. This can occur through better organized facilities, financial means, or the

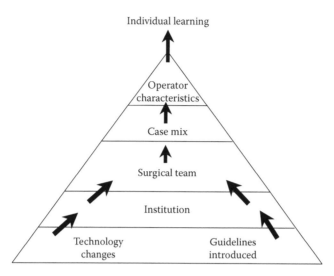

FIGURE 4.2
Pyramid of determinants of individual learning. (From Cook, J.A. et al., *Clin Trials*, 1, 421–427, 2004.)

training and supervision processes [18]. In particular, the makeup of the surgical team, roles, and experience will also contribute to the observed performance. If everyone in the team is relatively inexperienced and still learning their role in the operation, the performance is likely to be lower. Above this, we have "case mix," which denotes the clinical and patient characteristics of the person undergoing the procedure. There is often a process of loosening the "case mix" requirements for eligibility for surgery where the cases attempted become more difficult as the interventionalist becomes more experienced. As a consequence, outcomes may appear to be steady or even get worse if this is not taken into account. In principle, case mix should have less of an influence on the effect size of a randomized trial where inclusion criteria are defined a priori. The final influence in the pyramid is the characteristics of the interventionalist delivering the intervention. Factors such as attitude, natural abilities, capacity for acquiring new skills, and previous experience will all contribute as will personal circumstances to the performance in a particular case [3]. We note that as individuals have different characteristics and experience some degree of variation in learning between individuals should be anticipated.

On the basis of this conception of the learning process, it is useful to draw a distinction between two related though distinct learning curves: a community or technology learning curve and an individual learning curve; a community or technology learning curve, related to refinement of the new intervention (both technology and technique); second, personal learning of individual surgeons which, although impacted by the former, is mainly driven by their personal aptitude, training, and surgical experience. Further discussion of this can be found in Section 4.9.

4.7 How Are Trials Designed to Account for Learning Curves?

4.7.1 Traditional Approaches

The traditional approaches to designing trials of skill-based interventions has been either to provide intensive training and supervision [6], or to insist upon a minimum professional level (e.g., consultant) or to require participating interventionalists to perform a fixed number of procedures prior to participation in the trial [19]. For example, to participate in a trial comparing open and laparoscopic mesh repair of inguinal hernias, surgeons were required to have performed 10 laparoscopic interventions and to demonstrate their competence by undertaking another five with supervision from the trial collaborators. More recently, expertise-based trials have been proposed as another potential approach to account for differing levels of skill. Under this approach, the intervention will only be delivered by interventionalists who have expertise in that particular intervention. Therefore, a participant is randomized as usual, but the procedure is undertaken by an interventionalist skilled in only the randomized procedure. This potentially offers many benefits including interventionalists who are fully comfortable with the intervention they are delivering. However, some mechanism for determining those with sufficient "expertise" to deliver a particular intervention is still needed. Unlike in a conventional design, it is no longer possible to statistically control the interventionalist across groups. This leads to the potential for a differential in the interventionalists and also a clustering effect that requires an increase of sample size compared with the conventional trial [20]. Further discussion on this trial design can be found elsewhere in this book in Chapter 9. All these approaches attempt to measure the effect size toward the end of learning (the right-hand side of the graphs in Figure 4.1) and they do not provide information on the impact of learning per se.

4.7.2 Other Approaches

With regard to the design of trials, two possible options present themselves for protecting the trial from criticism. One approach would be to conduct a formal review of the literature to enable quantification of the learning curve for a new intervention and thereby inform trial design [21]. For example, a literature review was conducted on fiber-optic intubation where learning processes were widely acknowledged. A substantial number of studies were identified, which assessed this issue to some degree, though the practical value was limited by the general reporting—such as vague statements like "skilled" or "experienced." Nevertheless, it was possible to statistically quantify the main features of the learning curve for this procedure. An alternative complementary and more readily applied approach would be to explicitly record information on the expertise of interventionalists for the interventions under assessment [22,23]. RCTs conducted at different points in the evolution of an intervention may reasonably represent different levels of expertise. For example, a pragmatic trial would seek participating interventionalists who are representatives of the community in which the intervention will be used. As a consequence, this may require a relatively low level of procedural expertise. Conversely, an explanatory trial could seek to largely exclude learning by only including highly experienced surgeons. The results of trials where the intervention is conducted by the most experienced interventionalists in the high-caseload centers may not transfer readily to other settings.

Some criticisms of RCTs assessing skill-based interventions are ill-founded. Providing explicit information clarifies the applicability of the data and the likelihood or otherwise of routine clinical practice matching such a level. We recommend that RCT (and other study designs assessing learning) should report explicitly and informatively on expertise of participating interventionalists. Such information could be captured very readily through a questionnaire to participating interventionalist or through interventionalist participation forms at various stages of the RCT. Such an approach, potentially coupled with a statistical description of any learning curve effect within a trial, is a better way to address this problem [24].

4.8 Which Statistical Analyses Are Appropriate for Assessing Learning Curves in Trials?

Having recognized that the design of an RCT with learning effects requires careful consideration, we now consider the concomitant statistical analysis of such trials has until recently received little attention [14]. A systematic review, conducted in 2000, identified 272 studies that investigated learning curves in health technologies [1]. Only six studies (2%) used data from an RCT. The overall conclusions were that the statistical methods used were suboptimal and needed improvement. Several statistical approaches have subsequently been used which recognized that learning curve data is hierarchical in nature (e.g., cases nested within surgeon). A hierarchical model was used for a large trial assessing suturing material and technique for perineal repair found evidence of a learning effect in some outcomes [16]. A large surgical trial of over 2000 participants, which compared laparoscopic and open tension-free mesh repairs of inguinal hernias, found, in a post hoc analysis, a reduction in recurrence in the laparoscopic group with surgeon expertise [25]. Methodology for (observational) monitoring the introduction of a new surgical procedure has also been proposed [26]. A reduction in the 5-year recurrence rate of prostate cancer after radical prostatectomy was observed in a cohort of over 7000 cases. Similarly for laparoscopic radical prostatectomy, a sustained learning process was observed [27,28]. This reduction in recurrence seemed to continue even after a surgeon had performed a large number of procedures (250 cases).

In general, statistically identifying a learning curve has a high data requirement. To investigate learning in a trial requires systematic data collection of factors known to influence the learning curve. Such factors include performance measures for learning (process and/or outcome); the number and order of cases that an interventionalist performs in the study (i.e., the operative date and unique interventionalist identifier); and ideally the number of cases undertaken by each interventionalist before entering any patient into the trial and out with the study but during the study period. A hierarchical statistical model should be used to assess the impact of learning upon the trial effect size.

4.9 What Are the Implications for Technology Assessment of the Learning Curve?

Poor usage of the term "learning curve" in the clinical literature has in our opinion led to confusion on how "learning curve" interacts with "timing of technology assessment." Often "learning curves" are used to describe the poorer outcomes associated with an interventionalist's first cases, but little or no reference is made to the wider context of the diffusion of the technology [29]. Learning curves of interventionalists will be different depending upon where about on the technology diffusion curve an assessment has been undertaken. For example, we would expect the observed learning for an interventionalist to be greatest in the early stage of an intervention's development (early in the community learning curve), whereas in the later period such an individual will likely benefit from equipment developments, a more refined technique, and training support. Therefore, the interpretation of learning curve effects has to be made within the context that the evaluation takes place. It is only once the intervention technique and any associated technology, training, and other community factors have been well established, can a final assessment of the influence of learning as it pertains to routine clinical practice be made. Assessment prior to this point will be a "snapshot" of the current state and an RCT will likely be more explanatory than pragmatic in design. Whilst it is preferable to have RCTs at all stages of the technology development, this will often not be possible. When designing an RCT, it is therefore important to consider available learning curve data pertains to the stage of innovation under evaluation or an earlier stage. It should be noted that a learning effect is likely to influence cost as well as effectiveness [23,30].

4.10 Conclusion

Learning curves will continue to pose a challenge for rigorous assessments of skill-dependent interventions with no easy solution to address this issue. RCTs assessing skill-dependent interventions should be designed in light of potential learning effects. How this should be done will depend upon the research question being addressed. The impact of the learning process is likely to vary as an intervention becomes more established, which suggests two complementary evaluations. The first would focus on the benefits and costs of introducing the new intervention and the second on the benefits and costs of the new technology in a steady state. The likelihood of a learning curve should not prevent early and rigorous assessment of a new intervention.

References

1. Ramsay, C.R., Grant, A.M., Wallace, S.A., Garthwaite, P.H., Monk, A.F., Russell, I.T. Assessment of the learning curve in health technologies: A systematic review. *Int J Technol Assess Health Care*, 16, 1095–1108, 2000.
2. Smith, J.E., Jackson, A.P., Hurdley, J., Clifton, P.J. Learning curves for fibreoptic nasotracheal intubation when using the endoscopic video camera. *Anaesthesia*, 52, 101–106, 1997.
3. Cuschieri, A. Whither minimal access surgery: Tribulations and expectations. *Am J Surg*, 169, 9–19, 1995.
4. Russell, I. Evaluating new surgical interventions. *BMJ*, 311, 1243–1244, 1995.
5. Stirrat, G.M., Farrow, S.C., Farndon, J., Dwyer, N. The challenge of evaluating surgical interventions. *Ann R Coll Surg Engl*, 74, 80–84, 1992.
6. Bonenkamp, J.J., Songun, I., Hermans, J., et al. Randomized comparison of morbidity after D1 and D2 dissection for gastric cancer in 996 Dutch patients. *Lancet*, 345, 745–748, 1995.
7. Sakamoto, J., Yasue, M. Extensive lymphadenectomy for gastric cancer patients: What can the results of one trial tell us? *Lancet*, 345, 742–743, 1995.
8. Sue-Ling, H.M., Johnston, D. D1 versus D2 dissection for gastric cancer. *Lancet*, 345, 1515–1516, 1995.
9. McCulloch, P. D1 versus D2 dissection for gastric cancer. *Lancet*, 345, 1516–1517, 1995.

10. Guadagni, S., Catarci, M., de Manzoni, G., Kinoshita, T. D1 versus D2 dissection for gastric cancer. *Lancet*, 345, 1517, 1995.

11. Parikh, D., Johnson, M., Chagla, L., Lowe, D., McCulloch, P. D2 gastrectomy: Lessons from a prospective audit of the learning curve. *Br J Surg*, 83, 1595–1599, 1996.

12. Darzi, A., Smith, S., Taffinder, N. Assessing operative skill. Needs to become more objective. *BMJ*, 318, 887–888, 1999.

13. Okiishi, J.C., Lambert, M.J., Eggett, D., Nielsen, L., Dayton, D.D., Vermeersch, D.A. An analysis of therapist treatment effects: Toward providing feedback to individual therapists on their clients' psychotherapy outcome. *J Clin Psychol*, 62(9), 1157–1172, 2006.

14. Ramsay, C.R., Grant, A.M., Wallace, S.A., Garthwaite, P.H., Monk, A.F., Russell, I.T. Statistical assessment of the learning curves of health technologies. *Health Technol Assess*, 5, 1–79, 2001.

15. Yule, S., Flin, R., Paterson-Brown, S., Maran, N., Rowley, D. Development of a rating system for surgeons' non-technical skills. *Med Educ*, 40, 1098–1104, 2006.

16. Cook, J.A., Ramsay, C.R., Fayers, P. Statistical evaluation of learning curve effects in surgical trials. *Clin Trials*, 1, 421–427, 2004.

17. Huesch, M.D., Sakakibara, M. Forgetting the learning curve for a moment: How much performance is unrelated to own experience? *Health Econ*, 18, 855–862, 2008.

18. Bull, C., Yates, R., Sarkar, D., Deanfield, J., de Leval, M. Scientific, ethical, and logistical considerations in introducing a new operation: A retrospective cohort study from paediatric cardiac surgery. *BMJ*, 320, 1168–1173, 2000.

19. Wellwood, J., Sculpher, M.J., Stoker, D., et al. Randomised controlled trial of laparoscopic versus open mesh repair for inguinal hernia: Outcome and cost. *BMJ*, 317, 103–110, 1998.

20. Roberts, C. The implications of variation in outcome between health professionals for the design and analysis of randomized controlled trials. *Stat Med*, 18(19), 2605–2615, 1999.

21. Cook, J.A., Ramsay, C.R., Fayers, P. Using the literature to quantify the learning curve. *Int J Technol Assess Health Care*, 23, 255–260, 2007.

22. Devereaux, P.J., Bhandari, M., Clarke, M., et al. Need for expertise based randomised controlled trials. *BMJ*, 330, 88, 2005.

23. Cook, J.A. The design, conduct and analysis of surgical trials. *Trials*, 10, 9, 2009.

24. Medical Research Council. *Health Technology Assessment in Surgery: The Role of the Randomized Controlled Trial*. London, U.K.: Medical Research Council, 1994.

25. Neumayer, L., Giobbie-Hurder, A., Jonasson, O., et al. for the Veterans Affairs Cooperative Studies Program. Open mesh versus laparoscopic mesh repair of inguinal hernia. *N Engl J Med*, 350, 1819–1827, 2004.

26. Gorst-Rasmussen, A., Spiegelhalter, D.J., Bull, C. Monitoring the introduction of a surgical intervention with long-term consequences. *Stat Med*, 26(3), 512–531, 2007.

27. Vickers, A.J., Bianco, F.J., Serio, A.M., et al. The surgical learning curve for prostate cancer control after radical prostatectomy. *J Natl Cancer Inst*, 99, 1171–1177, 2007.

28. Vickers, A.J., Savage, C.J., Hruza, M., et al. The surgical learning curve for laparoscopic radical prostatectomy: A retrospective cohort study. *Lancet Oncol*, 10, 475–480, 2009.

29. Rogers, E.M. *Diffusion of Innovations*. Glencoe, Scotland: Free Press, 1962.

30. Bonastre, J., Noel, E., Chevalier, J., et al. Implications of learning effects for hospital costs of new health technologies: The case of intensity modulated radiation therapy. *Int J Technol Assess Health Care*, 23(2), 248–254, 2007.

5

Clustering Effects in RCTs
of Nonpharmacological Interventions

Jonathan A. Cook and Marion K. Campbell

University of Aberdeen

CONTENTS

5.1 Introduction

Most randomized controlled trials assume that the outcomes for each participant are independent—that is, it is unrelated to the outcome for any other participant. For a trial involving nonpharmacological treatments (NPTs), this assumption is at least open to debate and in some cases untenable. For trials involving skill-dependent interventions such as those delivered by a therapist or surgeon, it is realistic to expect participants under the care of the same health professional to be influenced in a similar manner as a result of the health professional practice, skill, and experience in delivering the intervention (see Chapter 4 for further discussion of the impact of experience). As a consequence, participants under the treatment of the same health professional are more likely to have similar outcomes than those under the care of another health professional; their outcomes can no longer be assumed to be independent. This phenomenon is referred to as the *clustering effect* [1]. Such an effect is also common in multicenter trials where the outcome for those who receive an intervention within the same center (e.g., hospital) is more similar than those in another center due to the policy and practices of the center and/or the interactions between individuals within the same center (e.g., for group therapy or knowledge-based interventions) [1–3]. The presence of a clustering effect has direct consequences for the design and analysis of trials, as the correlation between individual participant outcomes has to be accounted for. Both cluster randomized and expertise-based trials (see Chapters 8 and 9) explicitly incorporate clustering into the trial design though in different ways and for different reasons. The issue of clustering can also impact upon individually randomized trials [3,4]. In this chapter, we will describe how to measure the strength of the similarity (clustering) between individuals, give examples of its magnitude, and show how to account for its effect in the planning and analysis of NPT trials. Its impact in an NPT trial will be presented followed by a conclusion.

5.2 How to Measure the Clustering Effect

5.2.1 Definition of a Clustering Effect

The statistical measure of the clustering between participants under the care of a therapist/surgeon is known as the *intracluster correlation coefficient* (commonly described by the Greek letter rho or ρ) [5], the ICC.

The ICC can be defined as the proportion of the total variation in the participant outcome that can be attributed to the difference between clusters (e.g., health professional), and is described mathematically as

$$\rho = \frac{\sigma_b^2}{\sigma_b^2 + \sigma_w^2}$$

where
   σ_b^2 is the between-cluster (e.g., between-health professional) variance component
   σ_w^2 is the within-cluster (e.g., within health professional) variance component [6]

Using this definition, the ICC takes a value of between 0 and 1; the nearer the value is to 1, the greater the clustering effect is. A value of zero indicates no correlation (i.e., no clustering) and independent of observations.

If clustering exists, this has direct implications for the sample size calculations and the analysis approach that requires to be adopted. This is true whether individual or cluster randomization is adopted. The primary reason for this is that standard sample size calculations and analysis techniques also assume that outcomes for individual participants will be independent and consequently they will incorrectly estimate (typically underestimate) the true sample size required to detect a pre-specified difference with the desired accuracy and power.

5.2.2 Calculating the Intracluster Correlation Coefficient

There have been a variety of methods put forward to calculate the ICC. When the ICC is to be estimated from continuous data, the most common method adopted is the analysis of variance (ANOVA) approach [6]. This method calculates the ICC directly from the two components of variation in outcome in a clustered trial—the variation among individuals within a cluster (e.g., surgeon or therapist) and the variation between clusters. These two sources of variation can be separately identified using a standard one-way ANOVA, with the cluster level variable (the variable identifying the cluster an observation belong to) as a random factor [7]. The following variance component estimates can then be identified from the ANOVA table:

$$\hat{\sigma}_b^2 = \frac{(\text{MSB} - \text{MSW})}{n_0}$$

and

$$\hat{\sigma}_w^2 = \text{MSW}$$

where
   MSB is the between-cluster mean square
   MSW is the within-cluster mean square and
   n_0 is the "average" cluster size

The 'average' cluster size is calculated using

$$n_0 = \frac{1}{J-1}\left(N - \sum_J n_j^2\right)$$

where
   J is the number of clusters
   N is the total number of individuals and
   n_j is the number of individuals in the jth cluster

The ICC can then be estimated as

$$\hat{\rho} = \frac{\hat{\sigma}_b^2}{\hat{\sigma}_b^2 + \hat{\sigma}_w^2}$$

Random effects models can also be used to produce estimates for these between-cluster, σ_b^2, and within-cluster, σ_w^2, components of variation, and hence the ICC. Additionally, they enable adjustment of covariates that might improve estimation of the ICC.

For binary data, while a range of methods have also been suggested [8–10], the ANOVA approach is again commonly used to provide a point estimate as the ANOVA approach has been shown not to require any strict distributional assumptions [7]. While the ANOVA estimate of the ICC may be used with dichotomous data, standard methods to calculate confidence intervals around this estimate may not be reliable in the dichotomous case.

5.2.3 Calculating Confidence Intervals for an ICC

There are a number of methods available for calculating confidence intervals around an ICC for a continuous outcome. Donner and Wells [11] reviewed a range of different approaches for use when clusters are small and recommended the method described by Smith [12] as providing the best approach. This method is based upon a large sample approximation to the variance of the ICC, and can be implemented through many standard statistical packages such as Stata [13]. For cases where cluster sizes are larger, Ukuomunne [14] showed that the methods based on the variance ratio statistics, such as Searle's method [15], provided the most accurate coverage.

No corresponding exact formulae exist for the calculation of confidence intervals for ICCs calculated from binary data. The use of methods that make a normal distribution assumption (such as Searle's) is likely to lead to misleading CI values. Computational techniques, such as "bootstrapping," which use the study data to generate empirical estimates of the appropriate confidence limits, have, however, been advocated for confidence interval generation in other fields where similar problems have arisen [16,17] and been used by a number of authors to calculate confidence limits for ICCs. It is important that the bootstrap sampling method acknowledges the cluster structure of the data. Field and Welsh [18] give a technical assessment of different bootstrap sampling options for clustered data. Ukuomunne compared a range of possible bootstrap methods to generate confidence intervals for ICCs generated from binary data and showed that a particular variation (the bootstrap-*t* method) provided the best coverage [19] for non-normal continuous outcome. More recently, the use of parametric and nonparametric bootstrap methods were compared for hierarchical binary outcomes [20]. The estimation for multiple levels of clustering was investigated.

The nonparametric bootstrap method performed better that the parametric approach as did bootstrap sampling at the highest level of clustering rather than lower level. Bootstrap techniques can also be implemented through many standard statistical packages.

5.3 Typical ICC Values

ICC values are rarely reported in trial and observation publications. The size of dataset needed to calculate individual with a high level of precision is very large [7]. Table 5.1 gives ICC values for a number of variables from observational and randomized controlled trials where an NPT was used.

Basic information (population, setting, level, number, and mean size of clusters) is given on the measure and the dataset used to generate the ICC estimate. For each ICC, a 95% CI was generated (using bias correction bootstrap method unless otherwise stated in the table) were generated from individual patient data. It [7,21] has been shown that lower ICC values were higher for secondary over primary care and process measure over patient or clinical outcomes. The longer the time after the intervention was delivered the smaller the ICC is likely to become from the intervention would also seem to lead to a larger ICC as suggested by the STARS trial data where the point estimate of ICC reduces over time. Of particular note is the work by Roberts and Roberts where evidence for a different ICC by intervention is presented and a large ICC value was observed for the group therapy intervention as perhaps might be expected [22]. Reporting of ICC by intervention group for a range of measures for various NPT is needed.

5.4 Adjusting for the Clustering Effect in Sample Size Calculations

5.4.1 General Comment

If clustering is known or expected to be present in a trial, it should also be accounted for in the sample size calculations. As mentioned earlier, the primary reason for this is that standard sample size calculations assume that outcomes for individual participants will be independent of their setting (or health professional delivering their intervention). If a standard sample size calculation is used in the presence of clustering, it will generally (see the following for an exception) underestimate the true sample size required

TABLE 5.1

Intracluster Correlations Coefficients (ICCs) for NPT Interventions

Study	Setting/Population	Measure (Binary/Continuous)	Cluster Details — No., Mean (Range)	Cluster	ICC (95% CI)*	Data Source (Follow-Up)
COGENT	Primary care/asthma patients	Asthma symptom score (C)	62, 47.7 (32, 69)	Practice	0.069 (0.039, 0.099)	Pre-intervention (NA)
TEMPEST study	Secondary care	Length of stay (C)	20 (14, 30)	Hospital directorates	0.065 (0.016, 0.159)	Audit data (NA)
Group therapy trial	Secondary care/patients with schizophrenia and related psychological disorders	PANSS score (C)	10, 5.6 (5, 6)	Therapy group	+0.268 (0.022, 0.372)	Intervention group (12 months)
Nurse practitioner trial	Primary care/same day appointment patients	MISS/PMISS satisfaction score (C)	20, 30 (26, 36) [20 General practices]	Nurse practitioners	+0.044 (0.004, 0.096)	Intervention group (post-consultation)
Nurse practitioner trial	Primary care/same day appointment patients	MISS/PMISS satisfaction score (C)	71, 8 (2, 29) [20 General practices]	General practitioners	+0.116 (0.050, 0.186)	Intervention group (post-consultation)
Scottish Hip Fracture Audit	Secondary care/patients with hip fractures	Death (B)	12, 187.3 (151, 818)	Hospital	0.005 (0.001, 0.014)	Audit data (4 months)
Scottish Hip Fracture Audit	Secondary care/patients with hip fractures	Pain rating (B)	12, 199.6 (85, 567)	Hospital	0.061 (0.036, 0.104)	Audit data (4 months)
STARS trial	Secondary care/patients with displaced subcapital hip fractures	Operation time (C)	49, 6.1 (1, 96)	Randomizing surgeon	0.093 (0.003, 0.222)	Intervention group (NA)
STARS trial	Secondary care/patients with displaced subcapital hip fractures	Eq5d (C)	48, 5.8 (1, 92)	Randomizing surgeon	0.015 (0.000, 0.109)	Intervention groups (4 months)

Notes: B and C stands for binary and continuous respective; * ICC calculated using ANOVA method and 95% bias corrected bootstrap 95% confidence interval generated unless otherwise stated; + ICC and 95% were calculated using heteroscedastic model (see Ref [22] for further detail).

to detect a pre-specified difference with the desired accuracy and power. Cluster randomized and individually randomized trial designs will be considered separately.

5.4.2 Cluster Randomized Trials

We will focus upon the completely randomized case for cluster randomized trials. To accommodate for the clustering effect, standard sample size estimates require being inflated by a factor:

$$1 + (\bar{n} - 1)\rho$$

where

$\bar{n}$ is the average cluster size (e.g., average number of individuals per health professional) and

ρ is the estimated ICC (implicitly assuming the ICC will be the similar in both groups) [23]

This factor is commonly known as the "design effect" or the "variance inflation factor" [1]. Minor variations in the sample calculation exist for specific analyzes and to allow for uneven cluster sizes [24], which will generally result in a larger total sample size beyond the aforementioned formula. Formulas for stratified and matched–paired designs also exist [25,26].

It should be noted that the impact of clustering on the required sample size can be substantial even when the ICC is numerically small (this is because *both* the ICC and the cluster size influence the calculation). For example, even assuming a small ICC (e.g., 0.05), the sample size would have to be doubled for a cluster size as low as 20. Sample size calculators and published ICCs are available to aid sample size calculations in the presence of clustering [7,27,28].

5.4.3 Individually Randomized Trials

Where the clustering can be reasonable assumed to be consistent across interventions (i.e., similar ICC the use of stratification or minisation mininisation by clustering factor (e.g., health professional) in randomization will maximize the efficiency by maintaining even group sizes [22]), the same design factor can be used to adjust the sample size of individually randomized trials for clustering for binary outcome. The use of stratification or minimization by clustering factor (e.g., health professional) in randomization will maximize the efficiency by maintaining even group sizes. As derived by Vierron and Giraudeau [29], for a continuous outcome with even group sizes the design effect becomes

$$1 - \rho$$

ρ is the estimated ICC.

This leads to less uncertainty around the treatment effect in comparison with a standard analysis. This is unlikely to hold for a binary or ordinal outcome and only for a continuous outcome if the analysis appropriately accounts for the clustering [30]. This result holds when the group sizes are uneven if they are consistent across clusters (e.g., number of participants for one intervention is always twice the number for the other). Where there is variation in the proportion of individuals in an intervention group, this formula does not hold. Instead, the following formula is needed [29]:

$$1+(n_0-1)\rho$$

where $n_0 = \dfrac{n_1 n_2}{N} \displaystyle\sum_j \left(\dfrac{m_{1j}}{n_1} + \dfrac{m_{2j}}{n_2} \right)^2$

where m_{1j}, m_{2j}, n_1, and n_2 are the number of individuals in groups 1 and 2 respectively that are in cluster j and, J, P, and N are defined as previously.

As this variation increases, the true design effect increases toward the cluster randomization design effect and therefore can also result in a larger sample size (design effect greater than one) than with a conventional (non-clustered) design. Under an expertise-based trial design where the number of clusters and average cluster size are similar across treatment groups, the design effect given earlier for a cluster randomized trial can be used. Walter et al. suggest that it may be reasonable to expect a larger treatment effect in an expertise-based trial [31].

The aforementioned sample size calculations presuppose that the clustering effect is similar for both intervention groups. However, for an individually randomized trial comparing substantially different interventions (e.g., group psychotherapy versus individual psychotherapy), the intracorrelation and/or cluster size is likely to vary between interventions. It may be reasonable to expect clustering for only one intervention (e.g., surgery versus medical intervention). Alternatively, it may not be feasible to control cluster sizes. Such scenarios leads to different variances between intervention groups and as a result uneven randomization may provide improved efficiency and therefore smaller sample size [22]. Models which allow for the treatment effect to vary by cluster have also been proposed which would typically lead to an inflation in the sample size [30].

5.5 Adjusting for the Clustering Effect in the Trial Analysis

As with the sample size calculations, the analysis should also take account of any clustering. If the clustering is ignored, the analysis will return confidence intervals and corresponding *p*-values that are artificially small thus increasing the changes spuriously concluding that an intervention is significantly different to a comparator (when, in truth, it is not). The general principle of analyzing as randomizing (and vice versa in terms of planning) is important though some exceptions exist [32,33]. The analysis approach depends upon a number of factors including the level at which inference will be made, the study design, the type of outcome (whether binary or continuous), and the number of clusters in the trial and how many participants are linked to each.

There are three main approaches for the analysis of data arising from a clustered design—analysis at the cluster level, adjustment of standard tests to accommodate for the clustering in the data or more advanced analysis which uses the data from the individual level while accounting for the clustering of data. The simplest approach is to adopt a cluster level analysis, which essentially involves calculating one summary measure for each cluster such as a cluster mean (e.g., average length of stay per center) or proportion (e.g., proportion of successes per surgeon), and conducting an analysis using standard statistical tests. This approach is appropriate when the inference is intended to be at the cluster level. However, it is more often the case that inference is intended to be at the individual participant level. Specialist statistical techniques have been developed to allow this—techniques such as random effects (or multilevel) modeling and generalized estimating equations, which allow the hierarchical nature of the data to be incorporated directly into the analysis [7,34]. Alternatively, the estimate of the standard error from a standard statistical technique can be modified in order to compensate for the impact of cluster. Models where clustering is allowed to differ between treatments have also been suggested [22,30,35]. It is beyond the scope of this chapter to outline the specifics of each of these methods in detail, but a worked example of the methods is provided using an example trial—the KAT trial [36] to illustrate the impact adjusting for clustering.

5.6 Examples of the Clustering Effect Observed in Previous NPT Trials

The Knee Arthroplasty Trial (KAT) evaluated the effectiveness of three treatment options for knee replacement surgery: metal backing, patellar resurfacing, and mobile bearing [35]. KAT was an individually randomized trial with three randomized comparisons comprising 409, 1715, and 539 participants, respectively. Randomization was performed using a minimization algorithm, which included the recruiting surgeon. Data were collected 2 years after the operation is presented in Table 5.2 for an adjusted and unadjusted analysis for the Oxford Knee Score, the primary outcome for functional status.

TABLE 5.2

Comparison of Statistical Analysis Adjusted and Unadjusted for Clustering by Surgeon—Oxford Knee Score at 2 Years from KAT Trial

Comparison	Statistical Analysis	Treatment Effect (95% CI)
Metal backing (14 surgeons)		
	Unadjusted	1.33 (−0.91, 3.57)
	Adjusted for clustering	1.33 (−1.44, 4.10)
Patellar resurfacing (83 surgeons)		
	Unadjusted	0.27 (−0.86, 1.39)
	Adjusted for clustering	0.27 (−1.00, 1.53)
Mobile bearing (23 surgeons)		
	Unadjusted	0.13 (−2.20, 1.93)
	Adjusted for clustering	0.13 (−1.88, 1.61)

Two analyzes are presented for each of the three comparison: a linear regression model adjusted for the minimization factors bar surgeon (unadjusted) and an adjusted analysis where clustering by surgeon is addressed by using Huber–White standard errors. Analyzes were performed using Stata software. As expected using this approach, the adjusted and unadjusted treatment effects vary only in the confidence intervals. Interestingly, for the mobile-bearing comparison, the confidence interval for the adjusted analysis is narrower than the corresponding interval for the unadjusted analysis. For the other two comparisons this is not the case. In this example, clustering had little impact upon the estimated treatment effect though this may not be the case for other outcome or settings.

5.7 Conclusion

The potential impact of clustering can be substantial and should be given consideration in the design, conduct, and analysis of NPT trials. We have highlighted how sample size and the analysis approach can be modified to account for clustering. Clustering will also have an impact upon the conduct of trial regarding recruitment of cluster units (e.g., school or health professional) and data collection. While the impact on cluster randomized trial has been more widely recognized, there is a need for similar refinement of approach for individually randomized trials (including expertise-based trials) where clustering exists. Planning of NPT trials is hindered by a lack of reported ICCs particularly for individually randomized trials. The issue of uneven clustering between interventions has recently been highlighted and is a particular concern for some comparisons.

References

1. Donner, A., Klar, N. *Design and Analysis of Cluster Randomized Trials in Health Research*, 1st edn. London, U.K.: Arnold, 2000.
2. Senn, S. Some controversies in planning and analysing multi-centre trials. *Stat Med*, 17, 1753–1765, 1998.
3. Roberts, C. The implications of variation in outcome between health professionals for the design and analysis of randomized controlled trials. *Stat Med*, 18(19), 2605–2615, 1999.
4. Lee, K.J., Thompson, S.G. Clustering by health professional in individually randomised trials. *BMJ*, 330, 142–144, 2005.
5. Campbell, M.K., Grimshaw, J.M., Piaggio, G. Intraclass correlation coefficient, in D'Agostino, R.B., Sullivan, L., Massaro, J. (eds.), *Wiley Encyclopaedia of Clinical Trials*. Chichester, U.K.: Wiley, 2008.
6. Donner, A., Koval, J.J. The estimation of intraclass correlation in the analysis of family data. *Biometrics*, 36, 19–25, 1980.
7. Ukoumunne, O.C., Gulliford, M.C., Chinn, S., Sterne, J.A.C., Burney, P.G.J. Methods for evaluating area-wide and organization based interventions in health and health care: A systematic review. *Health Technol Assess*, 3(5), iii-92, 2000.
8. Moore, D.F., Tsiatis, A. Robust estimation of the variance in moment methods for extra-binomial and extra-Poisson variation. *Biometrics*, 47, 383–401, 1991.
9. Feng, Z., Grizzle, J.E. Correlated binomial variates: Properties of estimator of intraclass correlation and its effect on sample size calculation. *Stat Med*, 11, 1607–1614, 1992.
10. Ridout, M.S., Demetrio, C.G.B., Firth, D. Estimating intraclass correlation for binary data. *Biometrics*, 55, 137–148, 1999.
11. Donner, A., Wells, G. A comparison of confidence interval methods for the intraclass correlation coefficient. *Biometrics*, 42, 401–412, 1986.
12. Smith, C.A.B. On the estimation of intraclass correlation. *Ann Hum Genet*, 21, 363–373, 1956.
13. StataCorp. *Stata Statistical Software: Release 9*. College Station, TX: StataCorp LP, 2005.
14. Ukoumunne, O.C. A comparison of confidence interval methods for the intraclass correlation coefficient in cluster randomized trials. *Stat Med*, 21, 3757–3774, 2002.
15. Searle, S.R. *Linear Models*. New York: Wiley, 1971.
16. Davision, A.C., Hinkley, D.V. *Bootstrap Methods and Their Application*. Cambridge, U.K.: Cambridge University Press, 1997.
17. Campbell, M.K., Torgerson, D.J. Bootstrapping: Estimating confidence intervals for cost-effectiveness ratios. *QJM*, 92, 177–182, 1999.
18. Field, C.A., Welsh, A.H. Bootstrapping clustered data. *J R Stat Soc B*, 69(Part 3), 369–390, 2007.
19. Ukoumunne, O.C., Davison, A.C., Gulliford, M.C., Chinn, S. Non-parametric bootstrap confidence intervals for the intraclass correlation coefficient. *Stat Med*, 22, 3805–3821, 2003.

20. Ren, S., Yang, S., Lai, S. Intraclass correlation coefficients and bootstrap methods of hierarchical binary outcomes. *Stat Med*, 25, 3576–3588, 2006.

21. Campbell, M.K., Fayers, P.M., Grimshaw, J.M. Determinants of the intracluster correlation coefficient in cluster randomized trials: The case of implementation research. *Clin Trials*, 2, 99, 2005.

22. Roberts, C., Roberts, S.A. Design and analysis of clinical trials with clustering effects due to treatment. *Clin Trials*, 2, 152, 2005.

23. Donner, A., Birkett, N., Buck, C. Randomization by cluster. Sample size requirements and analysis. *Am J Epidemiol*, 114(6), 906–914, 1981.

24. Eldridge, S.M., Ashby, D., Kerry, S. Sample size for cluster randomized trials: Effect of coefficient of variation of cluster size and analysis method. *Int J Epidemiol*, 35, 1292–1300, 2006.

25. Hsieh, F.Y. Sample size formulae for intervention studies with the cluster as unit of randomisation. *Stat Med*, 7, 1195–1201, 1988.

26. Shipley, M.J., Smith, P.G., Dramaix, M. Calculation of power for matched pair studies when randomization is by group. *Int J Epidemiol*, 18, 457–461, 1989.

27. Pinol, A.P.Y., Piaggio, G. *ACLUSTER Version 2.0.* Geneva, Switzerland: World Health Organisation, 2000. http://www.abdn.ac.uk/hsru/research/research-tools/study-design (accessed on 12/08/2011).

28. Campbell, M.K., Thomson, S., Ramsay, C.R., MacLennan, G.S., Grimshaw, J.M. Sample size calculator for cluster randomized trials. *Comput Biol Med*, 34, 113–125, 2004.

29. Vierron, E., Giraudeau, B. Design effect in multicenter studies: Gain or loss of power? *BMC Med Res Methodol*, 9, 39, 2009.

30. Lee, K.J., Thompson, S.G. The use of random effects models to allow for clustering in individually randomized trials. *Clin Trials*, 2, 163, 2005.

31. Walter, S.D., Ismaila, A.S., Devereaux for the SPRINT study investigators. Statistical issues in the design and analysis of expertise-based randomized clinical trials. *Stat Med*, 27, 6583–6596, 2008.

32. Donner, A., Taljaard, M., Klar, N. The merits of breaking the matches: A cautionary tale. *Stat Med*, 26, 2036–2051, 2007.

33. Pocock, S.J., Assman, S.E., Enos, L.E., Kasten, L.E. Subgroup analysis, covariate adjustment and baseline comparisons in clinical trial reporting: Current practice and problems. *Stat Med*, 21, 2917–2930, 2002.

34. Biau, D.J., Porcher, R., Boutron, I. The account for provider and center effects in multicenter interventional and surgical randomized controlled trials is in need of improvement: A review. *J Clin Epidemiol*, 61(5), 435–439, 2008.

35. Walwyn, R., Roberts, C. Therapist variation within randomised trials of psychotherapy: Implications for precision, internal and external validity. *Stat Methods Med Res* 19, 291–315, 1–25, 2009.

36. KAT Trial Group. The Knee Arthroplasty Trial (KAT) design features, baseline characteristics, and two-year functional outcomes after alternative approaches to knee replacement. *J Bone Joint Surg Am*, 91, 134–141, 2009.

6

Assessment of Harm

Panagiotis N. Papanikolaou
University of Ioannina School of Medicine

John P.A. Ioannidis
University of Ioannina School of Medicine and
Foundation for Research and Technology–Hellas

CONTENTS

In this chapter, we deal with the following topics:

1. Randomized controlled trials (RCTs) vs. other sources of harms for nonpharmacological interventions: Here, we review examples where both randomized and nonrandomized evidence exists on specific harms for nonpharmacological interventions and examples where major harms have been recognized for nonpharmacological interventions and how the relevant evidence was obtained (RCTs or other means).

2. Suboptimal reporting: There is evidence that RCTs focus mainly on efficacy issues and that reporting of harms in RCTs is poor and often neglected. We provide examples.

3. Collection of harms data not incorporated in the design of RCTs of pharmacological interventions: Many trials of nonpharmacological interventions apparently do not collect any data on harms.

Definitions of harms and prestudy anticipation may not be straight-forward, but there are several efforts to improve the standardization of such data collection. We describe some examples of major harms related to nonpharmacological interventions.

4. Synthesis of data from harms from multiple trials—meta-analysis: Here, we discuss examples where data have been synthesized on harms from multiple trials in formal meta-analyses. Focus is placed on the lack of compatible data across studies, lack of any data, and other shortcomings of meta-analysis of harms.

5. Special considerations for harms of specific interventions: We discuss briefly special considerations for harms related to surgical interventions, psychological and behavioral interventions, and devices.

6. Suggestions for improvement: We discuss issues of improvement of design of trials, definition of harms, collection of harms information, reporting, integration of harms from many trials, and application of risk-benefit evaluations and decision-making approaches in RCTs of nonpharmacological interventions.

6.1 Randomized Controlled Trials vs. Other Sources of Harms for Nonpharmacological Interventions

RCTs are a potentially important source of information on adverse events associated with medical interventions [1]. RCTs of medical interventions may address both efficacy and safety issues, though traditionally RCTs focus mainly on the efficacy of treatment. Moreover, harms data from randomized trials tend to be poorly collected, analyzed, and reported [2–5]. Recording and reporting of both efficacy and safety issues in RCTs is of a great importance, directly affecting the quality of these clinical trials, and if this is done in an adequate way, then we may appraise the benefits and harms of each intervention in such a way that may affect clinical practice. Evaluation of harms has traditionally relied on nonrandomized studies, in particular surveillance, but this leaves considerable room for missing important harms. On the other hand, RCTs may have better experimental control and protection against bias than observational studies have, and they may use more active surveillance for recording adverse events, thus recognizing important harms. If evidence on adverse events is available from large-scale randomized trials, then we may also address serious, yet uncommon harms.

In a recent study of Papanikolaou et al. [6], where the aim was to determine whether large-scale RCTs and large observational studies (>4000 subjects) give different estimates of risk for important harms of medical interventions, specific harms of various medical interventions were targeted.

In 7 out of 15 harms where both randomized and nonrandomized evidence existed (eligible data), the interventions studied were nonpharmacological. The interventions concerned vaccines, surgical procedures, and vitamins intake. Specifically, in vaccination against pertussis with acellular vs. whole-cell pertussis vaccine, eligible data were available for encephalopathy, convulsions, and hypotonic hyporesponsiveness. Both randomized and nonrandomized evidence was available for visceral or vascular injury comparing laparoscopic and open surgery for inguinal hernia repair; for wound infection comparing laparoscopic and open surgery for acute appendicitis; and for spontaneous miscarriage and multiple gestation with folate supplementation during periconceptional period or pregnancy. In general, there was no clear predilection for randomized or nonrandomized studies to estimate greater relative risks, but usually (75%) the randomized trials estimated larger absolute excess risks of harm than the nonrandomized studies did.

Major harms have been recognized not only for pharmacological but also for nonpharmacological interventions. In another study by Papanikolaou and Ioannidis [7], where the *Cochrane Database of Systematic Reviews* was searched for reviews containing quantitative data on specific, well-defined harms for at least 4000 randomized subjects, and for which a formal meta-analysis of these data had been performed, nearly all nonpharmacological interventions addressed major harms. Reviews dealing with vaccinations with acellular vs. whole-cell pertussis vaccines for preventing whooping cough in children addressed encephalopathy, convulsions, and hypotonic hyporesponsiveness. The same major harms were targeted in studies about vaccinations with acellular vaccines vs. placebo/diphtheria–tetanus vaccines for preventing whooping cough in children. Systematic reviews in gynecology and obstetrics, examining mid-trimester amniocentesis vs. control and early vs. mid-trimester amniocentesis for prenatal diagnosis, addressed neonatal respiratory distress syndrome and spontaneous miscarriage, respectively. In the same medical field, blood loss >500 mL and second-degree perineal tears were outcome measures in reviews comparing upright or lateral vs. supine position/lithotomy during the second stage of labor. Miscarriage, ectopic pregnancy, and stillbirth were targeted by systematic reviews comparing periconceptional supplementation with folate or multivitamins vs. control for prevention of neural tube defects, whereas cesarean delivery and operative vaginal delivery were addressed in reviews studying continuous electronic fetal heart rate monitoring vs. intermittent auscultation during labor. Another major harm, 30-day stroke or death, was recognized in studies comparing carotid endarterectomy vs. control for symptomatic carotid stenosis. And finally, harms including vascular and visceral injury, deep and superficial infections, or wound infections were outcome measures in systematic reviews of laparoscopic vs. open surgery for inguinal hernia repair, or laparoscopic vs. open appendectomy in adults, respectively.

Another study of Chou et al. [8], in which a systematic review of RCTs comparing nonpharmacological interventions for the management of low back pain was performed, addressed several major but on the other hand rare harms, such as worsening of lumbar disc herniation or cauda equina syndrome with spinal manipulation, pneumothorax with acupuncture, and aggravation of neurological signs and symptoms requiring surgery and severe dermatitis with transcutaneous electrical nerve stimulation.

6.2 Suboptimal Reporting

There is evidence that RCTs focus mainly on efficacy issues and that reporting of harms in RCTs is poor and often neglected. In some medical areas that have been studied so far, reporting of harms in RCTs of nonpharmacological treatment is inadequate, whereas in pharmacological trials reporting of harms is better but still needs improvement. Nonpharmacological trials are the minority of RCTs, but they are of great significance, too, since for some situations only nonpharmacological interventions are available.

In the field of mental health, a study by Papanikolaou et al. [3] evaluated safety reporting in a large random sample of RCTs (132 eligible reports with 142 separate randomized trials involving a total of 11,939 subjects) on various mental health–related interventions derived from the PsiTri registry. Nonpharmacological trials (e.g., psychological, behavioral, and social interventions), concerning about one-quarter of the studied trials, failed to report safety data. Practically none of the nondrug trials had adequate reporting of clinical adverse events and laboratory-determined toxicity. The space given to safety information was little and less than the space given for the names of authors and their affiliations.

Ethgen et al. [9] assessed the reporting of harm in RCTs of pharmacological and nonpharmacological treatment for rheumatoid arthritis and hip or knee osteoarthritis during a 6 year period and found that harm was described less in reports of nonpharmacological treatment trials than in reports of pharmacological treatment trials, even after adjustment for confounding factors. Fewer than half of the nonpharmacological treatment trials reported any harm data and only 2 out of 74 (2.7%) reports involving nonpharmacological interventions adequately defined severity of adverse events according to the Ioannidis and Lau classification. Little space (6.6%) was dedicated to safety in the results section of the nonpharmacological treatment reports.

Martin et al. [10] evaluated the quality of complication reporting in the surgical literature. They analyzed 119 articles reporting outcomes in 22,530 patients. Of them, 42 articles pertained to RCTs and 77 articles pertained to retrospective series. They set 10 criteria relating to completeness of reporting surgical complications. No article met all reporting criteria.

The median score was 5 and the mean score 6, including both random-ized trials and retrospective studies. RCTs were no more thorough in reporting complication data but had higher-quality reports when com-pared with retrospective studies (31% vs. 26% scoring 7–9). Only 24 stud-ies (20%) reported severity grading, whereas 41 studies (34%) provided definitions of complications.

In the study by Papanikolaou and Ioannidis [7] mentioned earlier, where the *Cochrane Database of Systematic Reviews* was searched for reviews con-taining quantitative data on specific, well-defined harms for at least 4000 randomized subjects, and for which a formal meta-analysis of these data had been performed, only seven systematic reviews in the entire *Cochrane Database of Systematic Reviews* provided eligible data on specific harms from nonpharmacological interventions. The Cochrane Library is known for its high quality of its reviews and comprehensive approach to outcomes, so it seems unlikely that other systematic reviews would fare better.

Bibawy et al. [11] evaluated reporting of harms and adverse events in oto-laryngology journals. They found 576 studies (10% of which were RCTs) making therapeutic recommendations (70% surgical–nonpharmacological recommendations, 30% medical–pharmacological recommendations) and 377 (65%) mentioned harms or adverse events anywhere in the manuscript. Only 7% of the studies reported withdrawals due to harms per arm, whereas 29% of the studies used tables and/or figures containing harms-related data.

6.3 Collection of Harms Data Not Incorporated in the Design of RCTs of Nonpharmacological Interventions

Many trials of nonpharmacological interventions apparently do not col-lect any data on harms. For example, trials of mental health interventions not only do not report harms data, but probably do not even collect such data. Definitions of harms and prestudy anticipation may not be straightfor-ward, but there are several efforts to improve the standardization of such data collection. Overall, there is a misconception that many nonpharmaco-logical treatments may have minor adverse events, and that could explain a lower interest in the assessment of harms. However, this is not true. We will describe some examples where the nonpharmacological interventions may be followed by serious or disabling or life-threatening adverse events, even death. The following data come from both RCTs and nonrandomized studies.

Stapes surgery for the treatment of otosclerosis may result in deafness in the operated ear in about 1% [12]; radiation therapy for head and neck cancer may result in various severe adverse events [13] such as (a) severe or life-threatening edema of the larynx requiring urgent tracheotomy, (b) necrosis of laryngeal tissues, which may require a total laryngectomy,

(c) severe fibrosis of the esophagus causing severe dysphagia, which may require a permanent gastrostomy, and (d) panophthalmitis or blindness; death after pancreatectomy due to pancreatic fistula/leak, death after hepatectomy due to hepatic failure, and death after esophagectomy due to lower respiratory tract infections/pneumonia [11]; hepatitis B or C or human immunodeficiency virus infection after blood transfusion (as a result of careful donor selection and the use of advanced tests, including nucleic acid testing [NAT], the risk of transmission of human immunodeficiency virus and hepatitis C virus has been reduced to about 1 in 1.5 million donations in the United States) [14,15]; perioperative stroke or death after carotid endarterectomy in about 3% [16]; atrial fibrillation after cardiac surgery, occurring in 25%–40% of patients, and possible consequent stroke [17]; postoperative hemorrhage after adenotonsillar surgery (0.1%–8.1%) [18–20] with a bleeding mortality rate of about 0.002% [21,22]; death in psychiatric patients treated with electroconvulsive therapy and other serious complications, including critical cardiac events (arrhythmias, cardiac arrest) and critical respiratory events (bronchospasm, respiratory arrest, apnea) (the estimated mortality rate with electroconvulsive therapy is between 2 and 10 per 100,000, about 0.002% per treatment, and 0.01% for each patient) [23–25]; death due to malfunction of pacemakers and implantable cardioverter defibrillators (ICD) (device malfunction was directly responsible for 61 confirmed deaths during the study period (30 pacemaker patients, 31 ICD patients), according to Maisel et al. [26]; severe infection following cochlear implantation, eventually requiring explantation (in 7 out of 500 cochlear implantations, 1.4%) [27].

6.4 Synthesis of Data from Harms from Multiple Trials: Meta-Analysis

Since there is poor reporting of adverse events in randomized trials, systematic reviews and meta-analyses may provide inadequate information about safety issues, or they may even abandon all effort to present or synthesize information if the data are of poor quality or highly heterogeneous. In the previously mentioned study by Papanikolaou and Ioannidis [7] evaluating the availability of large-scale evidence on specific harms from systematic reviews of RCTs, it was found that when appropriate information on harms was missing from the systematic reviews, this information often was also missing or not presented in an appropriate way in the reports of the largest randomized trials that were included in these reviews. It was also identified that there were several instances where the largest RCTs presented detailed enough data on specific harms so as to qualify for the definition of large-scale evidence on specific harms, which were not conveyed appropriately in the systematic reviews. The systematic reviews included either no data on these harms (e.g., for the topic of heptavalent pneumococcal vaccine) or nonspecific information even though the original reports provided specific data on harms (e.g., for the topic of cholera vaccine). Thus, the lack of information on safety data in systematic reviews reflects both the poor quality of safety data presented in primary trial reports and the further loss of important information in the generation of systematic reviews of evidence from randomized trials.

6.5 Special Considerations for Harms of Specific Interventions

There are some issues about nonpharmacological trials that pose special challenges in their reporting. We will present some examples.

The adverse effects of nonpharmacological interventions in the field of mental health are difficult to document and attribute to the tested intervention. However, attribution and causality is a problem even in pharmacological trials [28]. This should not be an excuse for not recording and reporting adverse events. Investigators should bear in mind that almost every intervention, even the nonpharmacological ones (e.g., psychosocial interventions), may have adverse events, sometimes serious ones, and they should be prepared to collect any data related to harms, so as to recognize potentially harmful interventions, preventing them from becoming established into mental health clinical practice. Also, some information on safety outcomes may be impossible to disentangle from efficacy outcomes. Outcomes in mental health are often the integral composite of benefit and harm. Nevertheless, serious and life-threatening adverse events should be possible to record separately in most situations.

Special considerations also apply to the surgical medical field. Surgical interventions are multifactorial as they involve preoperative and postoperative care, anesthesia, and the intervention itself. This fact and several other factors such as experience of care providers and centers' volume, and blinding can influence surgical outcomes [29]. It is difficult indeed to attribute the harm of an intervention solely to the intervention itself, since the harm may be the result of a complication during anesthesia or the complication of a bad preoperative or postoperative care. On the other hand, there is also vast heterogeneity in the definition of various adverse events related to surgical procedures and in their classification by severity [11,30,31]. All these matters result in problematic reporting of RCTs studying surgical interventions and make it very difficult to evaluate a large number of RCTs in a systematic approach.

6.6 Suggestions for Improvement

As mentioned earlier, RCTs appear to be problematic as far as reporting of harms data is concerned. Though they could be a great tool for collecting, recording, and reporting harms data, randomized trials focus mainly on efficacy outcomes. There is accumulating evidence of the efficacy and effectiveness of various medical interventions, but that is not enough. What we should also know if the intervention proved to be effective is whether it is harmful or not. Only then, we may appraise the risk-benefit ratio of the studied intervention and decide whether that intervention would be beneficial into clinical practice.

In order to make the best out of the advantages that RCTs provide to clinical investigators, the design of the trials should be improved. In that direction, the Consolidated Standards of Reporting Trials (CONSORT) statement, a set of recommendations for the standardization, and thus improvement, of reports of RCTs was first published in 1996 [32]. It was revised 5 years later [33] and one of the additions to the checklist flowchart was an item about reporting adverse events. Given the importance of harms-related issues, an extension of the CONSORT statement describing recommendations on the appropriate reporting of harms in RCTs was published in 2004 [34]. Ten new recommendations apply to the 22-item checklist in the extended CONSORT statement. Authors are encouraged to use the term "harms" instead of "safety," because "safety" is a reassuring term that may obscure potentially major "harms" of medical interventions. Collected data on harms and benefits should be clearly stated in the Title–Abstract and addressed harms and benefits in the Introduction. In the Methods section, addressed adverse events should be listed with definitions for each as to grading, expected vs. unexpected events, reference to standardized and validated definitions, and descriptions of new definitions. Clarifications on how harms-related information was collected should be stated in the same section. Investigators should describe plans for presenting and analyzing information on harms, including coding, handling of recurrent events, specification of timing issues, handling of continuous measures, and any statistical analyses. In the Results section, authors should describe for each arm the participant withdrawals that are due to harms and the experience with the allocated treatment. Authors should also provide the denominators for analyses on harms, with the "intention-to-treat" being the preferred analysis both for efficacy and harms. The absolute risk of each adverse event (type, grade, and seriousness per arm should be specified) and appropriate metrics for recurrent events, continuous variables and scale variables, whenever pertinent, should be presented. Any subgroup analyses and exploratory analyses for harms should also be described. In the Discussion section, authors should provide a balanced discussion of benefits and harms with emphasis on study limitations, generalizability, and other sources of information on harms.

A new extension of the CONSORT statement for trials of nonpharma-cological treatment was developed in 2006 [35,36], taking into account specific issues concerning nonpharmacological interventions. In this extension, no change or addition was made for harms; what was done was an extension of 11 items from the CONSORT statement, the addition of one new item and the development of a modified flow diagram, all aiming at better reporting of randomized trials of nonpharmacological treatment. All this work is very promising for better quality of reporting of RCTs of nonpharmacological treatment, should investigators adopt these guide-lines in their work.

Definition of harms across different medical fields is often a challenge. Several investigators have proposed various definitions and classification systems for harms addressed in their medical field. Dindo et al. [31] pro-posed a classification of surgical complications and a definition of negative surgical outcomes (complications, failure to cure, sequelae) that practically modified a previous classification by Clavien et al. [30]. The proposed clas-sification consists of five severity grades: Grade I (any deviation from the normal postoperative course without the need for pharmacological treat-ment other than antiemetics, antipyretics, analgesics, diuretics, electro-lytes, and physiotherapy, or surgical other than wound infections opened at bedside, endoscopic, and radiological interventions), Grade II (compli-cations requiring pharmacological treatment with different drugs, blood transfusions, and total parenteral nutrition), Grade III (complications requiring surgical, endoscopic, or radiological intervention not under general anesthesia—Grade IIIa or under general anesthesia—Grade IIIb), Grade IV (life-threatening complications including central nervous system complications requiring intermediate care/intensive care unit manage-ment regarding single organ dysfunction including dialysis—Grade IVa or multiorgan dysfunction—Grade IVb) and Grade V (death). The suffix "d" (for "disability") is added to the respective grade of complication if the patient suffers from a complication at the time of discharge, indicating the need for a follow-up. The authors validated that modified classification in a cohort of 6336 patients and assessed acceptability and reproducibility of the classification by contacting an international survey. The proposed clas-sification has several advantages, since it may provide increased unifor-mity in reporting results, make comparisons of surgical outcomes between time periods and between different centers feasible, or may be the basis to perform adequate meta-analysis.

In the direction of increased unanimity in terms of adverse events definitions and severity among reports of all oncology clinical trials, the National Cancer Institute of the U.S. National Institutes of Health devel-oped and published in 2003 the Common Terminology Criteria for Adverse Events (CTCAE) v3.0 (a revised version called "CTCAE v3.0 Notice of Modifications" was published in 2006), which is a descriptive terminology that can be utilized for adverse event reporting [13]. A grading (severity) scale is provided for each adverse event term. As it was reported earlier, the CTCAE v3.0 is applicable to all oncology clinical trials; therefore, it refers not only to chemotherapy-induced harms but also to harms related to surgery and radiation, which are nonpharmacological interventions for cancer treatment.

It would be a mistake if we did not consider other trial designs when generating evidence for harms. Observational studies to date have been the standard for identifying serious, yet uncommon harms of medical interventions [37–39]. One of the reasons for that is that they can be larger and may have longer follow-up. Compared with randomized trials, observational studies are more prone to bias and often present suboptimal reporting when published. In the direction of better reporting of observational studies, the STROBE Initiative developed the "Strengthening the reporting of observational studies in epidemiology" (STROBE) statement and published it in 2007 [40]. The STROBE statement is a checklist of items that should be addressed in articles reporting on the three main study designs of analytical epidemiology: cohort, case-control, and cross-sectional studies. The intention is solely to provide guidance on how to report observational research well. Data from observational studies cannot be dismissed simply because the study was not randomized [41]. Evidence on harms from RCTs and observational studies should complement each other and be integrated in systematic reviews to have the best result in evaluating the existing evidence on harms from various medical interventions.

Although the need to incorporate, synthesize, and weigh data from different types of studies for systematic reviews of harms has become widely recognized [42–45], methods to combine data from different sources are just starting to be developed [46]. Since confounding and selection bias can distort findings from observational studies, it is especially important that researchers who include such studies avoid inappropriate statistical combination of data, carefully describe the characteristics and quality of included studies [47], and thoroughly explore potential sources of heterogeneity [48,49].

RCTs of nonpharmacological therapies account for about 25% of all published RCTs, according to a study by Chan et al. [50] that assessed all published RCTs in 2000. Therefore, they represent a significant portion of randomized trials and they deserve similar attention as the pharmacological trials. There is a need for better reporting of harms data induced by nonpharmacological interventions and the first steps in this difficult undertaking are done with the CONSORT statement and its extension for nonpharmacological trials. If the reporting of harms data in RCTs of nonpharmacological interventions becomes better, then evaluations of risk and benefit conferred by the studied interventions will be more precise. This will result in easier and better decision making, influencing clinical practice toward the right direction.

References

1. Ioannidis, J.P., Lau, J. Improving safety reporting from randomised trials. *Drug Saf*, 25, 77–84, 2002.
2. Ioannidis, J.P., Lau, J. Completeness of safety reporting in randomized trials: An evaluation of 7 medical areas. *JAMA*, 285, 437–443, 2001.
3. Papanikolaou, P.N., Churchill, R., Wahlbeck, K., Ioannidis, J.P. Safety reporting in randomized trials of mental health interventions. *Am J Psychiatry*, 161, 1692–1697, 2004.
4. Ioannidis, J.P., Contopoulos-Ioannidis, D.G. Reporting of safety data from randomised trials. *Lancet*, 352, 1752–1753, 1998.
5. Loke, Y.K., Derry, S. Reporting of adverse drug reactions in randomised controlled trials—A systematic survey. *BMC Clin Pharmacol*, 1, 3, 2001.
6. Papanikolaou, P.N., Christidi, G.D., Ioannidis, J.P. Comparison of evidence on harms of medical interventions in randomized and nonrandomized studies. *CMAJ*, 174(5), 635–641, 2006.
7. Papanikolaou, P.N., Ioannidis, J.P. Availability of large-scale evidence on specific harms from systematic reviews of randomized trials. *Am J Med*, 117, 582–589, 2004.
8. Chou, R., Huffman, L.H.; American Pain Society; American College of Physicians. Nonpharmacological therapies for acute and chronic low back pain: A review of the evidence for an American Pain Society/American College of Physicians clinical practice guideline. *Ann Intern Med*, 147:492–504, 2007.
9. Ethgen, M., Boutron, I., Baron, G., Giraudeau, B., Sibilia, J., Ravaud, P. Reporting of harm in randomized, controlled trials of nonpharmacological treatment for rheumatic disease. *Ann Intern Med*, 143, 20–25, 2005.
10. Martin, R.C., 2nd, Brennan, M.F., Jaques, D.P. Quality of complication reporting in the surgical literature. *Ann Surg*, 235, 803–813, 2002.
11. Bibawy, H., Cossu, A., Cogan, S., Rosenfeld, R. Reporting of harms and adverse events in otolaryngology journals. *Otolaryngol Head Neck Surg*, 140, 241–244, 2009.
12. Fisch, U. *Tympanoplasty, Mastoidectomy and Stapes Surgery. Part 3: Stapes Surgery.* Stuttgart, New York: Thieme, 2008, p. 285.
13. Common Terminology Criteria for Adverse Events, Version 3.0, Cancer Therapy Evaluation Program, http://ctep.cancer.gov, Published date: August 9, 2006.
14. Dodd, R.Y. Current safety of the blood supply in the United States. *Int J Hematol*, 80, 301–305, 2004.
15. Glynn, S.A., Kleinman, S.H., Schreiber, G.B., et al.; for the Retrovirus Epidemiology Donor Study (REDS). Trends in incidence and prevalence of major transfusion-transmissible viral infections in US blood donors, 1991 to 1996. *JAMA*, 284, 229–235, 2000.
16. Chambers, B.R., Donnan, G.A. Carotid endarterectomy for asymptomatic carotid stenosis. *Cochrane Database Syst Rev*, (4), CD001923, 2005.
17. Crystal, E., Connolly, S.J., Sleik, K., Ginger, T.J., Yusuf, S. Interventions on prevention of postoperative atrial fibrillation in patients undergoing heart surgery: A meta-analysis. *Circulation*, 106, 75–80, 2002.
18. Carmody, D., Vamadevan, T., Cooper, S.M. Post tonsillectomy haemorrhage. *J Laryngol Otol*, 96, 635–638, 1982.

19. Chowdhury, K., Tewfik, T.L., Schloss, M.D. Post-tonsillectomy and adenoidectomy hemorrhage. *J Otolaryngol,* 17, 46–49, 1988.
20. Tami, T.A., Parker, G.S., Taylor, R.E. Post-tonsillectomy bleeding: An evaluation of risk factors. *Laryngoscope,* 97, 1307–1311, 1987.
21. Pratt, L.W., Gallagher, R.A. Tonsillectomy and adenoidectomy: Incidence and mortality, 1968–1972. *Otolaryngol Head Neck Surg,* 87, 159–166, 1979.
22. Kristensen, S., Tveterås, K. Post-tonsillectomy haemorrhage. A retrospective study of 1150 operations. *Clin Otolaryngol Allied Sci,* 9, 347–350, 1984.
23. Shiwach, R.S., Reid, W.H., Carmody, T.J. An analysis of reported deaths following electroconvulsive therapy in Texas, 1993–1998. *Psychiatr Serv,* 52, 1095–1097, 2001.
24. Nuttall, G.A., Bowersox, M.R., Douglass, S.B., et al. Morbidity and mortality in the use of electroconvulsive therapy. *J ECT,* 20, 237–241, 2004.
25. UK ECT Review Group. Efficacy and safety of electroconvulsive therapy in depressive disorders: A systematic review and meta-analysis. *Lancet,* 361, 799–808, 2003.
26. Maisel, W.H., Moynahan, M., Zuckerman, B.D., et al. Pacemaker and ICD generator malfunctions: Analysis of Food and Drug Administration annual reports. *JAMA,* 295, 1901–1906, 2006.
27. Venail, F., Sicard, M., Piron, J.P., et al. Reliability and complications of 500 consecutive cochlear implantations. *Arch Otolaryngol Head Neck Surg,* 134, 1276–1281, 2008.
28. Naranjo, C.A., Busto, U., Sellers, E.M. Difficulties in assessing adverse drug reactions in clinical trials. *Prog Neuropsychopharmacol Biol Psychiatry,* 6, 651–657, 1982.
29. McCulloch, P., Taylor, I., Sasako, M., Lovett, B., Griffin, D. Randomised trials in surgery: Problems and possible solutions. *BMJ,* 324, 1448–1451, 2002.
30. Clavien, P.A., Sanabria, J.R., Strasberg, S.M. Proposed classification of complications of surgery with examples of utility in cholecystectomy. *Surgery,* 111, 518–526, 1992.
31. Dindo, D., Demartines, N., Clavien, P.A. Classification of surgical complications: A new proposal with evaluation in a cohort of 6336 patients and results of a survey. *Ann Surg,* 240, 205–213, 2004.
32. Begg, C., Cho, M., Eastwood, S., et al. Improving the quality of reporting of randomized controlled trials. The CONSORT statement. *JAMA,* 276, 637–639, 1996.
33. Altman, D.G., Schulz, K.F., Moher, D., et al.; CONSORT Group (Consolidated Standards of Reporting Trials). The revised CONSORT statement for reporting randomized trials: Explanation and elaboration. *Ann Intern Med,* 134, 663–694, 2001.
34. Ioannidis, J.P., Evans, S.J., Gøtzsche, P.C., et al.; CONSORT Group. Better reporting of harms in randomized trials: An extension of the CONSORT statement. *Ann Intern Med,* 141, 781–788, 2004.
35. Boutron, I., Moher, D., Altman, D.G., Schulz, K.F., Ravaud, P.; CONSORT Group. Methods and processes of the CONSORT Group: Example of an extension for trials assessing nonpharmacological treatments. *Ann Intern Med,* 148, W60–W66, 2008.
36. Boutron, I., Moher, D., Altman, D.G., Schulz, K.F., Ravaud, P.; CONSORT Group. Extending the CONSORT statement to randomized trials of nonpharmacological treatment: Explanation and elaboration. *Ann Intern Med,* 148, 295–309, 2008.
37. Jick, H. The discovery of drug-induced illness. *N Engl J Med,* 296, 481–485, 1977.

38. Edwards, I.R., Aronson, J.K. Adverse drug reactions: Definitions, diagnosis, and management. *Lancet*, 356, 1255–1259, 2000.
39. Stephens, M.D., Talbot, J.C., Routledge, P.A. (eds.). *The Detection of New Adverse Reactions*, 4th edn. London, U.K.: Macmillan Reference, 1998.
40. von Elm, E., Altman, D.G., Egger, M., Pocock, S.J., Gøtzsche, P.C., Vandenbroucke, J.P.; STROBE Initiative. Strengthening the reporting of observational studies in epidemiology (STROBE) statement: Guidelines for reporting observational studies. *BMJ*, 335, 806–808, 2007.
41. Vandenbroucke, J.P. What is the best evidence for determining harms of medical treatment? *CMAJ*, 174, 645–646, 2006.
42. Glasziou, P., Vandenbroucke, J.P., Chalmers, I. Assessing the quality of research. *BMJ*, 328, 39–41, 2004.
43. Cuervo, L.G., Aronson, J.K. The road to health care. *BMJ*, 329, 1–2, 2004.
44. Smith, L.A., Moore, R.A., McQuay, H.J., Gavaghan, D. Using evidence from different sources: An example using paracetamol 1000 mg plus codeine 60 mg. *BMC Med Res Methodol*, 1, 1, 2001.
45. Rosendaal, F.R. Bridging case–control studies and randomized trials. *Curr Control Trials Cardiovasc Med*, 2, 109–110, 2001.
46. Wald, N.J., Morris, J.K. Teleoanalysis: Combining data from different types of study. *BMJ*, 327, 616–618, 2003.
47. Stroup, D.F., Berlin, J.A., Morton, S.C., et al. Meta-analysis of observational studies in epidemiology: A proposal for reporting. Meta-analysis of observational studies in epidemiology (MOOSE) group. *JAMA*, 283, 2008–2012, 2000.
48. Egger, M., Schneider, M., Davey Smith, G. Spurious precision? Meta-analysis of observational studies. *BMJ*, 316, 140–144, 1998.
49. Berlin, J.A., Colditz, G.A. The role of meta-analysis in the regulatory process for foods, drugs, and devices. *JAMA*, 281, 830–834, 1999.
50. Chan, A.W., Altman, D.G. Epidemiology and reporting of randomised trials published in PubMed journals. *Lancet*, 365, 1159–1162, 2005.

7

External Validity and Applicability of Nonpharmacological Trials

Isabelle Boutron and Philippe Ravaud

Paris Descartes University and
Public Assistance—Hospitals of Paris

CONTENTS

7.1 General Principles

Randomized controlled trials are considered the gold standard for therapeutic evaluation. For results to be clinically useful, randomized controlled trials must take into account the internal validity of findings—the extent to which systematic errors or bias has been avoided—as well as their external validity, sometimes called applicability, or whether the results of a trial can be reasonably applied or generalized to a definable group of patients in a particular setting in routine practice.

Historically, internal validity has been considered a priority for clinical research. Several meta-epidemiological studies have identified the different dimensions of internal validity [1–6]. The concept and terminology of internal validity have been clarified [5]. The Cochrane collaboration now recommends evaluating the risk of bias in a trial instead of the quality of the trial [5]. The Consolidated Standards of Reporting Trials (CONSORT) statements, endorsed by many editors, improved the reporting of data related to internal validity [4,7]. Finally, several tools have been developed to evaluate the internal validity of trials included in systematic reviews [5,8–10].

Contrary to internal validity, external validity has been neglected [11,12]. Funding agencies, researchers, reviewers, and editors tend to be more concerned with the scientific rigor of interventions studied than with the applicability of the trial results. Nevertheless, as highlighted by Hill, "At its best a trial shows what can be accomplished with a medicine under careful observation and certain restricted conditions. The same results will not invariably or necessarily be observed when the medicine passes into general use ..." [13]. This neglect has probably contributed to the failure to transpose research into clinical practice. In fact, lack of external validity is frequently advocated as the reason why interventions found to be effective in clinical trials are underused in clinical practice [14].

Recently, a definition and conceptual framework has been proposed for the concept of external validity. Dekkers and colleagues performed a methodological systematic review to address the concept and distinguished two aspects: external validity and applicability of the trial results [15]. External validity, also named generalizability, applies to patients and disease characteristics, that is, whether the results of the study are valid for patients with the same disease other than those in the original study population in a treatment setting that is in all respects equal to the treatment setting of the original study. In contrast, applicability refers to the treatment setting, or whether study results are valid in a different treatment setting than that in the original study. This distinction is not always clear-cut and has been criticized [16] but will be useful for the purpose of this chapter.

7.2 Criteria for Assessing External Validity and Applicability

Several criteria can potentially influence external validity and the applicability of the trial results. These criteria concern the patient and disease characteristics; the setting, centers, and care providers; the complexity of the intervention; and the temporal aspects of the trial [14,17].

7.2.1 Patient and Disease Characteristics

Several studies demonstrated important differences between the characteristics of patients participating in randomized controlled trials and those undergoing treatment in clinical practice [18–20]. These differences are probably a consequence of the planning and conduct of the trial that could affect the sample representativeness.

Eligibility criteria and specifically exclusion criteria have an important impact. These criteria may be excessive and may exclude some patients that should be considered the target population [20]. A systematic review evaluated the nature and extent of exclusion criteria on the representation of certain patient populations. Exclusion criteria were classified as being strongly, potentially, or poorly justified according to a pre-specified classification. Patients were excluded in 81% of trials because of common medical comorbidities, in 51% because of commonly prescribed medications, and in 72% because of age. Reasons for exclusion were frequently not justified. In 84% of trials, at least one exclusion criteria was considered poorly justified [21]. Similarly, Coca et al. showed that more than half of randomized controlled trials of cardiovascular disease excluded patients with renal disease [22].

However, exclusion criteria are not the only factor influencing the definition of the target population. In fact, physicians may decide not to propose the trial to some eligible patients for several reasons and this decision could exclude elderly or patients with comorbidities [23]. Further, invited patients may refuse to participate, and this decision could be related to patients' prognostic factors [24]. Steg et al. compared the characteristics and event rates for patients who were eligible but were not included in a randomized controlled trial with those who were included [25]. Patients enrolled in the trial were younger and had less comorbidity. Further, nonparticipating eligible patients had a higher risk of death (7.1%) than participants (3.6%).

The recruitment modalities also have an important impact. For example, use of a run-in period that aims to exclude patients with low treatment compliance or with adverse effects will affect the external validity of the sample. The exclusion of these patients may lead to an underestimation of adverse effects and an overestimation of the beneficial effect.

Consequently, to evaluate external validity in published results of a randomized controlled trial, readers should focus on eligibility criteria but also on the recruitment modalities, the proportion of eligible patients who agreed to participate and finally the baseline characteristics of patients. However, some authors consider that the problem of external validity and generalizability has been exaggerated and that lack of external validity is often a good excuse for clinicians to not change their practice according to research results. For example, Sheldon and colleagues argued that "it is probably more appropriate to assume that research findings are generalizable across patients unless there is a strong theoretical or empirical evidence to suggest that a particular group of patients will respond differently" [26].

7.2.2 Setting, Centers, and Care Providers

Several factors related to the setting, centers, and care providers participating in the trial have an important impact on the applicability of trial results. Differences in the healthcare system, the diagnostic process, the speed of consultation, and cultural barriers could affect applicability. For example, in the field of education, cultural background may have an important effect. In the field of surgery, evidence demonstrates the role of center volume and care provider expertise on the success of the intervention. Halm et al., in a systematic review of research into center volume and outcomes [27], showed high center volume associated with better outcomes across a wide range of procedures and conditions, but the magnitude of the association varied greatly. In fact, 71% of all studies of hospital volume and 69% of studies of physician volume reported statistically significant associations between high volume and better outcomes [27]. Consequently, when planning a trial, investigators must evaluate the context in which the trial results will be applied. Similarly, when reporting the trial results or interpreting the findings of a systematic review, researchers must consider the context in which the intervention should be performed. Otherwise, a surgical procedure performed in high-volume center by very skilled surgeons might be better than the medical treatment, but in a low-volume center, the medical treatment might be better than the surgical procedure. For example, results from the asymptomatic carotid artery surgery (ACAS) trial showed that endarterectomy for asymptomatic carotid stenosis reduced the 5-year absolute risk of stroke by about 5% [28]. However, the trial accepted only surgeons with a good safety record, initially rejecting 40% of applicants and subsequently barring those with adverse operative outcomes from further participation in the trial. The benefit from surgery in the ACAS trial was due largely to the consequently low operative risk. In surgical case series, operative mortality was found to be eightfold higher than in the ACAS trial [14]. Trials should not include centers that do not have the competence to treat patients safely, but selection should not be so exclusive that the results cannot be generalized to clinical practice. For example, surgeons rejected by the ACAS trial are unlikely to have stopped performing surgery outside of the trial.

7.2.3 Complexity of the Intervention

Nonpharmacological treatments are usually complex interventions of several components, each having a potential impact on the success of the intervention. For example, surgery for arterial endarterectomy and reconstruction during carotid endarterectomy can be performed in various ways—standard endarterectomy with primary closure, standard endarterectomy with patch angioplasty, and eversion endarterectomy. Endarterectomy with patch angioplasty and eversion endarterectomy were found associated with significantly better perioperative outcomes than was primary closure, which highly affected the success rate [29]. Similarly, rehabilitation programs or education programs may be administered in many different ways that could influence their effectiveness. For example, supervised rehabilitation programs can be efficient but not unsupervised rehabilitation programs [30].

The complexity of the intervention is also an important barrier to the translation of trial results into clinical practice and could be responsible for important waste if clinicians are unable to reproduce the intervention in practice. A review of 80 studies selected by the journal *Evidence-based Medicine* as being both valid and important for clinical practice found that clinicians could replicate the intervention described in less than one-third of the reports evaluating nonpharmacological treatments [31]. Interventions may be used incorrectly or not used at all if the intervention is inadequately described. A review of 158 reports of randomized controlled trials of surgery showed that important components of the intervention such as anesthesia protocol or perioperative management were not described in more than half of the reports.

The development of reporting guidelines [7,32–35], research into this topic, and innovative communication methods to disseminate and allow for the replication of the intervention should improve this issue in the future. For example, graphic techniques [36], video [37], and audio could be used to improve the reproducibility of interventions.

7.2.4 Temporal Aspect

The delay between the conduct of the trial and the use of the trial results from a published report or in a meta-analysis may be very long [38]. This delay could translate into important changes in clinical practice, particularly for complex interventions such as surgical procedures, for which the success of the treatment depends highly on factors other than the surgical procedure, mainly those related to the quality of critical care provided.

7.3 Criteria for Pharmacological and Nonpharmacological Treatments

For trials assessing pharmacological treatments, the main questions relate to the external validity of the trial results (i.e., whether the results can be applied to other populations). The clinician wants to know "whether there are other groups of individuals to whom the results might apply" [39]. This implies that assessment of external validity focuses on the patients included in trials and discusses the validity of inferences as they pertain to future patients rather than the specific trial participants [39]. For nonpharmacological treatment, assessing external validity and applicability is probably more complex, as shown in Table 7.1. In fact, results will depend on the setting, treatments, and complexity of the intervention [14,17].

TABLE 7.1

External Validity Criteria for Pharmacological
and Nonpharmacological Interventions

	Pharmacological Treatment	Nonpharmacological Treatment
Patients and disease characteristics	+++	+++
Setting, centers, and care providers	+	+++
Complexity of the intervention	+	+++
Temporal aspect	+++	+++

7.4 Assessment of External Validity and Applicability

7.4.1 Reporting of External Validity and Applicability Data

Assessing the external validity and applicability of trial results to turn research into action supposes that information is adequately reported in published articles. However, some evidence of inadequate reporting exists. In a sample of reports of randomized controlled trials assessing pharmacological and nonpharmacological treatments for hip and knee OA, the country was clearly reported in 21% reports, the setting in one-third, and the number of centers in less than half, with details about the centers (volume of care) given in one-fifth of the reports [12]. One study of surgical reports found that the setting and center volume of activity was reported in only 7% and 3% of articles, respectively; selection criteria for care providers were reported in 41% of the articles, and the number of care providers performing the intervention was reported in 32% [40]. Some evidence suggests an important disparity between the information provided in published reports of randomized controlled trials and the information required by clinicians to be able to apply trial results in practice [31].

The issue of inadequate reporting was embraced in the mid-1990s by an international group of clinical epidemiologists, statisticians, and biomedical editors. The first reporting guidelines for randomized controlled trials were developed and published—the Consolidated Standards of Reporting Trials (CONSORT) statement. These guidelines are regularly updated, the most recent publication in 2010 [7]. Recently, to take into account specific issues relating to the planning and reporting of the results of surgical trials, the CONSORT group published an extension to the statement for nonpharmacological treatment [33,34]. The CONSORT statement, as well as the extension for trials assessing nonpharmacological treatment, dedicated a specific item to the discussion of the "Generalizability (external validity, applicability) of the trial findings" [7,32]. The essential data necessary to evaluate the external validity and applicability of a trial's results that need to be reported were detailed in the extension of the CONSORT for nonpharmacological treatment [33,34] with specific items dedicated to centers, care providers and describing the intervention.

7.4.2 Specific Issues When Assessing External Validity and Applicability

Assessing internal validity and external validity raises some similar issues. In fact, assessing both issues depends on the quality of reporting. However, external validity and applicability cannot be assessed by analogy to internal validity [15]. In fact, the internal validity of a trial depends on the planning, conduct, and analysis of the trial. Consequently, the assessment of internal validity can be performed once and will be "immutable." In contrast, as illustrated in Figure 7.1, assessing external validity depends on the planning and conduct of the trial and on the consumers who are willing to use the trial

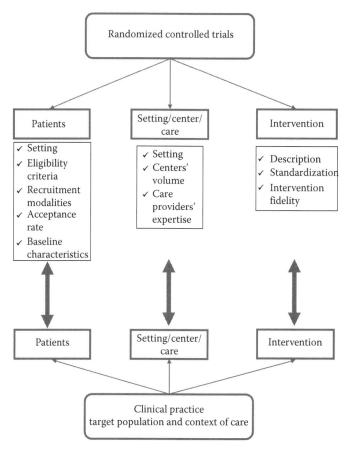

FIGURE 7.1
Assessing external validity depends on the planning and conduct of the trial but also on the consumers who are willing to use the trial results, that is, the target populatal results.

results (i.e., the target population and context where one wants to apply the trial results). For example, if the trial included patients recruited and treated in a high-volume university hospital, external validity and applicability will be considered adequate for a clinician working in a high-volume university hospital but inadequate for a clinician working in a low-volume center. Consequently, assessing the external validity of trial results implies defining the target population and context.

Most of the tools, users' guides, and recommendations related to the assessment of external validity and applicability clearly indicate that (1) internal validity is a clear prerequisite for external validity and applicability and (2) the assessment must determine whether the trial results are generalizable, that is, "whether the patients are adequately representative of the patients to be encountered in normal practice" [17]; "whether the treatment under consideration […] is appropriate for normal clinical practice" [17]; and "[t]he extent to which the results of a study provide a correct basis for applicability to other clinical circumstances" [41]. All these conceptual considerations for assessing external validity imply comparing how the trial was planned and conducted to the clinical practice situation. However, a major issue is defining "clinical practice." In fact, the definition of clinical practice will vary greatly among different countries with different patient characteristics (comorbidities, socioeconomic status) and context of care (e.g., co-interventions). For example, the patients and context of care and, consequently, "clinical practice" will differ greatly between New York and a rural town in Africa.

Some authors argue for assessing the main restrictions in external validity and applicability of the trial results, that is, to whom and in which context the study results should *not* be applied [42]. They argue that the study design may call for restricting the inclusion of subjects to those with a narrow range of characteristics rather than making a futile attempt to make the subjects representative, in a sample sense, of the target population [42].

7.4.3 Balance between Internal and External Validity and Applicability

Some evidence shows that randomized controlled trials often lack external validity [14], particularly those in which methodological constraints are used to improve the internal validity of the trial. For example, in a trial evaluating arthroscopic surgery for osteoarthritis of the knee, a placebo to arthroscopic surgery was used to blind patients to the treatment received [43]. This procedure allowed for increasing internal validity because of the limited risk of performance and evaluation bias. However, the use of a placebo was also a barrier to patients' participation in the trial. In fact, nearly half of the patients declined participation. Further, in most of these trials, patients with strong preferences will not agree to participate [44,45]. These limitations do not invalidate the results of trials assessing nonpharmacological treatment, but the external validity and applicability of the results may be limited. In the field of nonpharmacological trials, some authors argue that nonrandomized studies are essential because they improve external validity at the expense of internal validity [46,47]. Nevertheless, one study comparing the context of care (number of centers and surgeons involved) in published reports of randomized controlled trials and nonrandomized studies of surgery [48] found that the results did not support the hypothesis that, in general, results of nonrandomized studies have better applicability than do those of randomized controlled trials. Both study types were mainly of single-center studies with one or two participating surgeons. Other factors potentially affecting applicability, such as the relevance of a radiographic primary outcome and duration of follow-up of <1 year, also did not differ by study design and limited the applicability of the results of the selected studies.

7.5 Conclusion and Perspectives

To be clinically useful, the results of a trial must be relevant to clinical practice, that is, be reasonably likely to be replicated when applied to a definable group of patients in a particular clinical setting [16]. External validity and applicability of trial results are difficult to conceptualize because they depend on both the trial itself and on where and to whom the trial results should be applied. However, data necessary to evaluate the external validity and the applicability of the treatment evaluated in randomized trials should systematically be reported. Further, research should focus on the development of specific methods for representing external validity and applicability and favoring the dissemination of research results in clinical practice.

References

1. Schulz, K.F., Chalmers, I., Hayes, R.J., Altman, D.G. Empirical evidence of bias. Dimensions of methodological quality associated with estimates of treatment effects in controlled trials. *JAMA*, 273(5), 408–412, 1995.
2. Moher, D., Pham, B., Jones, A., et al. Does quality of reports of randomised trials affect estimates of intervention efficacy reported in meta-analyses? *Lancet*, 352(9128), 609–613, 1998.
3. Wood, L., Egger, M., Gluud, L.L., et al. Empirical evidence of bias in treatment effect estimates in controlled trials with different interventions and outcomes: Meta-epidemiological study. *BMJ*, 336(7644), 601–605, 2008.
4. Plint, A.C., Moher, D., Morrison, A., et al. Does the CONSORT checklist improve the quality of reports of randomised controlled trials? A systematic review. *Med J Aust*, 185(5), 263–267, 2006.

5. Higgins, J.P.T., Green, S. *Cochrane Handbook for Systematic Reviews of Interventions.* Chichester, U.K.: Wiley-Blackwell, 2008.
6. Pildal, J., Chan, A.W., Hrobjartsson, A., Forfang, E., Altman, D.G., Gotzsche, P.C. Comparison of descriptions of allocation concealment in trial protocols and the published reports: Cohort study. *BMJ*, 330(7499), 1049, 2005.
7. Moher, D., Hopewell, S., Schulz, K.F., et al. CONSORT 2010 explanation and elaboration: Updated guidelines for reporting parallel group randomised trials. *BMJ*, 340, c869, 2010.
8. Boutron, I., Moher, D., Tugwell, P., et al. A checklist to evaluate a report of a nonpharmacological trial (CLEAR NPT) was developed using consensus. *J Clin Epidemiol*, 58(12), 1233–1240, 2005.
9. Jadad, A.R., Moore, R.A., Carroll, D., et al. Assessing the quality of reports of randomized clinical trials: Is blinding necessary? *Control Clin Trials*, 17(1), 1–12, 1996.
10. Verhagen, A.P., de Vet, H.C., de Bie, R.A., et al. The Delphi list: A criteria list for quality assessment of randomized clinical trials for conducting systematic reviews developed by Delphi consensus. *J Clin Epidemiol*, 51(12), 1235–1241, 1998.
11. Glasgow, R.E., Green, L.W., Klesges, L.M., et al. External validity: We need to do more. *Ann Behav Med*, 31(2), 105–108, 2006.
12. Ahmad, N., Boutron, I., Moher, D., Pitrou, I., Roy, C., Ravaud, P. Neglected external validity in reports of randomized trials: The example of hip and knee osteoarthritis. *Arthritis Rheum*, 61(3), 361–369, 2009.
13. Lancet, T. (ed.). *Bradford Hill's Principles of Medical Statistics*, 11th edn. London, U.K.: Hodder Arnold Publication, 1937.
14. Rothwell, P.M. External validity of randomised controlled trials: "To whom do the results of this trial apply?" *Lancet*, 365(9453), 82–93, 2005.
15. Dekkers, O.M., von Elm, E., Algra, A., Romijn, J.A., Vandenbroucke, J.P. How to assess the external validity of therapeutic trials: A conceptual approach. *Int J Epidemiol*, 39(1), 89–94, 2009.
16. Rothwell, P.M. Commentary: External validity of results of randomized trials: Disentangling a complex concept. *Int J Epidemiol*, 39(1), 94–96, 2010.
17. Julian, D.G., Pocock, S.J. *Interpreting a Trial Report*. London, U.K.: WB Saunders, 1997.
18. Wei, L., Ebrahim, S., Bartlett, C., Davey, P.D., Sullivan, F.M., MacDonald, T.M. Statin use in the secondary prevention of coronary heart disease in primary care: Cohort study and comparison of inclusion and outcome with patients in randomised trials. *BMJ*, 330(7495), 821, 2005.
19. Travers, J., Marsh, S., Caldwell, B., et al. External validity of randomized controlled trials in COPD. *Respir Med*, 101(6), 1313–1320, 2007.
20. Travers, J., Marsh, S., Williams, M., et al. External validity of randomised controlled trials in asthma: To whom do the results of the trials apply? *Thorax*, 62(3), 219–223, 2007.
21. Van Spall, H.G., Toren, A., Kiss, A., Fowler, R.A. Eligibility criteria of randomized controlled trials published in high-impact general medical journals: A systematic sampling review. *JAMA*, 297(11), 1233–1240, 2007.
22. Coca, S.G., Krumholz, H.M., Garg, A.X., Parikh, C.R. Underrepresentation of renal disease in randomized controlled trials of cardiovascular disease. *JAMA*, 296(11), 1377–1384, 2006.
23. Amiel, P., Moreau, D., Vincent-Genod, C., et al. Noninvitation of eligible individuals to participate in pediatric studies: A qualitative study. *Arch Pediatr Adolesc Med*, 161(5), 446–450, 2007.

24. Petersen, M.K., Andersen, K.V., Andersen, N.T., Soballe, K. "To whom do the results of this trial apply?" External validity of a randomized controlled trial involving 130 patients scheduled for primary total hip replacement. *Acta Orthop*, 78(1), 12–18, 2007.
25. Steg, P.G., Lopez-Sendon, J., Lopez de Sa, E., et al. External validity of clinical trials in acute myocardial infarction. *Arch Intern Med*, 167(1), 68–73, 2007.
26. Sheldon, T.A., Guyatt, G.H., Haines, A. Getting research findings into practice. When to act on the evidence. *BMJ*, 317(7151), 139–142, 1998.
27. Halm, E.A., Lee, C., Chassin, M.R. Is volume related to outcome in health care? A systematic review and methodologic critique of the literature. *Ann Intern Med*, 137(6), 511–520, 2002.
28. Executive Committee for the Asymptomatic Carotid Atherosclerosis Study. Endarterectomy for asymptomatic carotid artery stenosis. *JAMA*, 273(18), 1421–1428, 1995.
29. Rockman, C.B., Halm, E.A., Wang, J.J., et al. Primary closure of the carotid artery is associated with poorer outcomes during carotid endarterectomy. *J Vasc Surg*, 42(5), 870–877, 2005.
30. Herbert, R.D., Bo, K. Analysis of quality of interventions in systematic reviews. *BMJ*, 331(7515), 507–509, 2005.
31. Glasziou, P., Meats, E., Heneghan, C., Shepperd, S. What is missing from descriptions of treatment in trials and reviews? *BMJ*, 336(7659), 1472–1474, 2008.
32. Schulz, K.F., Altman, D.G., Moher, D. CONSORT 2010 statement: Updated guidelines for reporting parallel group randomised trials. *PLoS Med*, 7(3), e1000251, 2010.
33. Boutron, I., Moher, D., Altman, D.G., Schulz, K.F., Ravaud, P. Extending the CONSORT statement to randomized trials of nonpharmacological treatment: Explanation and elaboration. *Ann Intern Med*, 148(4), 295–309, 2008.
34. Boutron, I., Moher, D., Altman, D.G., Schulz, K.F., Ravaud, P. Methods and processes of the CONSORT Group: Example of an extension for trials assessing nonpharmacological treatments. *Ann Intern Med*, 148(4), W60–W66, 2008.
35. Glasziou, P., Chalmers, I., Altman, D.G., et al. Taking healthcare interventions from trial to practice. *BMJ*, 341, c3852, 2010.
36. Perera, R., Heneghan, C., Yudkin, P. Graphical method for depicting randomised trials of complex interventions. *BMJ*, 334(7585), 127–129, 2007.
37. Haynes, A.B., Weiser, T.G., Berry, W.R., et al. A surgical safety checklist to reduce morbidity and mortality in a global population. *N Engl J Med*, 360(5), 491–499, 2009.
38. Patsopoulos, N.A., Ioannidis, J.P. The use of older studies in meta-analyses of medical interventions: A survey. *Open Med*, 3(2), e62–e68, 2009.
39. Horton, R. Common sense and figures: The rhetoric of validity in medicine (Bradford Hill Memorial Lecture 1999). *Stat Med*, 19(23), 3149–3164, 2000.
40. Jacquier, I., Boutron, I., Moher, D., Roy, C., Ravaud, P. The reporting of randomized clinical trials using a surgical intervention is in need of immediate improvement: A systematic review. *Ann Surg*, 244(5), 677–683, 2006.
41. Juni, P., Altman, D.G., Egger, M. Systematic reviews in health care: Assessing the quality of controlled clinical trials. *BMJ*, 323(7303), 42–46, 2001.
42. Rothman, K.J. *Modern Epidemiology*. Boston, MA: Little Brown, 1986.
43. Moseley, J.B., O'Malley, K., Petersen, N.J., et al. A controlled trial of arthroscopic surgery for osteoarthritis of the knee. *N Engl J Med*, 347(2), 81–88, 2002.

44. Torgerson, D., Moffett, J.K. Patient preference and validity of randomized controlled trials. *JAMA*, 294(1), 41–42, author reply 42, 2005.
45. Torgerson, D.J., Sibbald, B. Understanding controlled trials. What is a patient preference trial? *BMJ*, 316(7128), 360, 1998.
46. Godwin, M., Ruhland, L., Casson, I., et al. Pragmatic controlled clinical trials in primary care: The struggle between external and internal validity. *BMC Med Res Methodol*, 3(1), 28, 2003.
47. Persaud, N., Mamdani, M.M. External validity: The neglected dimension in evidence ranking. *J Eval Clin Pract*, 12(4), 450–453, 2006.
48. Pibouleau, L., Boutron, I., Reeves, B.C., Nizard, R., Ravaud, P. Applicability and generalisability of published results of randomised controlled trials and nonrandomised studies evaluating four orthopaedic procedures: Methodological systematic review. *BMJ*, 339, b4538, 2009.

8

Assessing Nonpharmacological Interventions in Cluster Randomized Trials

Bruno Giraudeau
University Hospital of Tours

Philippe Ravaud
Paris Descartes University and
Public Assistance—Hospitals of Paris

CONTENTS

8.1 Introduction

A cluster randomized trial is defined as a trial in which "intact social units, or clusters of individuals, rather than individuals themselves, are randomized" [1]. Such a design is well adapted to assess organizational and behavioral interventions implemented at the level of health organizational units or geographical areas [2], as well as interventions that apply at the patient level such as therapeutic education programs [3], interventions aimed at curing or preventing transmission of contagious diseases [4], or global care plans [5]. The cluster randomized trial, also named "real-world trial" [6], is now considered a well-adapted design for pragmatic trials [7] as "a way to allow for real-world practice within study centers while addressing inter-center bias by randomizing those to the study interventions." The use of such trials has greatly increased over the past 15 years [8] and even motivated an extension of the CONSORT statement [9].

8.2 Why Randomize Clusters?

There exist several motivations to randomize clusters. First, some interventions apply at the cluster level, even though outcomes and expected benefits are measured at the patient level. As an example, Althabe et al. [2] demonstrated that implementation of evidence-based obstetrical practices at the hospital level (i.e., for the whole team of birth attendants) increased the prophylactic use of oxytocin during the third stage of labor and reduced the use of episiotomy. Such interventions are out of the scope of the present book (focusing on nonpharmacological treatment aimed at the patient level) and are therefore not discussed.

Second, if the disease of interest is contagious, there is a natural risk of contamination between individuals, in the medical sense of the term, and therefore also from a statistical point of view. As explained by Hayes et al. [10], "some interventions are expected to influence a person's infectiousness to others rather than, or as well as, decreasing their own susceptibility to infection." Thus, group contamination exists bilaterally. In one view, infectious agents can be spread among subjects within a cluster, and in another, people living within a cluster of noncontaminated people are not susceptible to infection. Therefore, cluster randomization is highly appropriate to study interventions aimed at curing contagious diseases or preventing their transmission. Most but not all treatments of contagious infectious diseases are pharmacological treatments. In a study of head lice, the effect of the Bug Buster kit [11] was assessed with cluster randomization (the immediate family was given the same treatment as was allocated to the family member with confirmed head–louse infestation, if necessary), although only the index case was taken into account in the analysis [12]. In the same way, for interventions aimed at preventing disease transmission, a household randomization, for example, is appropriate. As an example, Cowling et al. demonstrated that using surgical facemasks plus hand hygiene for all household members seemed to prevent household transmission of influenza virus when implemented within 36 h of symptom onset of the index patient [4].

Third, in some situations, although individual randomization is theoretically possible (i.e., there is no risk of contamination), cluster randomization is performed for logistic convenience or pragmatism [6,7]. Mason et al. [13], who evaluated the benefits of paramedic practitioners assessing and, when possible, treating older people in the community after minor injury or illness, randomized weeks so that in some periods, the paramedic practitioner service was active and in other periods inactive. In the same way, Plaisance et al. [14] allocated patients with cardiac arrest to receive active compression–decompression cardiopulmonary resuscitation or standard cardiopulmonary resuscitation according to whether their arrest occurred on an odd or even day of the month, respectively. Obviously, in this latter emergency situation, there was no other solution than considering time units as randomization units.

Finally, the use of cluster randomization is necessary in cases of some form of "herd effect" or "mass effect" because of interactions between members of clusters. Campbell et al. [15] assessed the effect of a peer-led intervention for smoking prevention in adolescence by randomizing schools and training influential students to act as peer supporters. Other examples are randomization of football clubs to assess the effect of a comprehensive warm-up program to improve strength, awareness, and neuromuscular control during static and dynamic movements to prevent injuries of female players [16], south African villages to assess the impact of an HIV prevention program on the incidence of HIV [17], or work units within companies to evaluate the effectiveness of a prevention program for low back pain [17]. Actually, for these interventions, despite addressing individuals, they may affect others, or compliance may be enhanced by cluster implementation. Hahn et al. [18], in discussing a Cochrane review of hip protectors, observed a large positive effect with cluster randomized trials but no clear benefit with individually randomized trials. To explain such a difference, one possibility was the existence of a "herd effect" in cluster randomized trials, "which were often performed in nursing homes, where compliance with using the protectors may have been enhanced" because of some form of collective compliance [19].

8.3 Which Randomization Unit?

Donner and Klar [1] define randomization units as "social units" or "clusters of individuals." In 89.9% of reports of 199 cluster randomized trials reviewed by Eldridge et al. [20], randomization units were health professionals, general practices, or other organizations such as clinics. Social units may also correspond to schools, households, companies, or, in a more anecdotal way, sport clubs, churches, prisons, or YMCAs. Two other types of clusters are geographic and temporal. For instance, sub-Saharan cluster randomized trials commonly randomize villages or residential areas as units [21]. Time units are of great interest in emergency situations (with lack of time to perform individual randomization) as was done by Plaisance et al. [14]. Such a design is sometimes named an "alternate month design," even though the unit of time is not the month. In the latter examples, there is no randomization (just alternation), but nevertheless, this design shares great similarities with the cluster randomization design. A recent, more simple form of the "alternate month design," is the cluster randomized crossover design [22,23], whereby each cluster receives both intervention and control interventions consecutively, in separate time periods, and only once. Randomization then determines the order in which interventions are administered. This design was used by Marks et al. [24], who demonstrated that classroom exposure to low nitrogen oxide unflued gas heaters caused increased respiratory symptoms. This latter design is of great interest when few clusters are included because it allows for limiting the imbalance between groups, especially if the case mix of patients differs among centers or clusters. The design is also of statistical interest, because the interperiod correlation increases power, in the same way as in individual randomized crossover trials [22]. However, it can be used only to study interventions with transient effect, to avoid risk of carryover effect.

8.4 Which Type of Intervention?

In cluster randomized trials, because of the two units (individuals embedded in clusters), we may differentiate two types of interventions, whether they apply at the cluster level or the individual level. Indeed, Eldridge et al. [25] defined a four-class typology of trial interventions based on the primary reason for adopting a cluster randomization design. First, the "individual-cluster" type refers to interventions that apply at the individual level, such as treatments or therapeutic educations. These interventions include "professional-cluster" interventions (interventions that apply directly to health professionals who are then in charge of recruiting patients) and "cluster–cluster" interventions (interventions also implemented at the cluster level but which correspond to changes in organizations rather than interventions for health professionals). Finally, "external-cluster" interventions refer to additional staff, as in the Wood et al. study [26], in which nurses were in charge of a multidisciplinary family-based cardiovascular disease prevention program. This latter type of intervention can motivate the use of cluster randomization for convenience in the conduct of the trial, although an individually randomized trial is theoretically often conceivable.

A major distinction between cluster randomized trials and individually randomized trials is that in some situations, participants, embedded in clusters, cannot opt out of the intervention. For instance, when Marks et al. [24] randomized time periods during which classrooms were exposed to low-nitrogen oxide unflued gas heaters or to non-indoor-air-emitting flued gas heaters, students could not opt out of the intervention. Other examples are the implementation of a community-wide intervention for overall antibiotic uses for young children by randomizing 16 communities in Massachusetts [27] or adjunction of ferrous sulfate in school meals to reduce anemia in Brazil [28].

8.5 Which Type of Consent? Which Information?

The latter examples raise an ethical issue specific to cluster randomized trials [29]. Indeed, in some situations, the consent at the individual patient level is not feasible, as in the Finkelstein et al. study [27], in which geographical areas were randomized and the intervention consisted in information directed at physicians and patients by mail, pharmacies, child centers, and so on. Eldridge et al. [25] estimated that only 40.2% of 199 reports of cluster randomized trials described obtaining consent for participation, but the authors found no evidence of underreporting of consent procedures in publications. Hutton [29] asserted that "consent can be sought, or not sought" and actually defines three levels of consent: (1) the use of routinely held data, (2) the collection of additional data (with or without invasive procedures), and (3) the offer or administration of an intervention. Interestingly, the 1996 version of the Declaration of Helsinki described situations in which informed consent is not to be obtained: "if the physician considers it essential not to obtain informed consent, the specific reasons for this proposal should be stated in the experimental protocol for transmission to the independent committee." However, in the current version of the declaration (Tokyo, 2004), this notion no longer appears.

In some situations, although patients consent, giving complete information is not desirable because it may lead to a contamination, which the cluster randomization design tends to avoid. In the Leonhardt et al. [30] study, which assessed the impact of a motivational counseling approach to promote physical activity in low back pain patients, patient information was partial: patients received no information about their practice allocation and the intervention in other study arms. Ethical issues in cluster randomized trials (among them patient information and individual-level consent) may then have methodological impacts and therefore must be handled considering patient protection and the internal validity of the trial. As advocated by Eldridge et al. [25], a public debate regarding the acceptability of not obtaining consent in certain situations is necessary, and some work on this topic is in progress [31,32].

8.6 Which Statistical Specificities?

The hierarchical nature of the cluster randomization design implies that for patients or participants embedded in clusters, clusters are the randomization unit and targets of inference are usually individuals. Such a design implies correlation in collected data: two individuals from the same cluster have more similar outcomes than two individuals from different clusters. As an example, Nourhashemi et al. [5] assessed whether a global care plan decreased the rate of functional decline in patients with mild to moderate Alzheimer's disease. Thus, not surprisingly, patients cared for in the same memory clinics, by a common multidisciplinary team (which may vary in its effectiveness), are more prone to exhibit similar decline than if they were cared for by distinct teams. In this example, the underlying concept is "clustering by health professional" [33], as described in another section of this book. The same concept also holds when therapeutic education programs are provided by health professionals or when guidelines or organizational interventions are implemented at the physician level. Otherwise, correlation in outcomes is naturally expected if we are interested in an infectious disease [10] but also in case of the "herd effect," as previously described. On the contrary, we may expect low correlation or even no correlation if cluster randomization has been performed for logistic convenience or pragmatism, although, to our knowledge, no data confirm such a viewpoint.

The within-cluster similarity is generally assessed by the intraclass correlation coefficient (ICC), defined as the proportion of variation in the outcome that can be explained by the variation between clusters. This coefficient is usually of small magnitude (median value of 0.048 in the Campbell et al. study [34]), and known to be higher for process variables than for outcome variables. Indeed, "there is potential for greater variability in measures of patient outcome at the cluster level compared with measures of process, such as physician behavior" [34]. From a statistical point of view, this correlation must be taken into account both for the sample size calculation (cluster randomized trials are known to be less efficient than individually randomized trials) and for data analysis. This situation is presently well known, and, as stated by Murray et al. [35], "there are valid methods that are readily available and well documented for the design and analysis of group randomized trials." Nevertheless, improvements are needed: a recent review illustrated that less than one trial in four was correctly planned (i.e., no sample size inflation because of intracluster correlation) and less than one in two reported correct analyses (i.e., took into account clustering in data) [36].

8.7 What Are the Limits to Cluster Randomization?

To date, the major methodological challenge cluster randomized trialists face is the risk of lack of internal validity because of selection bias. Puffer et al. [37] were the first to point out this potential problem in cluster randomized trials, then Hahn et al. [18], Eldridge et al. [38,39], and Giraudeau and Ravaud [40]. Eldridge et al. [38] concluded in their review that "about a quarter [of CRTs] were potentially biased because of procedures surrounding recruitment and identification of patients." As emphasized by Torgerson and Torgerson, the point is often that "cluster randomized trials first recruit the clusters, then randomize and finally recruit the participants: such an approach invites bias" [41]. As an example, in the UK BEAM trial, which assessed three treatments for back pain in primary care [42], 71.4% of recruited participants were in the intervention group, even though the randomization was 1:1, and groups were not comparable on several clinically important criteria. Selection bias may thus be quantitative but also qualitative.

To prevent selection bias, Puffer et al. [37] proposed the identification and complete inclusion of participants before the randomization of a cluster, which allows for maintaining the usual chronology of a randomized trial: inclusion before randomization. If this solution is not possible, the authors also advise blinding-independent recruiters to the allocation group, so that recruitment can be performed independently of the randomization result. However, such solutions are not always possible.

Finally, the intention-to-treat (ITT) principle is highly challenged in cluster randomized trials. Indeed, in individually randomized trials "we know which participants were randomized and must be taken into account in the ITT analysis, whereas in cluster randomized trials, we know which clusters have been randomized but we frequently do not know exactly which participants within clusters should have been included in the trial" [40]. Moreover, empty clusters are discarded before the statistical analysis and therefore randomized units are discarded, which violates the ITT principle. Therefore, we may question analyzing the results of cluster randomized trials as randomized trials, and some authors have used some form of adjustment (as in observational studies), such as propensity scores [43,44].

8.8 Conclusion

The cluster randomized design is well adapted to situations that may involve intergroup contamination, whatever the reason (contagiousness, herd effect, contamination induced by health professionals, etc.). Some authors even consider this design "a preferred option" for real-world trials [7] as a way to allow for real-world practice. Nevertheless, cluster randomization is not a panacea: it requires more patients and may also incur important biases. Opting for cluster randomization may therefore be justified by methodological, scientific, or feasibility reasons, and the rationale for such a choice must be reported, as advised by the CONSORT Statement extension for cluster randomized trials [9].

References

1. Donner, A., Klar, N. *Design and Analysis of Cluster Randomization Trials*. London, U.K.: Arnold, 2000.
2. Althabe, F., Buekens, P., Bergel, E., et al. A behavioral intervention to improve obstetrical care. *N Engl J Med*, 358(18), 1929–1940, 2008.

3. Davies, M.J., Heller, S., Skinner, T.C., et al. Effectiveness of the diabetes education and self management for ongoing and newly diagnosed (DESMOND) programme for people with newly diagnosed type 2 diabetes: Cluster randomised controlled trial. *BMJ*, 336(7642), 491–495, 2008.
4. Cowling, B.J., Chan, K.H., Fang, V.J., et al. Facemasks and hand hygiene to prevent influenza transmission in households: A cluster randomized trial. *Ann Intern Med*, 151(7), 437–446, 2009.
5. Nourhashemi, F., Andrieu, S., Gillette-Guyonnet, S., et al. Effectiveness of a specific care plan in patients with Alzheimer's disease: Cluster randomised trial (PLASA study). *BMJ*, 340, c2466, 2010.
6. Freemantle, N., Strack, T. Real-world effectiveness of new medicines should be evaluated by appropriately designed clinical trials. *J Clin Epidemiol*, 63(10), 1053–1058, 2010.
7. Schwartz, D., Lellouch, J. Explanatory and pragmatic attitudes in therapeutical trials. *J Chronic Dis*, 20(8), 637–648, 1967.
8. Bland, J.M. Cluster randomised trials in the medical literature: Two bibliometric surveys. *BMC Med Res Methodol*, 4, 21, 2004.
9. Campbell, M.K., Elbourne, D.R., Altman, D.G. CONSORT statement: Extension to cluster randomised trials. *BMJ*, 328(7441), 702–708, 2004.
10. Hayes, R.J., Alexander, N.D., Bennett, S., Cousens, S.N. Design and analysis issues in cluster-randomized trials of interventions against infectious diseases. *Stat Methods Med Res*, 9(2), 95–116, 2000.
11. Hill, N., Moor, G., Cameron, M.M., et al. Single blind, randomised, comparative study of the Bug Buster kit and over the counter pediculicide treatments against head lice in the United Kingdom. *BMJ*, 331(7513), 384–387, 2005.
12. Chosidow, O. Bug Buster for head lice: Is it effective? *Arch Dermatol*, 142(12), 1635–1637, 2006.
13. Mason, S., Knowles, E., Colwell, B., et al. Effectiveness of paramedic practitioners in attending 999 calls from elderly people in the community: Cluster randomised controlled trial. *BMJ*, 335(7626), 919, 2007.
14. Plaisance, P., Lurie, K.G., Vicaut, E., et al. A comparison of standard cardiopulmonary resuscitation and active compression–decompression resuscitation for out-of-hospital cardiac arrest. French Active Compression-Decompression Cardiopulmonary Resuscitation Study Group. *N Engl J Med*, 341(8), 569–575, 1999.
15. Campbell, R., Starkey, F., Holliday, J., et al. An informal school-based peer-led intervention for smoking prevention in adolescence (ASSIST): A cluster randomised trial. *Lancet*, 371(9624), 1595–1602, 2008.
16. Soligard, T., Myklebust, G., Steffen, K., et al. Comprehensive warm-up programme to prevent injuries in young female footballers: Cluster randomised controlled trial. *BMJ*, 337, a2469, 2008.
17. Jewkes, R., Nduna, M., Levin, J., et al. Impact of stepping stones on incidence of HIV and HSV-2 and sexual behaviour in rural South Africa: Cluster randomised controlled trial. *BMJ*, 337, a506, 2008.
18. Hahn, S., Puffer, S., Torgerson, D.J., Watson, J. Methodological bias in cluster randomised trials. *BMC Med Res Methodol*, 5(1), 10, 2005.
19. Higgins, J.P.T., Green, S. *Cochrane Handbook for Systematic Reviews of Interventions*. Chichester, U.K.: Wiley-Blackwell, 2008.

20. Eldridge, S.M., Ashby, D., Feder, G.S., Rudnicka, A.R., Ukoumunne, O.C. Lessons for cluster randomized trials in the twenty-first century: A systematic review of trials in primary care. *Clin Trials*, 1(1), 80–90, 2004.

21. Isaakidis, P., Ioannidis, J.P. Evaluation of cluster randomized controlled trials in sub-Saharan Africa. *Am J Epidemiol*, 158(9), 921–926, 2003.

22. Giraudeau, B., Ravaud, P., Donner, A. Sample size calculation for cluster randomized cross-over trials. *Stat Med*, 27(27), 5578–5585, 2008.

23. Turner, R.M., White, I.R., Croudace, T. Analysis of cluster randomized cross-over trial data: A comparison of methods. *Stat Med*, 26(2), 274–289, 2007.

24. Marks, G.B., Ezz, W., Aust, N., et al. Respiratory health effects of exposure to low-NOx unflued gas heaters in the classroom: A double-blind, cluster-randomized, crossover study. *Environ Health Perspect*, 118(10), 1476–1482, 2010.

25. Eldridge, S.M., Ashby, D., Feder, G.S. Informed patient consent to participation in cluster randomized trials: An empirical exploration of trials in primary care. *Clin Trials*, 2(2), 91–98, 2005.

26. Wood, D.A., Kotseva, K., Connolly, S., et al. Nurse-coordinated multidisciplinary, family-based cardiovascular disease prevention programme (EUROACTION) for patients with coronary heart disease and asymptomatic individuals at high risk of cardiovascular disease: A paired, cluster-randomised controlled trial. *Lancet*, 371(9629), 1999–2012, 2008.

27. Finkelstein, J.A., Huang, S.S., Kleinman, K., et al. Impact of a 16-community trial to promote judicious antibiotic use in Massachusetts. *Pediatrics*, 121(1), e15–e23, 2008.

28. Arcanjo, F.P., Pinto, V.P., Coelho, M.R., Amancio, O.M., Magalhaes, S.M. Anemia reduction in preschool children with the addition of low doses of iron to school meals. *J Trop Pediatr*, 54(4), 243–247, 2008.

29. Hutton, J.L. Are distinctive ethical principles required for cluster randomized controlled trials? *Stat Med*, 20(3), 473–488, 2001.

30. Leonhardt, C., Keller, S., Chenot, J.F., et al. TTM-based motivational counselling does not increase physical activity of low back pain patients in a primary care setting—A cluster-randomized controlled trial. *Patient Educ Couns*, 70(1), 50–60, 2008.

31. Taljaard, M., Weijer, C., Grimshaw, J.M., et al. Ethical and policy issues in cluster randomized trials: Rationale and design of a mixed methods research study. *Trials*, 10, 61, 2009.

32. http://www.sante.gouv.fr/IMG/pdf/resultats_PHRC_2008_hors_cancer.pdf.

33. Lee, K.J., Thompson, S.G. Clustering by health professional in individually randomised trials. *BMJ*, 330(7483), 142–144, 2005.

34. Campbell, M.K., Fayers, P.M., Grimshaw, J.M. Determinants of the intracluster correlation coefficient in cluster randomized trials: The case of implementation research. *Clin Trials*, 2(2), 99–107, 2005.

35. Murray, D.M., Varnell, S.P., Blitstein, J.L. Design and analysis of group-randomized trials: A review of recent methodological developments. *Am J Public Health*, 94(3), 423–432, 2004.

36. Murray, D.M., Pals, S.L., Blitstein, J.L., Alfano, C.M., Lehman, J. Design and analysis of group-randomized trials in cancer: A review of current practices. *J Natl Cancer Inst*, 100(7), 483–491, 2008.

37. Puffer, S., Torgerson, D., Watson, J. Evidence for risk of bias in cluster randomised trials: Review of recent trials published in three general medical journals. *BMJ*, 327(7418), 785–789, 2003.
38. Eldridge, S., Ashby, D., Bennett, C., Wakelin, M., Feder, G. Internal and external validity of cluster randomised trials: Systematic review of recent trials. *BMJ*, 336(7649), 876–880, 2008.
39. Eldridge, S., Kerry, S., Torgerson, D.J. Bias in identifying and recruiting participants in cluster randomised trials: What can be done? *BMJ*, 339, b4006, 2009.
40. Giraudeau, B., Ravaud, P. Preventing bias in cluster randomised trials. *PLoS Med*, 6(5), e1000065, 2009.
41. Torgerson, D.J., Torgerson, C.J. *Designing Randomised Trials in Health, Education and the Social Sciences. An Introduction*. New York: Palgrave Macmillan, 2008.
42. Farrin, A., Russell, I., Torgerson, D., Underwood, M. Differential recruitment in a cluster randomized trial in primary care: The experience of the UK back pain, exercise, active management and manipulation (UK BEAM) feasibility study. *Clin Trials*, 2(2), 119–124, 2005.
43. Ravaud, P., Flipo, R.M., Boutron, I., et al. ARTIST (osteoarthritis intervention standardized) study of standardised consultation versus usual care for patients with osteoarthritis of the knee in primary care in France: Pragmatic randomised controlled trial. *BMJ*, 338, b421, 2009.
44. van Marwijk, H.W., Ader, H., de Haan, M., Beekman, A. Primary care management of major depression in patients aged > or = 55 years: Outcome of a randomised clinical trial. *Br J Gen Pract*, 58(555), 680–686, I–II, discussion 7, 2008.

9

Expertise-Based Trials

Jonathan A. Cook

University of Aberdeen

CONTENTS

9.1 Background

There is a lack of trials of interventional procedures, such as surgery, with many authors issuing repeated call for more studies [1,2]. However, the level has remained low in comparison with other medical areas and specifically pharmacological-based interventions. As noted elsewhere in this book, a randomized evaluation of nonpharmacological interventions raises distinct issues that require careful consideration. This is particularly the case for those that are skill-dependent. An alternative trial design, the expertise-based (EB) trial design, has been proposed for evaluations of such interventions [3,4]. This trial design potentially overcomes a number of the objections to a conventional design [5].

Under a conventional trial design, a participant is randomized to receive one or the other of the interventions. In a classic placebo-controlled drug trial, the physician delivers both interventions with participants, physicians, and any other outcome assessors blinded to the treatment allocation. The earliest surgical trials followed this design, bar blinding, with the participant randomized to the surgical procedures under evaluation with participating surgeons required to perform both type of procedures (say A and B). In an EB design, the participant is randomized to a procedure as before; however, this procedure will be delivered by a surgeon with expertise in the randomized procedure. Surgeons are no longer required to undertake both. A diagrammatical representation of a conventional and an EB design is shown in Figure 9.1a and b, respectively. Under both approaches, an eligible patient is randomly allocated to an intervention. The key distinction is who the allocated intervention is delivered by. Under an EB design, intervention A is delivered by a surgeon with expertise in delivering intervention A, and correspondingly, intervention B is delivered by a surgeon with expertise in delivering intervention B. This contrasts with a conventional design where every participating surgeon will be expected to be able to offer both interventions A and B.

The use of an EB for the evaluation of surgical interventions was proposed in 1980 as a solution to a number of issues that were seen to hinder the conduct of surgical trials [3]. Despite having been proposed over 20 years ago, this design has rarely been utilized and it has been neglected until more recently [5,6]. However, a parallel can be seen in trials where the interventions under evaluation differ substantially. Medical vs. surgical are commonly performed in this manner where it is both unrealistic and undesirable to have both medical and surgical treatment delivered by the same physician. This type of trial, though not common, has become more common recently and is universally acknowledged to require an EB approach. An example is

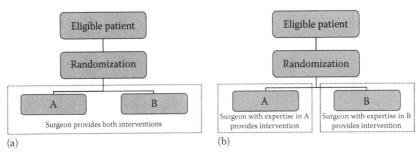

(a)   (b)

FIGURE 9.1
The two design options. (a) Conventional trial design. (b) Expertise-based trial design.

the REFLUX trial which compared a policy of relatively early laparoscopic surgery with a continued best medical management policy for people with more severe gastroesophageal reflux disease [7].

More recently, it has been proposed for more widespread usage due to potential advantages over the conventional design [5]. We will consider the merits of this approach in contrast to a conventional randomized controlled trial (RCT) design highlighting the challenges for both and the need for further research. As noted earlier, this approach has relevance beyond to the surgical situation to other areas, which requires specialist knowledge or skills where knowledge of both interventions is not common; for example, the delivery of psychotherapy interventions [8]. It is, however, in the surgical field where it has received most attention to date and from which we will draw our examples.

9.2 Consideration of the Potential Benefits of Expertise-Based Trial Design

We now consider the potential benefits of adopting an EB trial design over the conventional design.

9.2.1 Differential Expertise

In a conventional RCT, a participating surgeon is expected to have similar expertise in both procedures in practice, which is unlikely to be the case for all surgeons in many situations. Whether due to having more exposure during training, positive personal experience, through colleagues and/or because they believe it to be more effective, surgeons will tend to have a preference between interventions and develop greater expertise in one than the other. It is therefore possible that there will be differential skill and experience with respect to the interventions. In some situations, a substantial proportion of surgeons may only routinely perform one or the other of the procedures. This differential expertise may lead to more rigorous and skilled performance when patients are randomized to the preferred intervention.

It is widely acknowledged that the adoption of a new intervention by a surgeon leads to an individual learning curve, whereby performance changes over time as experience with the new intervention is gained (see Chapter 4). In some cases, this learning process may continue over a large number of procedures [9]. A common criticism against trials with a conventional design, which compare a new vs. an established intervention, is that the participation surgeons have insufficient expertise in the new intervention and thus biasing the results against the new intervention (*differential expertise bias*) [10].

The EB design explicitly acknowledges the process of learning by not requiring participant surgeons to have expertise in both interventions and thereby can prevent such bias. The requirement of a sufficient number of surgeons to have acquired expertise in the new intervention in addition to having expertise in the established intervention may result in a narrower timeframe before widespread adoption of the new intervention and the potential loss of equipoise, which may prejudice the undertaking of an RCT evaluation with a conventional design being more susceptible.

9.2.2 Protection from Other Potential Biases

Unlike the delivery of pharmacological interventions, blinding of the surgeon to the allocated surgical intervention being delivered is not possible. This inevitably leads to greater potential for the introduction of bias into the evaluation. This can take two main forms whereby the relative performance of the interventions is artificially modified. First, bias may occur through the application of a co-intervention. Second, knowledge of the allocation can also result in bias being introduced when an outcome is assessed. It is not uncommon in surgical trials for the surgeon who delivered the intervention also to undertake the assessment of outcome. By virtue of only surgeons with appropriate expertise (and possibly matching their preference) delivering an intervention, the EB approach may provide some protection against both of these forms of bias beyond that of a conventional trial design. For situations where contamination between interventions is a concern, the EB design would provide additional protection. Surgical treatment crossovers can occur before, during, or after the scheduled procedure. An inevitable consequence of undertaking a less familiar intervention will be an increase in the likelihood of abandonment of the less familiar intervention for a more familiar one. For example, conversions between laparoscopic and open surgeries have been shown to reduce with experience. An EB design would, in principle, avoid crossovers due to unfamiliarity with an intervention.

9.2.3 Recruitment of Surgeons/Patients

In comparison with drug trials, an additional challenge facing trialists is the recruitment of surgeons who are in essence integral part of the proposed intervention [11]. As noted earlier, understandably surgeons often have strong preferences for a particular intervention. This can lead to situations where although the surgical community in general is uncertain about the relative merits of the interventions (community equipoise), most individual surgeons may have strong views that would preclude them from participation in a conventional trial. Under an EB design, such surgeons could participate in the arm which they are comfortable with provided they have sufficient expertise. We will consider some empirical evidence of this later (Section 9.4). For a participant, an EB design provides explicit assurance that they will receive a procedure delivery by someone with appropriate expertise who is comfortable undertaking this intervention and thereby increase the recruitment of eligible patients.

9.3 Challenges Faced in Conducting EB Trials

Having considered the potential benefits of adopting a EB trial design over the conventional design, we now consider the main questions trialist face in undertaking an EB trial: How can expertise be defined? How does the EB design differ in trial organization? What is the impact upon the required sample size? To whom are the results generalizable?

9.3.1 How Can Expertise Be Defined?

While the generic concept of an EB design is both attractive and intuitive, the defining of "expertise" for a particular intervention is remarkably difficult. Expertise cannot be measured directly and some form of proxy must be used. Commonly used proxies are the number of previously performed procedures (possibly including similar interventions), the years of operative experience the surgeon has, the surgeon's clinical grade, formal training (e.g., courses, simulations, and/or mentoring), a performance/outcome assessment, or some combination of them. All of these approaches are inhibited by the generally poor level of reporting in surgical studies and understanding of the learning process [12]. Further research in this area is needed along with better reporting of surgical (and other skill-dependent treatment) trials and a more formalized mechanism of training and recording experience to provide increased transparency and transferability.

However, a further and perhaps equally concerning issue is whether the inherent ability of the two "expertise" groups to deliver the respective intervention is equivalent. Where a more complex intervention is being compared with a less sophisticated or novel method those who have "expertise" in the complex intervention could be a more skilled cohort in comparison with the general surgical community [13]. Thus, an EB trial could conceivably have an reverse "differential expertise bias" favoring the complex intervention.

9.3.2 What Is the Impact on the Required Sample Size?

As a consequence of surgeons only delivering interventions in one arm, the EB design introduces extra variation in comparison with a conventional RCT and therefore all other factors being equal would require a larger sample size [14]. It is likely that an EB trial will require a larger sample size in comparison with a corresponding conventional trial. The relative loss of information is most pronounced, as in a cluster randomized trial when a relatively small number of surgeons (clusters) perform a large number of cases. It is conceivable that in some situation an expertise trial could have a large treatment difference than a conventional trial due to the aforementioned reduced crossover and perhaps more rigorous application of a familiar intervention. Also weighing in favor of an EB approach is the possible expectation of increased surgeon and patient participants. Generally, in terms of sample size, the EB design suits an evaluation of a common surgical intervention where a large pool of potential surgeons is available. It should be noted that conventional trials also suffer from a loss of information (compared with a constant pharmacological intervention) themselves even though this is commonly overlooked in the sample size calculation.

9.3.3 How Does the Expertise-Based Design Differ in Trial Organization?

A unique issue related to conducting a trial with an EB design over the conventional design is the need for two sets of expertise to be available. In a multicenter study, this requires dual expertise in both interventions to be available in every participating center (or at least a commutable distance to them), which may be difficult to achieve in smaller centers.

9.3.4 To Whom Are the Results Generalizable?

In contrast with a conventional trial, the results of EB trial need more careful generalization of the results. Whereas a conventional trial can, at least in theory, be applied to any surgeon, the results of an EB trial are directly applicable only to surgeons who have (or could acquire) the required expertise. It does not necessarily follow that any surgeon will achieve similar outcome. While, realistically, this will to some degree also be the case with conventional trials, it is explicitly part of the design of an EB trial and this difference should be recognized. A more "loose" definition of expertise can increase the applicability of the results as with a conventional trial.

9.4 Usage and Examples

As noted early while the EB design was proposed in 1980, its usage has been limited. A summary of example surgical EB trial designs and basic trial features is given in Table 9.1, though a number are currently underway which use this design. There is limited though increasing evidence with regards to the EB design's potential benefits. As many trials suffer from difficulty in achieving recruitment, the willingness of surgeons to participate in a potential EB and conventional design is of key importance. We briefly summarize the results of three surveys of surgeons practice and opinion.

TABLE 9.1

Example Surgical EB Trials for Further Information

Trial	Clinical Area	Interventions	No. of Participants	No. of Surgeons per Intervention
Finkemeier et al. [15]	Orthopedics	Nail insertion with reaming vs. nail insertion without reaming	94	2 vs. 3
Phillips et al. [16]	Orthopedics	Open reduction and internal fixation ASIF technique vs. closed cast treatment or open reduction and internal fixation of medial malleolus	138	1 vs. 1
Machler et al. [17]	Cardiac	Minimally invasive aortic valve surgery vs. conventional aortic valve surgery	120	2 vs. 2
Wihlborg et al. [18]	Orthopedics	Rydell four-flanged nail vs. Gouffon pins	200	4 vs. 3
Wyrsch et al. [19]	Orthopedics	Open reduction and internal fixation of the tibia and fibula vs. external fixation with or without limited internal fixation	39	2 vs. 4
CABRI [20]	Cardiac/ cardiology	Coronary angioplasty vs. bypass revascularization	1054	N/R

Notes: See references [5,20] for further information. N/R stands for not reported.

9.4.1 SPRINT Trial

The experience of surgeons participating in a large conventional RCT (SPRINT trial) [21] of two surgical interventions (reaming vs. no reaming) for treating a tibial shaft fracture was surveyed. More surgeons had no prior experience of the more complex non-ream intervention than the ream approach (26 (35%) vs. 7 (9%) of the 74 for which experience was known). The median difference in the number of procedures experience was 7, 95% CI (5–11). Over 80% of surgeons believed that the reamed procedure was superior prior to trial commencement.

9.4.2 Open versus EVAR Repair

Canadian vascular surgeons were surveyed to assess their attitudes and experience of endovascular aortic repair (EVAR) and traditional open repair and their perspective on the merits of the two designs options for a randomized comparison of elective surgery [22]. In total, 101 practicing surgeons responded. Based upon pre-specified expertise of 100 or more procedures' experience for open repair and 60 or more for EVAR, 80 (80%) and only 13 (13%), respectively, had appropriate expertise. Willingness to participate was similar between the designs with 54% and 51% for EB and conventional designs, respectively.

9.4.3 HTO versus UKA Repair

Members of the Canadian Orthopedic Association were asked on willingness to participate in an EB or conventional RCT and any preference between the two trial designs for an evaluation of high tibial osteotomy (HTO) and unicompartmental knee arthroplasty (UKA) [23]. Patients who had a varus alignment, a passively correctable deformity, and a clinically intact anterior cruciate ligament were eligible. Two hundred and one practicing surgeons responded. Expertise in a procedure was defined as having performed 10 or more procedures. One hundred and two (53%) were willing to participate in an EB RCT compared with 35 (18%) for a conventional RCT. Ninety-seven surgeons (52%) stated a strong or moderate preference for EB design compared with 25 (14%) for conventional and 63 (34%) who had mild or no preference. Surgeons' expertise was related to opinion about the relative merits of the procedures favoring the intervention in which they had expertise.

From these surveys, there would seem to be at least equal and in some cases substantially increased surgeon willingness to participate in an EB trial in comparison with a conventional trial design. There is supportive evidence for the potential impact of differential expertise bias in some conventional trials. However, the results varied across scenario suggest that any benefit is situation-specific and argue for the conduct of such a survey prior to the commencement of any large, multicenter surgical trials.

9.5 Conclusion

The EB trial design has been proposed to address concerns related to the conventional trial design. The EB design offers theoretical benefits over conventional approach in terms of ethics, increased participation of surgeons and patients, reduced occurrence of crossover between interventions, less scope for differential performance bias, and assurance of parity of expertise. The potential increased rate of participation of surgeons (and ultimately patients) is particularly of note. However, there are difficult methodological issues (such as defining expertise), and practical challenges that need to be considered. Additionally, there is also a likely loss of precision associated with the EB design, which could result in a substantially larger sample size being needed. It is uncertain how the respective benefits of the two designs are realized in practice and further exploration with the EB design is needed. In some scenarios, a hybrid of both designs (expertise and conventional options for participating surgeon) might be the optimal design. As with any trial, an EB trial design needs careful implementation.

References

1. Wente, M.N., Seiler, C.M., Uhl, W., Büchler, M.W. Perspectives of evidence-based surgery. *Dig Surg*, 20, 263–269, 2003.
2. Pollock, A.V. Surgical evaluation at the crossroads. *Br J Surg*, 80, 964–966, 1993.
3. Van der Linden, W. Pitfalls in randomized surgical trials. *Surgery*, 87, 258–262, 1980.
4. Rudicel, S., Esdaile, J. The randomized clinical trial in orthopaedics: Obligation or option? *J Bone Joint Surg Am*, 67, 1284–1293, 1985.
5. Devereaux, P.J., Bhandari, M., Clarke, M., et al. Need for expertise based randomised controlled trials. *BMJ*, 330, 88, 2005.
6. Johnston, B.C., da Costa, B.R., Devereaux, P.J., Akl, E.A., Busse, J.W.; Expertise-Based RCT Working Group. The use of expertise-based randomized controlled trials to assess spinal manipulation and acupuncture for low back pain: A systematic review. *Spine*, 33(8), 914–918, 2008.
7. Grant, A.M., Wileman, S.M., Ramsay, C.R., et al., and the REFLUX trial Group. Minimal access surgery compared with medical management for chronic gastro-oesophageal reflux disease: UK collaborative randomised trial. *BMJ*, 337, a2664, 2008.
8. Elkin, I., Parloff, M.B., Hadley, S.W., Autry, J.H. NIMH treatment of depression collaborative research program. Background and research plan. *Arch Gen Psychiatry*, 42(3), 305–316, 1985.
9. Vickers, A.J., Bianco, F.J., Serio, A.M., et al. The surgical learning curve for prostate cancer control after radical prostatectomy. *J Natl Cancer Inst*, 99, 1171–1177, 2007.
10. Cook, J.A., Ramsay, C.R., Fayers, P. Statistical evaluation of learning curve effects in surgical trials. *Clinical Trials*, 1, 421–427, 2004.
11. Ergina, P.L., Cook, J.A., Blazeby, J.M., et al., for the Balliol Collaboration. Surgical innovation and evaluation 2: Challenges in evaluating surgical innovation. *Lancet*, 374(9695), 1097–1104, 2009.
12. Cook, J.A., Ramsay, C.R., Fayers, P. Using the literature to quantify the learning curve. *Int J Technol Assess Health Care*, 23, 255–260, 2007.
13. Biau, D.J., Porcher, R. Letter to the editor re: Orthopaedic surgeons prefer to participate in expertise-based randomized trials. *Clin Orthop Relat Res*, 467, 298–300, 2009.
14. Walter, S.D., Ismaila, A.S., Devereaux, P.J., for the SPRINT study investigators. Statistical issues in the design and analysis of expertise-based randomized clinical trials. *Stat Med*, 27, 6583–6596, 2008.

15. Finkemeier, C.G., Schmidt, A.H., Kyle, R.F., Templeman, D.C., Varecka, T.F. A prospective, randomized study of intramedullary nails inserted with and without reaming for the treatment of open and closed fractures of the tibial shaft. *J Orthop Trauma*, 14, 187–193, 2000.

16. Phillips, W.A., Schwartz, H.S., Keller, C.S., et al. A prospective, randomized study of the management of severe ankle fractures. *J Bone Joint Surg Am*, 67, 67–78, 1985.

17. Machler, H.E., Bergmann, P., Anelli-Monti, M., et al. Minimally invasive versus conventional aortic valve operations: A prospective study in 120 patients. *Ann Thorac Surg*, 67, 1001–1005, 1999.

18. Wihlborg, O. Fixation of femoral neck fractures: A four-flanged nail versus threaded pins in 200 cases. *Acta Orthop Scand*, 61, 415–418, 1990.

19. Wyrsch, B., McFerran, M.A., McAndrew, M., et al. Operative treatment of fractures of the tibial plafond: A randomized, prospective study. *J Bone Joint Surg Am*, 78, 1646–1657, 1996.

20. CABRI Trial Participants. First-year results of CABRI (coronary angioplasty versus bypass revascularisation investigation). *Lancet*, 346, 1179–1184, 1995.

21. Devereaux, P.J., Bhandari, M., Walter, S., Sprague, S., Guyatt, G. Participating surgeons' experience with and beliefs in the procedures evaluated in a randomized controlled trial. *Clin Trials*, 1, 225, 2004.

22. Mastracci, T.M., Clase, C.M., Devereaux, P.J., Cinà, C.S. Open versus endovascular repair of abdominal aortic aneurysm: A survey of Canadian vascular surgeons. *Can J Surg*, 51, 142–148, 2008.

23. Bednarska, E., Bryant, D., Devereaux, P.J. for the Expertise-Based Working Group. Orthopaedic surgeons prefer to participate in expertise-based randomized trials. *Clin Orthop Relat Res*, 466, 1734–1744, 2008.

10

Pragmatic Trials and Nonpharmacological Evaluation

Merrick Zwarenstein

Sunnybrook Research Institute and
University of Toronto

CONTENTS

10.1 Introduction

Randomized controlled trials (RCTs) are the gold standard evaluation design for validly comparing the effects of alternative interventions for a problem. The first randomized trial ever published tested a drug, streptomycin, against tuberculosis in hospitalized patients, but increasingly, RCTs are also being used to test many other interventions besides pharmacotherapeutics, including, in order of increasing degree of complexity—implantable devices, surgical interventions, counseling, and other psychological interventions aimed at individuals, families, or groups; knowledge translation interventions aimed at providing post-professional training to clinicians; task-shifting interventions aimed at downloading physician or high skilled professional tasks to less scarce cadres of health workers, health financing changes; and interventions aimed at changing the financing, or the organization of healthcare.

The number of pragmatic trials has risen over time, but still remains <1% of all randomized trials. However, pragmatic approaches to randomized trial design seem to be particularly suited to nonpharmacological trials; indeed, almost all self-described pragmatic trials listed on PubMed for 2009 were nonpharmacological trials.

In this chapter, we review briefly the nature and history of pragmatic trials and contrast their designs with more traditional explanatory trials, explaining the benefits of the pragmatic approach, especially for nonpharmacological evaluations. We then go on to describe two instruments recently developed, one for guiding the design of a pragmatic trial, and one for guiding reporting. We end by exploring the range of nonpharmacological pragmatic trials and imagining a new role for pragmatic trials in future health and social care decision making, providing support for decisions on the implementation of nonpharmacological interventions.

10.2 Applicability: The Key to Appropriate Choice of Design

RCTs are used to assess the benefits and harms of interventions in health care. If conducted properly, they minimize the risk of bias (threats to internal validity), particularly selection bias [1]. But internal validity or lack of bias is not all that counts in judging the usefulness of an evaluation design: an RCT may be free of bias in its estimate of the effects of an intervention in the setting in which the trial was conducted but it may nevertheless have limited applicability (also known as generalizability or external validity) beyond this exact setting [2]. Such a trial would not be of much use to anyone except users of the intervention in the trial setting, and even then, only provided the site remained unchanged and they planned to implement the tested intervention in exactly the way it had been implemented during the trial [3].

The problem of applicability is not trivial, not rare, and not limited to differences between places and facilities. In order to adhere to regulatory requirements, and to maximize the chances that the effects of an experimental drug (in which billions of dollars might have been invested) are shown during the trial, most drug trials are conducted in very specialized settings, with extra resources to implement the intervention "properly". These resources cover more detailed follow-up of the recipients of the intervention to ensure full outcome information for the evaluation. This sometimes results in extra care being given to subjects in the trial as a result of this intensive follow-up that ordinary recipients of usual care would not receive. Resources are also added to ensure that the participants in the trial (both those who receive care and those who provide it) are highly selected to ensure optimal delivery and uptake of the intervention. Practitioners may be unusually well trained or highly skilled, patients may be selected to ensure that they are above average in their adherence to treatment regimens, and healthcare facilities may be selected for the trial on the basis that they are particularly good at sticking to guidelines or extra-careful in their dealings with patients. This means that many of the standard attributes of randomized trial evaluations are distinctly nonstandard, indeed, unrealistic in the real world of usual care delivery, where the tested intervention might be later used, if it shows positively in the trial. And this means that when applied in the real-world setting, an intervention that did well in its original trial might do much worse, as most of these differences between real world and trial setting favor a better outcome under idealized conditions—the results of widespread implementation may well be disappointing.

This question of applicability is central to those who have to choose between treatments and interventions for groups of patients (policy makers), for their own patients (clinicians), or for themselves (patients and families). How likely is it, these decision makers may ask, that this intervention (apparently successful in this randomized trial) will achieve important benefits in my context, administered to me by my clinicians, by me to my patients, or by clinicians to patients in my organization? In other words, "Are these published findings applicable to my decision?"

10.3 History of Pragmatic Trials

Two French statisticians were the first to publish on the distinction between trials aimed at supporting decision making in the real world, and trials aimed at answering a research question of a more academic character: Schwartz and Lellouch [4] coined the terms "pragmatic" to describe trials conducted under usual care conditions, and "explanatory" to describe trials conducted under idealized conditions. They pointed out that pragmatic trials were perfectly designed to help choose between options for care in the real world, whereas explanatory trials were perfectly designed to test causal research hypotheses, for example, that an intervention causes a particular biological change. Table 10.1 shows some key differences between explanatory and pragmatic trials.

There is a continuum rather than a dichotomy between explanatory and pragmatic trials. In fact, Schwartz and Lellouch characterized pragmatism as an attitude to trial design rather than a fixed, binary characteristic of the trial itself. The pragmatic attitude favors design choices that maximize applicability of the trial's results to usual care settings, rely on unarguably important outcomes such as mortality and severe morbidity, and are tested in a wide range of participants.

Do these differences in participants, in care delivery, and in the intervention matter? Yes, as the effects of an intervention, treatment, a counseling

TABLE 10.1

Key Differences between Trials with Explanatory and Pragmatic Attitudes

	Explanatory Attitude	Pragmatic Attitude
Question	Efficacy: can the intervention work?	Effectiveness: does the intervention work when used in normal practice?
Setting	Well-resourced, "ideal" setting	Normal healthcare setting—no extra resources
Participants	Highly selected; poorly adherent participants and those with conditions which might dilute the effect are often excluded	Little or no selection beyond the clinical indication of interest
Intervention	Strictly enforced and adherence is monitored closely	Applied flexibly as it would be in normal practice
Outcomes	Often short-term surrogates, or process measures	Important, often life-changing outcomes that are long term. Directly relevant to participants, funders, communities, and healthcare practitioners
Relevance to practice	Indirect: little effort is made to match the design of the trial to the decision-making needs of those in the usual setting in which the intervention will be implemented	Direct: the trial is designed to meet the needs of those making decisions about treatment options in the setting in which the intervention will be implemented

Source: From Schwartz, D. and Lellouch, J., *J Chron Dis*, 20, 637, 1967. [Reprinted in J Clin Epidemiol, 62, 499–505, 2009].

approach, an educational intervention, or a healthcare reorganization measured in a randomized trial conducted under idealized conditions may be very different in more ordinary care settings, either settings elsewhere than the trial site or even in the very same facility in which the trial was conducted, but absent the extra capacities added during the trial. And if the results of a trial are only valid in the setting and time in which it was conducted, what use are the results to anyone else? As Schwartz and Lellouch wrote: "Most trials done hitherto have adopted the explanatory approach without question; the pragmatic approach would often have been more justifiable."

10.4 Designing and Reporting Pragmatic Trials

A recent tool, the pragmatic–explanatory continuum indicator summary (PRECIS) [6] (Figure 10.1) is intended to be used by trialists designing a trial. It is intended to help them to assess the degree to which their design decisions align with the trial's stated purpose, whether that purpose is informing a decision between two interventions, or a more academic issue of confirming a causal relationship. This tool has 10 dimensions (see Table 10.2 that

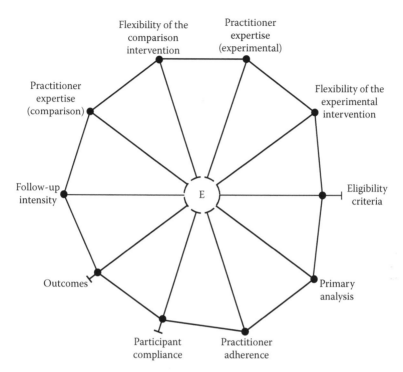

FIGURE 10.1
PRECIS diagram for DOTS trial.

describes the extremes of the scale at pragmatic and explanatory ends) based on trial design decisions (e.g., participant and practitioner expertise, flexibility with which the intervention can be delivered, and choice of comparator), and presents these on a graphical, 10-spoked "wheel." A highly pragmatic trial is out at the rim, whereas explanatory trials are nearer the hub. The advantage of this graph is that it quickly highlights inconsistencies in how the 10 dimensions will be managed in a trial. For example, if, in the trial, shown in Figure 10.1, of directly observed treatment for TB, we had intensely monitored compliance and intervened when it faltered, a single glance at the wheel would have immediately identified this inconsistency with the trial's otherwise pragmatic attitude. This allows trialists to make adjustments, if possible and appropriate, to the design to obtain greater consistency with their trial's purpose.

Decisions about applicability depend on readers being able to assess the feasibility of the intervention in their own context and a new tool is available to assist in this process [5] (Table 10.3). However, understanding what comprises the intervention (and often the comparator) is not always a simple matter of reading the trial report as trial reports currently stand [7]. Detailed reporting of the content of interventions, especially complex, nonpharmacological ones, is often poor [8]. In drafting a publication describing

TABLE 10.2

PRECIS Domains Illustrating the Extremes of Explanatory and Pragmatic Approaches to Each Domain

Domain	Pragmatic Trial	Explanatory Trial
Participants		
Participant eligibility criteria	All participants who have the condition of interest are enrolled, regardless of their anticipated risk, responsiveness, comorbidities, or past compliance	Stepwise selection criteria are applied that (a) restrict study individuals to those previously shown to be at highest risk of unfavorable outcomes, (b) further restrict these high-risk individuals to those who are thought likely to be highly responsive to the experimental intervention, and (c) include just those high-risk, highly responsive study individuals who demonstrate high compliance with pretrial appointment keeping and mock intervention
Interventions and expertise		
Experimental intervention—flexibility	Instructions on how to apply the experimental intervention are highly flexible, offering practitioners considerable leeway in deciding how to formulate and apply it	Inflexible experimental intervention, with strict instructions for every element
Experimental intervention— practitioner expertise	The experimental intervention typically is applied by the full range of practitioners and in the full range of clinical settings, regardless of their expertise, with only ordinary attention to dose setting and side effects	The experimental intervention is applied only by seasoned practitioners previously documented to have applied that intervention with high rates of success and low rates of complications, and in practice settings where the care delivery system and providers are highly experienced in managing the types of patients enrolled in the trial. The intervention often is closely monitored so that its "dose" can be optimized and its side effects treated; co-interventions against other disorders often are applied

TABLE 10.2 (continued)

PRECIS Domains Illustrating the Extremes of Explanatory and Pragmatic Approaches to Each Domain

Domain	Pragmatic Trial	Explanatory Trial
Comparison intervention—flexibility	"Usual practice" or the best alternative management strategy available, offering practitioners considerable leeway in deciding how to apply it	Restricted flexibility of the comparison intervention; may use a placebo rather than the best alternative management strategy as the comparator
Comparison intervention—practitioner expertise	The comparison intervention typically is applied by the full range of practitioners and in the full range of clinical settings, regardless of their expertise, with only ordinary attention to their training, experience, and performance	Practitioner expertise in applying the comparison intervention(s) is standardized to maximize the chances of detecting whatever comparative benefits the experimental intervention might have
Follow-up and outcomes		
Follow-up intensity	No formal follow-up visits of study individuals. Instead, administrative databases (e.g., mortality registries) are searched for the detection of outcomes	Study individuals are followed with many more frequent visits and more extensive data collection than would occur in routine practice, regardless of whether patients experienced any events
Primary trial outcome	The primary outcome is an objectively measured, clinically meaningful outcome to the study participants The outcome does not rely on central adjudication and is one that can be assessed under usual conditions (e.g., special tests or training are not required)	The outcome is known to be a direct and immediate consequence of the intervention. The outcome is often clinically meaningful but may sometimes (e.g., early dose-finding trials) be a surrogate marker of another downstream outcome of interest. It may also require specialized training or testing not normally used to determine outcome status or central adjudication

(continued)

TABLE 10.2 (continued)

PRECIS Domains Illustrating the Extremes of Explanatory and Pragmatic Approaches to Each Domain

Domain	Pragmatic Trial	Explanatory Trial
Compliance/adherence		
Participant compliance with "prescribed" intervention	There is unobtrusive (or no) measurement of participant compliance. No special strategies to maintain or improve compliance are used	Study participants' compliance with the intervention is monitored closely and may be a prerequisite for study entry. Both prophylactic strategies (to maintain) and "rescue" strategies (to regain) high compliance are used
Practitioner adherence to study protocol	There is unobtrusive (or no) measurement of practitioner adherence. No special strategies to maintain or improve adherence are used	There is close monitoring of how well the participating clinicians and centers are adhering to even the minute details in the trial protocol and "manual of procedures"
Analysis		
Analysis of primary outcome	The analysis includes all patients regardless of compliance, eligibility, and others (intention-to-treat analysis). In other words, the analysis attempts to see if the treatment works under the usual conditions, with all the noise inherent therein	An intention-to-treat analysis is usually performed However, this may be supplemented by a per-protocol analysis or an analysis restricted to "compliers" or other subgroups in order to estimate maximum achievable treatment effect. Analyses are conducted that attempt to answer the narrowest, "mechanistic" question (whether biological, educational, or organizational)

Source: From Thorpe, K.E. et al., *CMAJ*, 180, E47, 2009.
Note: PRECIS, pragmatic-explanatory continuum indicator summary.

TABLE 10.3

Extension of the CONSORT Statement for Pragmatic Trials

Section	Item	Standard CONSORT Description	Extension for Pragmatic Trials
Title and abstract	1	How participants were allocated to interventions (e.g., "random allocation," "randomized," or "randomly assigned")	
Introduction			
Background	2	Scientific background and explanation of rationale	Describe the health or health service problem that the intervention is intended to address and other interventions that may commonly be aimed at this problem
Methods			
Participants	3	Eligibility criteria for participants; settings and locations where the data were collected	Eligibility criteria should be explicitly framed to show the degree to which they include typical participants and/or where applicable, typical providers (e.g., nurses), institutions (e.g., hospitals), communities (or localities, e.g., towns), and settings of care (e.g., different healthcare financing systems)
Interventions	4	Precise details of the interventions intended for each group and how and when they were actually administered	Describe extra resources added to (or resources removed from) usual settings in order to implement intervention. Indicate if efforts were made to standardize the intervention or if the intervention and its delivery were allowed to vary between participants, practitioners, or study sites Describe the comparator in similar detail to the intervention

(continued)

TABLE 10.3 (continued)

Extension of the CONSORT Statement for Pragmatic Trials

Section	Item	Standard CONSORT Description	Extension for Pragmatic Trials
Objectives	5	Specific objectives and hypotheses	
Outcomes	6	Clearly defined primary and secondary outcome measures and, when applicable, any methods used to enhance the quality of measurements (e.g., multiple observations, training of assessors)	Explain why the chosen outcomes and, when relevant, the length of follow-up are considered important to those who will use the results of the trial
Sample size	7	How sample size was determined; explanation of any interim analyses and stopping rules when applicable	If calculated using the smallest difference considered important by the target decision-maker audience (the minimally important difference) then report where this difference was obtained
Randomization—sequence generation	8	Method used to generate the random allocation sequence, including details of any restriction (e.g., blocking, stratification)	
Randomization—allocation concealment	9	Method used to implement the random allocation sequence (e.g., numbered containers or central telephone), clarifying whether the sequence was concealed until interventions were assigned	
Randomization—implementation	10	Who generated the allocation sequence, who enrolled participants, and who assigned participants to their groups	

TABLE 10.3 (continued)

Extension of the CONSORT Statement for Pragmatic Trials

Section	Item	Standard CONSORT Description	Extension for Pragmatic Trials
Blinding (masking)	11	Whether participants, those administering the interventions, and those assessing the outcomes were blinded to group assignment	If blinding was not done, or was not possible, explain why
Statistical methods	12	Statistical methods used to compare groups for primary outcomes; methods for additional analyses, such as subgroup analyses and adjusted analyses	
Results			
Participant flow	13	Flow of participants through each stage (a diagram is strongly recommended)—specifically, for each group, report the numbers of participants randomly assigned, receiving intended treatment, completing the study protocol, and analyzed for the primary outcome; describe deviations from planned study protocol, together with reasons	The number of participants or units approached to take part in the trial, the number which were eligible, and reasons for nonparticipation should be reported
Recruitment	14	Dates defining the periods of recruitment and follow-up	
Baseline data	15	Baseline demographic and clinical characteristics of each group	

(continued)

TABLE 10.3 (continued)

Extension of the CONSORT Statement for Pragmatic Trials

Section	Item	Standard CONSORT Description	Extension for Pragmatic Trials
Numbers analyzed	16	Number of participants (denominator) in each group included in each analysis and whether analysis was by "intention-to-treat"; state the results in absolute numbers when feasible (e.g., 10/20, not 50%)	
Outcomes and estimation	17	For each primary and secondary outcome, a summary of results for each group and the estimated effect size and its precision (e.g., 95% CI)	
Ancillary analyses	18	Address multiplicity by reporting any other analyses performed, including subgroup analyses and adjusted analyses, indicating which are pre-specified and which are exploratory	
Adverse events	19	All important adverse events or side effects in each intervention group	
Discussion			
Interpretation	20	Interpretation of the results, taking into account study hypotheses, sources of potential bias or imprecision, and the dangers associated with multiplicity of analyses and outcomes	

TABLE 10.3 (continued)

Extension of the CONSORT Statement for Pragmatic Trials

Section	Item	Standard CONSORT Description	Extension for Pragmatic Trials
Generalizability	21	Generalizability (external validity) of the trial findings	Describe key aspects of the setting that determined the trial results. Discuss possible differences in other settings where clinical traditions, health service organization, staffing, or resources may vary from those of the trial
Overall evidence	22	General interpretation of the results in the context of current evidence	

Source: From Zwarenstein, et al., *BMJ*, 337, a2390, 2008.

a nonpharmacological trial, which has taken a pragmatic approach, both CONSORT statements for pragmatic trials and for nonpharmacological trials should be used in combination in guiding the writing, to ensure that all context- and intervention-related descriptions are complete.

10.5 Pragmatic Trials: Demand and Supply

Patients, advocacy groups, clinicians, systematic reviewers, funders, and policy makers want to use the results of RCTs. Calls have been made for more pragmatic trials in general [9], in relation to specific clinical problems [10,11] and in policy [12]. Large patient surveys have indicated the preference of those patients to participate in more pragmatic rather than less pragmatic trials of treatments for their conditions [13]. Researchers and methodologists are also calling for more pragmatic trials.

We are aware of only a single published study [14] that has attempted to identify pragmatic trials (identified using MeSH term "clinical trial," keyword "pragmatic" and authors' judgment to identify clinical trials with a pragmatic attitude) and it found just 95 published between 1976 and 2002. Since PubMed identifies over 168,000 RCTs for that period, trials with a pragmatic attitude are clearly the exception. The number of pragmatic trials has risen over time, but still remains <1% of all randomized trials. Our own broader search (using the keyword "pragmatic" and the PubMed narrow/specific reference search for rigorous studies of therapy (randomized controlled trial[Publication Type] OR (randomized[Title/Abstract] AND controlled[Title/Abstract] AND trial[Title/Abstract])) identified 46 randomized trials self-described as pragmatic in the period 1972–1996 (1.84 per annum), 84 in the period 1997–2002 (14 per annum), 22 in 2003, 37 in 2004, 44 in 2005, 42 in 2006, 58 in 2007, 51 in 2008, and 61 in 2009. The number of self-described pragmatic trials appears to be growing, especially in the last decade.

10.6 Pragmatic Trials: Almost Always Nonpharmacological

Although no formal systematic review has been done, my review of abstracts for self-described pragmatic trials in 2009 suggests that the pragmatic design is particularly suitable for trialists wishing to evaluate nonpharmacological interventions. In 2009, 58 of the 61 self-declared pragmatic trials were of nonpharmacological treatments. Since the vast majority of randomized trials are drug trials, this suggests that the pragmatic approach to trial design is particularly suited to nonpharmacological trials. And of the three pharmacological trials, only one was a simple head-to-head comparison of two drugs; another incorporated drug regimens into a complex multifaceted treatment approach that included nonpharmacological care [15], whereas the third was a trial of treatment changes, rather than of any specific drug itself. This trial was asking whether for patients with epilepsy, it was worth carefully tailoring drug regimens to reduce side effects, even when fits were controlled [16].

Many other nonpharmacological interventions have been evaluated in self-described pragmatic trials. These examples include nurse follow-up of critical care unit patients to promote post-discharge health [17], child support systems to prevent conduct disorder [18], group exercise programs for patients with breast cancer [19], homeopathy [20], and direct observation of tuberculosis treatment, a trial whose pragmatic attributes are displayed in the star diagram in Figure 10.1 [21]. Other nonpharmacological interventions include surgical treatment in ophthalmology [22], orthopedics [23], dentistry [24], and pediatric surgery [25]; knowledge translation to change clinician practices [26], interventions to promote public health [27], and even larval therapy in wound care [28].

10.7 What Is It about Nonpharmacological
Interventions That So Suits Pragmatic Trials?

The typical approach to interventions in explanatory designs is to standardize them, where it is possible to do so, and indeed, doing so assists in eliminating another source of variation, and thus easing the task of identifying the causal connection which is at the heart of the explanatory trial attitude. But the main characteristic of most nonpharmacological trials is that they are, even at their simplest, for example, implantable devices—very like complex interventions. Since they test complex interventions, standardization of the intervention is a very difficult challenge, and may even be antithetical to the treatment approach, since often tailoring of treatment is required, for example, in physical therapy, psychological treatments, or multifaceted interventions. This makes the pragmatic attitude of randomized trial design attractive, because it specifically accommodates the notion that care is variable and may be tailored between subjects.

A further distinguishing characteristic of nonpharmacological interventions is that, unlike drugs, they are more often developments from existing clinical approaches, by active clinicians near the frontline of care, than they are back room creations, designed and developed in laboratories far away from the clinical interface. This has several implications—one is that they are often similar enough to current care for the cost of the treatment itself to be borne by the health service as part of its normal offerings to patients, which means that it becomes much cheaper to run the trial as part of normal clinical care, and this in turn imposes certain pragmatic expectations of inclusive access, a contrast with what might otherwise be the tendency of trialists to exclude, control, or restrict.

Another characteristic of nonpharmacological interventions, which makes them particularly suitable for evaluation using pragmatic trial designs, is that there is seldom a regulatory process that must be gone through before the intervention is allowed to be implemented widely among real-world patients. In the absence of a compulsory prerelease testing phase, many nonpharmacological interventions are tested under real-world conditions, since there is no regulatory authority to distort the nature of the evaluation away from the real-world question that trials most naturally gravitate toward.

A further characteristic of nonpharmacological trials is that they are even more severely underfunded than are drug trials, most of which are conducted at enormous cost using resources from pharmacological companies. The low budget of most nonpharmacological trials creates substantial constraints that have led their designers to prefer a pragmatic approach. With no external pressure to collect more than clinical data cost constraints encourage clinical depth and quality of data collection and focus attention more on the simple question of effectiveness than the more data heavy questions of mechanisms of action.

With nonpharmacological trials, the developers of the intervention are very often the designers of the trial, with few extra resources in terms of outside skills or money to complicate the design. This encourages very precise mapping of the characteristics of the trial to the requirements of both the intervention, and of the real-world service in which it is tested and will later be used, if proven effective.

Many of the trials are in single or few sites—this reduces overhead costs, and allows a far greater reliance on "free" labor and nonfinancial relationships and exchanges to get collaboration, and data, recruit patients, and sites. Also, even in single site trials the need to get collaboration from a swathe of people in management, nursing, and from other clinicians may act as a reality check encouraging trial designs that fit more smoothly into normal clinical practice.

10.8 Future of Nonpharmacological Pragmatic Trials

At their most advanced, nonpharmacological trials are sometimes designed as early stages of a planned wide rollout of a new program, treatment, or policy, should it be shown to be effective under real-world conditions in the early phase of rollout in the target setting. This was the case for the recent randomized trial of Seguro Popular, the National Health Insurance Scheme in Mexico, which was initially tested in a massive randomized trial involving over half-a-million citizens in hundreds of communities [29]. This can only occur when the intervention, the trial, and the researcher are well integrated with policy and program decision making. If trialists wish to have an effect in the real world, they should set up their trials in such a way that they are most likely to directly influence decisions, that is, alongside decision makers, in the context of usual care delivery. This reaches toward a concept that Campbell, a social scientist and methodologist, described as the "experimental society" [30]. This would be an important achievement that would embed scientific thinking and randomized trials as an important and useful supporting actor in the advances of societies.

References

1. Altman, D.G., Bland, J.M. Statistics notes. Treatment allocation in controlled trials: Why randomise? *BMJ*, 318, 1209, 1999.
2. Rothwell, P.M. External validity of randomised controlled trials: "To whom do the results of this trial apply?" *Lancet*, 365, 82–93, 2005.
3. Cochrane, A.L. *Effectiveness and Efficiency. Random Reflections on Health Services.* London, U.K.: Nuffield Provincial Hospitals Trust, 1972.
4. Schwartz, D., Lellouch, J. Explanatory and pragmatic attitudes in therapeutic trials. *J Chron Dis*, 20, 637–648, 1967 [Reprinted in *J Clin Epidemiol*, 62, 499–505, 2009].
5. Zwarenstein, M., Treweek, S., Gagnier, J., et al. Improving the reporting of pragmatic trials: An extension of the CONSORT statement. *BMJ*, 337, a2390, 2008.

6. Thorpe, K.E., Zwarenstein, M., Oxman, A.D., et al. A pragmatic-explanatory continuum indicator summary (PRECIS): A tool to help trial designers. *CMAJ*, 180, E47–E57, 2009.

7. Glasziou, P., Meats, E., Heneghan, C., Shepperd, S. What is missing from descriptions of treatment in trials and reviews? *BMJ*, 336, 1472–1474, 2008.

8. Boutron, I., Moher, D., Altman, D.G., Schulz, K.F., Ravaud, P. Extending the CONSORT statement to randomized trials of nonpharmacologic treatment: Explanation and elaboration. *Ann Intern Med*, 148, W60–W66, 2008.

9. Tunis, S.R., Stryer, D.B., Clancy, C.M. Practical clinical trials: Increasing the value of clinical research for decision making in clinical and health policy. *JAMA*, 290, 1624–1632, 2003.

10. Marson, A., Kadir, Z., Chadwick, D. Large pragmatic randomized studies of new antiepileptic drugs are needed. *BMJ*, 314, 1764, 1997.

11. Hotopf, M., Churchill, R., Glyn, L. Pragmatic randomised controlled trials in psychiatry. *Br J Psychiatry*, 175, 217–223, 1999.

12. Lavis, J.N., Posada, F.B., Haines, A., Osei, E. Use of research to inform public policymaking. *Lancet*, 364, 1615–1621, 2004.

13. Murad, M.H., Shah, N.D., Van Houten, H.K., et al. Individuals with diabetes preferred that future trials use patient-important outcomes and provide pragmatic inferences. *J Clin Epidemiol*, 64(7), 743–748, 2011.

14. Vallvé, C. A critical review of the pragmatic clinical trial [in Spanish]. *Med Clin (Barc)*, 27, 384–388, 2003.

15. Kendrick, T., Chatwin, J., Dowrick, C., et al. Randomised controlled trial to determine the clinical effectiveness and cost-effectiveness of selective serotonin reuptake inhibitors plus supportive care, versus supportive care alone, for mild to moderate depression with somatic symptoms in primary care: The THREAD (THREshold for AntiDepressant response) study. *Health Technol Assess*, 13(22), iii–iv, ix–xi, 1–159, 2009.

16. Uijl, S.G., Uiterwaal, C.S., Aldenkamp, A.P., et al. Adjustment of treatment increases quality of life in patients with epilepsy: A randomized controlled pragmatic trial. *Eur J Neurol*, 16(11), 1173–1177, 2009.

17. Cuthbertson, B.H., Rattray, J., Campbell, M.K., et al.; PRaCTICaL study group. The PRaCTICaL study of nurse led, intensive care follow-up programmes for improving long term outcomes from critical illness: A pragmatic randomised controlled trial. *BMJ*, 339, b3723, 2009. Erratum in: *BMJ*, 339, 2009.

18. Hutchings, J., Bywater, T., Daley, D., et al. Parenting intervention in Sure Start services for children at risk of developing conduct disorder: Pragmatic randomised controlled trial. *BMJ*, 334, 678, 2007.

19. Mutrie, N., Campbell, A.M., Whyte, F., et al. Benefits of supervised group exercise programme for women being treated for early stage breast cancer: Pragmatic randomised controlled trial. *BMJ*, 334, 517, 2007.

20. Steinsbekk, A., Fonnebo, V., Lewith, G., Bentzen, N. Homeopathic care for the prevention of upper respiratory tract infections in children: A pragmatic, randomised, controlled trial comparing individualised homeopathic care and waiting-list controls. *Complement Ther Med*, 13, 231–238, 2005.

21. Zwarenstein, M., Schoeman, J.H., Vundule, C., Lombard, C.J., Tatley, M. Randomised controlled trial of self-supervised and directly observed treatment of tuberculosis. *Lancet*, 352, 1340–1343, 1998.

22. Lois, N., Burr, J.M., Norrie, J., et al. Internal limiting membrane peeling versus no peeling for idiopathic full thickness macular hole: A pragmatic randomised controlled trial. *Invest Ophthalmol Vis Sci*, 52(3), 1586–1592, 2011.

23. Handoll, H., Brealey, S., Rangan, A., et al. Protocol for the ProFHER (PROximal Fracture of the Humerus: Evaluation by Randomisation) trial: A pragmatic multi-centre randomised controlled trial of surgical versus non-surgical treatment for proximal fracture of the humerus in adults. *BMC Musculoskelet Disord*, 10, 140, 2009.

24. Cannizarro, G., Torchio, C., Felice, P., Leone, M., Esposito, M. Immediate occlusal versus non-occlusal loading of single zirconia implants. A multicentre pragmatic randomised clinical trial. *Eur J Oral Implantol*, 3(2), 111–120, 2010.

25. Lock, C., Wilson, J., Steen, N., et al. North of England and Scotland Study of Tonsillectomy and Adeno-tonsillectomy in Children (NESSTAC): A pragmatic randomised controlled trial with a parallel non-randomised preference study. *Health Technol Assess*, 14(13), 1–164, iii–iv, 2010.

26. Wolfe, C.D., Redfern, J., Rudd, A.G., Grieve, A.P., Heuschmann, P.U., McKevitt, C. Cluster randomized controlled trial of a patient and general practitioner intervention to improve the management of multiple risk factors after stroke: Stop stroke. *Stroke*, 41(11), 2470–2476, 2010.

27. Fiscella, K., Yosha, A., Hendren, S.K., et al. Get screened: A pragmatic randomized controlled trial to increase mammography and colorectal cancer screening in a large, safety net practice. *BMC Health Serv Res*, 10, 280, 2010.

28. Dumville, J.C., Worthy, G., Soares, M.O., et al.; VenUS II team. VenUS II: A randomised controlled trial of larval therapy in the management of leg ulcers. *Health Technol Assess*, 13(55), 1–182, iii–iv, 2009.

29. King, G., Gakidou, E., Imai, K., et al. Public policy for the poor? A randomised assessment of the Mexican universal health insurance programme. *Lancet*, 373, 1447–1454, 2009.

30. Campbell, D. Reforms as experiments. *Am Psychol*, 24, 409–429, 1969.

11

Preference Trials

Arthur Kang'ombe, Helen Tilbrook, and David Torgerson

University of York

CONTENTS

11.1 Introduction

The randomized controlled trial (RCT) is acknowledged as the most scientifically rigorous study design for evaluating medical interventions [1]. By randomly allocating patients to the different treatment arms of the trial, systematic biases or errors are minimized as unknown potential confounders are distributed equally between the different treatment arms [2]. Although random allocation results in an even distribution of the physical characteristics of participants, it may not deal with other potential biases. One of these is patients' preferences.

In an RCT, patients may have either a preference for the standard treatment or the new treatment being evaluated or may be indifferent to both treatments. If patients with preferences consent to be randomized, then some patients will get their preferred treatment and others will not. It is hypothesized that participants who receive their preferred treatment might be better motivated and comply better with the treatment programs and report better outcomes [3], whereas patients who do not receive their preferred treatment may experience "resentful demoralization" [4], may be less motivated, may not comply with the treatment program, may not report accurately during follow-up, and may even drop out of the trial [5]. Therefore, the motivation in patients may not be equally distributed following randomization across the different treatment arms [6,7]. For example, in a trial in which one of the treatments is preferred by the majority of patients, then those receiving the less favored treatment are likely to be less motivated than those receiving the more desirable one [7], and consequently the patients in the two treatment arms of the trial would be different. Hence, there would be a potential threat to the internal validity of the trial [6].

In trials in which patients have strong preferences for a treatment that is also available, the trial setting may decline to participate, and in a trial in which strong preferences exist and a high number of patients refuse randomization the external validity of the trial will be affected. Therefore, generalizing these results to a wider population will be limited [8].

As well as the direct effects of patients' preferences on compliance and motivation, there is also the "therapeutic effect" [9] of patients' preferences. These are the psychological effects that influence outcomes and are similar to the placebo effect [10]. The evidence for the placebo effect, both beneficial (placebo) and adverse effects (nocebo) [11], in medicine is widely supported [11,12]. The therapeutic effects of patients' preferences are problematic for trials because they seek to determine or assess the physiological or pharmacological effects of an intervention. Disentangling the psychological effects of patients' preferences from the physiological effects is complex [13].

11.2 Evidence for the Effects of Patients' Preferences in Randomized Controlled Trials

Trials do not tend to have sufficient power to detect the effects of patients' preferences, and therefore, the effect of patients' preferences on treatment outcomes in RCTs is uncertain. Recently, however, there have been some systematic reviews that have investigated the effects of preferences. One of the reviews found some evidence for preference effects.

The effects of participants' and professionals' preferences in RCTs were conducted by King et al. [5]. They investigated the effects of preferences on outcomes and trial recruitment, and the influence of preferences on attrition. The review identified 34 RCTs but only 2 of these [14,15] were conventional RCTs with preference data obtained before participants were randomized (a fully randomized patients' preference trial). They found no evidence that preferences influenced attrition. They found some evidence of an effect on outcomes in some of the preference trials and no effect on outcomes in the two conventional RCTs. The trials and studies included in the review were very heterogeneous and the authors reported that therefore they were unable to "reach definitive conclusions for any particular clinical field" [5].

A review by van Schaik et al. [16] was conducted to examine patients' preferences in the treatment of depressive disorder in primary care. The authors report that adherence to antidepressant medication is low, which may be caused by most patients preferring psychotherapy. They identified six studies, two of which were fully randomized patients' preference trials [17,18]. The authors found no evidence of the effect of preferences on outcomes in two partially randomized trials but a main finding of their report was that many of the participants would not have consented to randomization as most preferred psychotherapy to antidepressant medication.

More recently, a systematic review and patient level data meta-analysis was conducted to investigate the effects of patients' preferences on treatment outcomes and attrition (Preference Collaborative Review Group, 2008). They pooled and analyzed the data from eight trials that investigated treatments for musculoskeletal conditions. They found that patients who received their preferred treatment had better clinical outcomes than those with no preference; however, for those who did not receive their preferred treatment there was no evidence that they performed worse than those with no preference. Also, patients who did not receive their preferred treatment were *more* likely to have data at first follow-up than patients without a preference, which is the opposite of what was expected. They concluded that patients' preferences do affect outcomes and attrition in trials that investigate musculoskeletal conditions. To summarize, there were no statistically significant differences between the two preference groups (i.e., between those who were randomly allocated to their preferred treatment and those who were randomly allocated to the treatment they did not prefer). The only differences that were found were between patients with a preference (who did and did not receive their preference) and those who were indifferent (i.e., had no preference).

11.3 Fully Randomized Preference Trial

The fully randomized patients' preference trial is a conventional standard RCT in which all participants are randomly allocated to treatment (Figure 11.1). The only difference is that after consent has been obtained and before randomization patients are asked which treatment they would prefer or are they indifferent to the treatments, thereby providing the opportunity of taking preferences into account in the analysis [4] that can be included "as one of the potential modifiers (or 'covariates') of treatment effect in the analysis" [19].

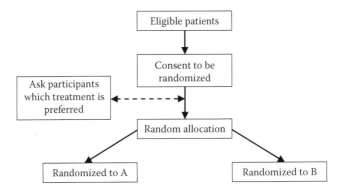

FIGURE 11.1
Fully randomized preference trial.

11.3.1 When to Use This Type of Design

Preference effects are likely to be more apparent in nonblinded trials where the patient is aware of the treatment they are receiving and where the outcome measure is self-reported by the patient. Indeed, Halpern [2] describes three conditions in which there will be bias due to patients' preferences: (1) the "outcome is sensitive to patients' treatment preferences" (p. 13) (i.e., subjective outcomes); (2) patients are able to determine which treatment they are receiving "at a rate greater than chance" (p. 14); and (3) the "majority of patients must prefer one of the treatments being compared at the trial's outset." (p. 14). Preference effects may also be more apparent in what Brewin and Bradley [6] describe as "participative interventions." These are interventions that rely on patients' active participation (e.g., self-monitoring, counseling) and require motivated patients to make them effective.

11.3.2 Collecting Preference Data

A single yes or no answer to questions about preference to the trial treatment can be provided on the eligibility questionnaire or baseline questionnaire (note however that initial preference data should be collected before randomization). Preference data could be collected at baseline and follow-up to determine if there has been a change in preferences. Also, a measure of strength of preference and exploration of reasons for preference might be obtained.

11.3.3 Advantages and Disadvantages of the Fully Randomized Preference Trial

The advantages of this design are that it follows the rigorous traditional design, analysis is straightforward, and patients' preferences and the characteristics of patients who have a preference can be explored. Additionally, patients' preferences can be used in the analysis on the trial. A disadvantage with this trial is that patients with strong preferences for a treatment may decide not to participate in the trial that could affect its external validity.

11.3.4 Statistical Analysis in Fully Randomized Preference Trials

In the following section, we will discuss how we can include patients' preferences within a statistical analysis. However, it is important to note that most trials are not powered to detect interactions between treatment effects and baseline covariates and this includes patients' preferences. Consequently, many analyses of patients' preference and treatment interactions are likely to find relatively important interactions as being not statistically significant. One way around this problem might be to specify—in advance of any analysis—an interaction effect that would be deemed to be important, and clinically significant, even if it were not statistically significant.

To estimate the effects of patients' preference on outcomes and attrition, three groups of patients are compared: patients who had a preference and were randomly allocated to their treatment; patients who had preference and were randomly allocated to the treatment they did not prefer; and patients who had no preference.

The two later groups can be combined to yield two groups overall: patients expressing preference and patients indifferent to treatment assignment. Klaber Moffett et al. [15] conducted a randomized trial of exercise for low back pain: clinical outcomes, costs, and preferences. The objective of the study was to evaluate effectiveness of an exercise program in a community setting for patients with low back pain to encourage a return to normal activities. One hundred and eighty-seven patients aged 18–60 years with mechanical low back pain of 4 weeks to 6 months duration were randomly assigned to a progressive exercise program (intervention group) or usual primary care management (control group). The main outcome variable was the Roland–Morris disability questionnaire (RMDQ) score at baseline, 6 weeks, 6 months, and 12 months post-randomization. Patients were asked what kind of treatment they would prefer before randomization.

The RMDQ is a list of 24 questions that describe how the patients feel about themselves. The minimum score is 0 and the maximum score is 24 with 24 the worst score corresponding to patients' characteristics fulfilling all the criteria.

To determine how patients' treatment preference affects the outcome and attrition preference need to be included in the model as a covariate. Candidate models can be considered. Since the RMDQ outcome is continuous, then a simple regression model of RMDQ against patient treatment preference adjusting for other covariates (treatment group, time of follow-up, and gender) could be considered. There is a problem with this approach in that follow-up measurements measured on a patient are not independent. It would therefore make more sense to consider a model that will use all the follow-up data taking into account the correlation between repeated measurements on an individual. One such model is the linear model for the continuous response RMDQ.

The probability that patient data are missing at any time during follow-up could be modeled using ordinary logistic regression. This also ignores the correlation between the repeated measurements on a subject because it treats all observations as independent. Therefore, models for discrete longitudinal data can be considered: generalized linear models. This can be implemented by fitting generalized estimating equations (GEEs). This will model the marginal mean structure, which is the marginal probability of missing as a function of covariates. If interest is placed on the inference of the association between two binary measurements on an individual, then this association can be modeled by fitting an alternating logistic regression (ALR) model [20,21].

The limitation of the correlations approach to models for binary data is that the correlations are constrained by the mean (probability of missing a visit). Hence, the odds ratio can be used to model the associations in the binary data since they are not constrained by the mean. In this respect, specific emphasis will be placed on regressing the outcome on the explanatory variables and describing the association between the outcomes. In the present analysis, this association was modeled by estimating the common odds ratio between two binary measurements in an individual using ALR [22].

Therefore, three candidate models will be fitted to the data: a linear model for the continuous outcome (RMDQ score) at each visit, a marginal model implemented by GEE and ALR for the longitudinal binary outcome (0 = missing, 1 = observed) at each visit [23].

The mean RMDQ scores at baseline seems to be higher for those who prefer intervention treatment compared with those patients who were indifferent to treatment assignment. From baseline (0 weeks) to 6 weeks, the mean RMDQ scores for both preference groups had decreased. After 6 weeks, the evolution of the mean RMDQ scores was almost constant in both preference groups but seems to be always higher in those patients who prefer intervention than those who were indifferent to treatment assignment. There seems no difference between the two preference groups at the end of the study (52 weeks). The mean profiles portray a quadratic trend against time of follow-up, which needs to be considered in the modeling by adding a quadratic time effect.

In Figure 11.2, there seems no difference between the proportions of respondents against each follow-up visit for the two preference groups. However, the proportion of respondents dropped between baseline and 26 weeks. It increased from 26 to 52 weeks. Hence, though the missing data vary among the different follow-up visits, it seems there are no differences between the two preference groups. Quadratic time effects will be included in the model.

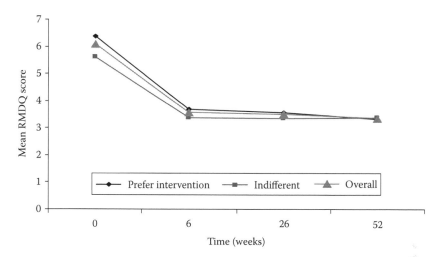

FIGURE 11.2
Mean RMDQ scores by treatment preference group against time of follow-up for the low back pain study.

11.3.4.1 Model Fitting to the Low Back Pain Study

The linear model was fitted using unstructured correlation of the repeated measurements. GEEs were fitted using the exchangeable working assumption of the two binary measurements in an individual. ALR was fitted using the exchangeable assumption of the log odds ratio between two binary measurements in an individual. The linear model fitted to the RMDQ outcome score was as follows:

$$RMDQ = bo + b1 \, gender + b2 \, treatment + b3 \, preference + b4 \, time$$
$$+ b5 \, time * treatment$$
$$+ b6 \, time^2$$
$$+ b7 * time^2 * treatment + error$$

where
bo is the average RMDQ score at baseline
b1 is the gender effect
b2 is the treatment effect
b3 is the effect of preference
b4 is the time effect
b5 is an interaction between time and treatment group
b6 is the quadratic effect of time
b7 is an interaction between the quadratic effect of time and treatment group

TABLE 11.1

Parameter Estimates (EST); Standard Error (SE), and p Values of a Linear Model Fitted to RMDQ Score, a Marginal Model by Generalized Estimating Equations and Alternating Logistic Regression Fitted to the Missing Data Indicator

Parameter	Linear Model		Generalized Estimating Equations		Alternating Logistic Regression	
	Est. (SE)	p	Est. (SE)	p	Est. (SE)	p
Intercept	5.94 (0.57)	<0.0001	−5.66 (1.10)	<0.0001	−5.64 (1.10)	<0.0001
Gender (female vs. male)	0.68 (0.47)	0.1535	−0.83 (0.43)	0.0533	−0.80 (0.43)	0.0616
Treatment (control vs. intervention)	−1.10 (0.58)	0.0574	0.04 (0.42)	0.9231	0.0037 (0.43)	0.9931
Preference (intervention vs. indifferent)	0.40 (0.49)	0.4133	0.08 (0.43)	0.8446	0.07 (0.43)	0.868
Time	−3.46 (0.47)	<0.0001	3.71(0.93)	<0.0001	3.68 (0.94)	<0.0001
Time*treatment (control vs. intervention)	1.79 (0.65)	0.0061	*__	*__	*__	*__
Time2	0.75 (0.13)	<0.0001	−0.83 (0.22)	0.0001	−0.82 (0.22)	0.0002
Time2*treatment (control vs. intervention)	−0.37 (0.18)	0.0411	*__	*__	*__	*__
Log odds ratio	—	—	—	—	4.56 (0.67)	<0.0001

*__ Implies that the parameter was nonsignificant and therefore was excluded from the model.
__ Implies that the log odds ratio parameter does not apply in the linear model and GEE.

From Table 11.1, we observe that the average RMDQ score at baseline was 5.94 points. There was no significant effect of preference on the RMDQ scores (0.40, p = 0.4133). There was a significant time effect in that the scores decreased with time. That decrease was more pronounced in the intervention group than the control group. There was also a significant quadratic effect of time indicating that scores did not decrease linearly with time. Therefore, we can conclude that patients' preference does not significantly influence patients' RMDQ scores after adjusting for other covariates in the model.

In the second part of the analysis, a marginal model for the missing data at each visit (0 = missing, 1 = observed) was modeled by GEE and ALR. The results are shown in Table 11.1. Under the GEE model, we are interested in finding out how preference affects the marginal probability of missing a visit adjusting for the other covariates in the model. In the ALR model, we are additionally interested in modeling how pairs of binary observations in an individual are associated on the log odds scale. Larger associations mean

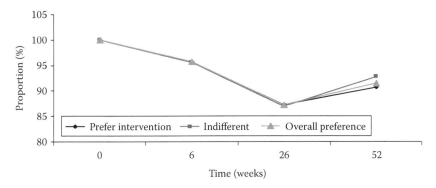

FIGURE 11.3
Proportion of respondents by treatment preference group against time of follow-up for the low back pain study.

that individuals who have a missed visit are very likely going to miss other visits. This may have implications in study design.

In both GEE and ALR, there were borderline gender effects, nonsignificant treatment, and preference effects. However, there were significant linear time and quadratic effects. This implies that the probability of missing was increasing with time of follow-up and was quadratic in the sense that it started to decrease at the end (quadratic time effect; -0.83, $p = 0.0001$ and -0.82, $p = 0.0002$ in the GEE and ALR, respectively; cf. Figure 11.3). The estimated log odds ratio association parameter between two binary measurements from an individual was estimated as 4.56, $p < 0.0001$.

This means that two binary measurements from an individual were highly associated implying individuals who missed a visit were more likely to miss other visits as well.

To conclude, it has been demonstrated in this chapter that patients' preference can be included in fully randomized clinical trials as a covariate in the model for the outcome of interest and also as a covariate in a model where the outcome is the data missing indicator. The study design influences what kind of statistical modeling framework to follow. And in the study by Klaber Moffett et al. [15], (Table 11.2) patients' preference did not influence the RMDQ scores and missing data.

TABLE 11.2

Experimental Design and Weekly Sample Sizes

Group	0 Weeks	6 Weeks	26 Weeks	52 Weeks
Control ($n = 98$)	98	94	86	88
Intervention ($n = 89$)	89	85	77	83
Missing	0	8	24	16

11.4 Conclusion

There is some evidence to suggest that patients' preferences do affect treatment outcomes in RCTs. The fully randomized patients' preference trials maintain all the rigor of a standard RCTs but collect preference data at baseline, after patient consent and before randomization, and therefore enable preferences to be taken into account in the analysis.

References

1. Pocock, S.J. *Clinical Trials: A Practical Approach.* Chichester, U.K.: John Wiley & Sons Ltd., 1983.
2. Halpern, S.D. Evaluating preference effects in partially unblinded, randomized clinical trials. *J Clin Epidemiol*, 56, 190–115, 2003.
3. King, M. The effects of patients' and practitioners' preferences on randomized clinical trials. *Palliat Med*, 14, 539–542, 2000.
4. Torgerson, D.J., Klaber-Moffett, J., Russell, I.T. Patient preferences in randomised trials: Threat or opportunity? *J Health Serv Res Policy*, 1(4), 194–197, 1996.
5. King, M., Nazareth, I., Lampe, F., et al. Impact of participants and physician intervention preferences on randomised trials. A systematic review. *JAMA*, 293(9), 1089–1099, 2005.
6. Brewin, C.R., Bradley, C. Patient preferences and randomised clinical trials. *BMJ*, 299, 313–315, 1989.
7. Janevic, M.R., Janz, N.K., Didge, J.A., et al. The role of choice in health education intervention trials: A review and case study. *Social Sci Med*, 56, 1581–1594, 2003.
8. Howard, L., Thornicroft, G. Patient preference randomised controlled trials in mental health research. *Br J Psychiatry*, 188, 303–304, 2006.
9. McPherson, K., Britton, A.R., Wennberg, J.E. Are randomized controlled trials controlled? Patient preferences and unblind trials. *J R Soc Med*, 90(12), 652–656, 1997.
10. McPherson, K., Britton, A.R., Wennberg, J.E. Are randomised trials controlled? Patient preferences and unblind trials. *J R Soc Med*, 90, 652–656, 1997.
11. Crow, R., Gage, H., Hampson, S., Hart, J., Kimber, A., Thomas, H. The role of expectancies in the placebo effect and their use in the delivery of health care: A systematic review. *Health Technol Assess*, 3(3), 1–96, 1999.
12. Silverman, W.A., Altman, D.G. Patients' preferences and randomised trials. *Lancet*, 347, 171–174, 1996.
13. McPherson, K., Chalmers, I. Incorporating patient preferences into clinical trials. Letters. *BMJ*, 317, 78, 1998.
14. Hardy, G.E., Barkham, M., Shapiro, D., Reynolds, S., Rees, A. Credibility and outcome of cognitive-behavioural and psychodynamic-interpersonal psychotherapy. *Br J Clin Psychol*, 34, 555–569, 1995.
15. Klaber Moffett, J., Torgerson, D., Bell-Syer, S., et al. Randomised controlled trial of exercise for low back pain: Clinical outcomes, costs, and preferences. *BMJ*, 319, 279–283, 1999.

16. van Schaik, D.J.F., Klijn, A.F.J., van Hout, H.P.J., et al. Patients' preferences in the treatment of depressive disorder in primary care. *Gen Hosp Psychiatry*, 26(3), 184–189, 2004.

17. Simpson, S., Corney, R., Fitzgerald, P., Beecham, J. A randomized controlled trial to evaluate the effectiveness and cost-effectiveness of counselling patients with chronic depression. *Health Technol Assess*, 4(36), 1–83, 2000.

18. Unützer, J., Katon, W., Callahan, C.M., et al. For the IMPACT Investigators. Collaborative care management of late-life depression in the primary care setting. A randomized controlled trial. *JAMA*, 288(22), 2836–2845, 2002.

19. Lambert, M.F., Wood, J. Incorporating patient preferences into randomized trials. *J Clin Epidemiol*, 53, 163–166, 2000.

20. Fitzmaurice, G.M., Laird, N.M., Ware, J.H. *Applied Longitudinal Analysis*. Hoboken, NJ: John Wiley & Sons, Inc., 2004.

21. Molenberghs, G., Verbeke, G. *Models for Discrete Longitudinal Data*. New York: Springer, 2005.

22. Verbeke, G., Molenberghs, G. *Linear Mixed Models for Longitudinal Data*. New York: Springer, 2000.

23. Preference Collaborative Review Group, Patients' preferences within randomised trials: Systematic review and patient level meta-analysis. *BMJ* 2008; doi: 10.1136/binj.a1864.

.

12

Nonrandomized Studies to Evaluate the Effects of a Nonpharmacological Intervention

Barnaby C. Reeves

University of Bristol

CONTENTS

12.1 General Principles and Aims of Evaluations of Effectiveness

From the point of view of a health policy maker, evaluation studies should have two key properties. First, they should be designed to protect against bias ("validity"). Second, they should be designed so that the findings are relevant to, and can be applied to, typical healthcare situations ("applicability" or "external validity") [1]. It can sometimes be difficult to design a study that is both highly valid and highly applicable and different study designs may be appropriate to evaluate subtly different research questions, defined by the population, intervention, comparator, and outcome(s) of interest ("PICO") [1].

Some evaluation studies carried out at an early stage of clinical translation [2] need to focus on understanding mechanisms or obtaining evidence of proof of concept in "ideal" circumstances [3]. For example, such studies help both researchers and potential funders to know whether a new technology at an early stage of development warrants further research investment. For pharmacological interventions and some nonpharmacological interventions, which can be easily randomized, such studies are typically small randomized controlled trials (RCTs). These RCTs often have limited applicability, recruiting participants who are more likely to benefit or least likely to have side effects, administering the intervention of interest in ideal circumstances (e.g., expert providers, specialist centers), and may be limited to investigating surrogate outcomes (e.g., serum levels of biomarkers or physiological measures like blood pressure) rather than the outcomes that an intervention is ultimately designed to influence (e.g., heart attack or stroke). Since they are small, they are likely to lack power to detect rare but potentially serious harms.

Other "pragmatic" evaluation studies, at a later stage of evaluation, are designed to inform health policy. Such studies aim to recruit participants who are representative of the kinds of patients who would typically be considered for the intervention under evaluation and to study the intervention as provided by usual healthcare services. They measure important clinical or patient-reported outcomes and, therefore, tend to be large in order to have adequate power. Many large, pragmatic RCTs of nonpharmacological interventions, intended to inform general clinical decisions and health policy, have been carried out [4–9]. However, RCTs of this kind are major undertakings, expensive, and often of no commercial interest given the current regulatory framework for approving many nonpharmacological interventions. A tendency for researchers to do small mechanistic RCTs has led to skepticism about applying the findings of RCTs to inform general clinical or health policy decisions [10].

High-quality evaluation studies should also, ideally, investigate all important effects of interventions, that is, their benefits, harms, and costs (and allow benefits and harms to be weighed up quantitatively, and the net effect balanced against costs). To facilitate the interpretation and synthesis of findings from evaluations of interventions for a particular health condition and to promote assessment of the risk of outcome reporting bias [11,12], the development of core outcome sets is being promoted [13].

RCTs of nonpharmacological interventions are sometimes perceived to be difficult to do. Nonrandomized study (NRS) designs are often thought to be an easier way to obtain more applicable evidence, studying more inclusive populations, with the intervention of interest provided by a multiple practitioners in a wide range of usual healthcare settings. In fact, there is not much evidence about whether NRS of nonpharmacological interventions do in fact have these characteristics; a review of evaluations of new orthopedic procedures did not find this to be the case [14].

RCTs of interventions, which depend on the skill of a practitioner for their provision (e.g., therapies of various kinds, psychotherapy, physiotherapy, occupational therapy, nursing and surgery, and interventional procedures), have also often been viewed with skepticism [15]. These perceptions warrant a close look at the special challenges of evaluating nonpharmacological interventions.

12.2 Why Are Nonpharmacological Treatments Challenging to Evaluate?

The complexity of nonpharmacological interventions, arising from the intervention itself and the need for a practitioner to deliver the intervention, has already been described. This complexity causes challenges when evaluating nonpharmacological treatments. These challenges can be considered under three headings, namely those arising from the practitioner, from the nonpharmacological treatment itself, and from the patient/participant.

12.2.1 Practitioner

Challenges relating to the practitioner arise because the practitioner is an intrinsic part of the intervention, bringing a particular level of skill and experience of the treatment being evaluated. Evaluations of pharmacological treatments have to deal only with individual differences among study participants. Evaluations of nonpharmacological treatments also have to deal with individual differences between the practitioners. An evaluation cannot include an infinite number of practitioners and needs to guard against obtaining results that apply only to the practitioners in the study. Because of individual differences between practitioners, practitioners sometimes question the relevance to their own practice of the findings of RCTs that represent an average effect of multiple [other] practitioners. Expertise-based RCTs [16] have been proposed as one way to allow practitioners who are expert in providing the conventional or control intervention to participate in an evaluation of new intervention and hence promote the likelihood of the findings being implemented.

One treatment may be effective when provided by one practitioner but not for another, an interaction of treatment and practitioner. Practitioners may have a strong preference for one treatment despite two or more treatments being in widespread use, that is, collective but not individual equipoise. There are different nuances of preference: personal investment in learning something new, perhaps getting a commercial edge on rivals when competing for patients in privately funded health services; a feeling that "I know best"; discomfort at acknowledging uncertainty, perhaps backed up by a feeling that an average effect achieved by other practitioners simply is not relevant to one's own practice.

These challenges are summarized in Table 12.1. Note that the first three of these practitioner-centered challenges apply equally when doing an NRS as when doing an RCT. Only the issues of practitioner preference and unease about discussing uncertainty are easier to accommodate in an NRS, where they can simply be avoided by deciding the choice of treatment in the usual way. Although this may allow a wide range of practitioners to participate in an evaluation study, it causes an additional challenge for an NRS, which does not arise in an RCT, namely allocation (or selection) bias. If a practitioner provides two or more treatments for a particular condition, he or she almost certainly recommends one or the other on the basis of some characteristics of the patients, causing imbalance between the groups of patients having one treatment or another (confounding by indication) [17,18]. If a practitioner always provides one treatment, then the evaluation must compare the practitioner's patients with those of another practitioner, resulting in the practitioner's characteristics (e.g., learning curve or innate skill) being inextricably confounded with the treatment being provided.

TABLE 12.1

Challenges Posed by the Practitioner

Challenges Posed by the Practitioner	Challenge for	
	RCT	NRS
1. Innate skill varies between practitioners	Yes	Yes
2. A practitioner may have a learning curve for a new intervention	Yes	Yes
3. The treatment that is most effective when provided by one practitioner may not be the most effective when provided by another	Yes	Yes
4. A practitioner may have a strong preference for one treatment rather than another	Yes	No
5. A practitioner unwilling to discuss uncertainty about alternative treatments with patient	Yes	No

12.2.2 Nonpharmacological Treatment

In contrast to a pharmacological treatment, a nonpharmacological treatment rarely consists of a single component. For example, a rehabilitation program for chronic knee pain may involve supervised exercise, progression tailored to the needs of patient, and counseling about self-management and coping [6]; a new surgical strategy such as off-pump coronary artery bypass surgery may evolve with increasing experience among the "pioneers" [19], a different phenomenon to a practice effect when learning an established technique [20]. In these circumstances, it can be challenging to know when to carry out an evaluation, and how to define and standardize the intervention of interest, that is, to decide which features are necessary and sufficient to define the intervention [21]. Nevertheless, it is essential for researchers to define these core features. Quality assurance to measure compliance by practitioners with the core features (i.e., "process evaluation") is an important aspect of an evaluation of a nonpharmacological intervention in order to interpret the findings fully [22].

The key elements of the new treatment may be difficult to define when the new treatment is a modification of a previous treatment. A new treatment may involve modification of one or more aspects of a complex treatment and it can be difficult to identify precisely what is new and whether it warrants evaluation. Consequently, in an evaluation study, it may be unclear which elements of a procedure need to be standardized and which can be provider at the provider's discretion. The issue of whether to standardize other elements of a complex treatment overlaps with the mechanistic/explanatory vs. pragmatic dimension since, in everyday practice, it may be difficult to control the practices of other professionals delivering aspects of a complex treatment [23]. An alternative, if standardization is not possible, is to measure important process features. For example, a recent NRS (using data collected in an RCT) evaluated minimally invasive vs. open surgical methods for obtaining saphenous vein from the leg to make coronary artery bypass grafts; in this study, the surgeons chose which method to use, so the analysis had to control for surgical features likely to affect the patency of the graft at 1 year, for example, graft and target artery quality and the use of a composite or non-composite graft [24].

A final challenge raised by nonpharmacological treatments is blinding. Awareness of the treatment received can lead practitioners to adopt health-related behaviors or co-interventions differentially, causing performance bias [25]. If outcome assessors are aware of the treatment received, then there is also the risk of detection bias [25,26]; therefore, researchers should either use objective outcomes or blind outcome assessment (at least for the primary outcome), if at all possible. Since practitioners are involved in delivering the intervention, it is almost always impossible to blind them to the allocated treatment. (There may be rare exceptions where a third party can prepare active or sham versions of key aspect of the treatment and give the "preparation" to the practitioner to administer.) It may also be difficult to blind study

TABLE 12.2

Challenges Posed by the Nonpharmacological Treatment

Challenges Posed by the Nonpharmacological Treatment	Challenge for	
	RCT	NRS
1. Treatments are complex and their core features difficult to define	Yes	Yes
2. There may be uncertainty about the extent to which non-core aspects of treatments should be standardized	Yes	Yes
3. Treatments may evolve over time	Yes	Yes
4. There may be uncertainty about whether a treatment has changed sufficiently to warrant evaluation	Yes	Yes
5. Practitioners cannot be blinded to treatment allocation	Yes	Yes

participants to the treatment they have received, although sham interventions are sometimes devised [27–29]. Alternative surgical interventions delivered under general anesthetics through the same incision are exceptions [30,31].

These challenges are summarized in Table 12.2. Note that all five of these treatment-related challenges apply equally when doing an NRS as when doing an RCT.

12.2.3 Patients/Participants

Patients as well as surgeons may have strong preferences. It can be frustrating to researchers that patients have preferences despite accepting that no one knows whether the intervention or comparator is more effective in terms of the outcome of primary interest, e.g. survival after cancer surgery. In such circumstances, may value different effects (including ones not consider by the researchers) in highly individual ways, for example, avoid serious complications at all costs, or accept almost any complication to prevent recurrence/symptoms. Patients may simply not accept that there is uncertainty between alternatives that are extremely different, for example, major surgery such as CABG or medical treatment.

The difficulty of carrying out RCTs when potential participants are likely to have preferences has been extensively researched over recent years [32–34]. This body of research has shown that, in many circumstances, patients' preferences can be modified by changing the precise way in which information is given so that they are prepared to accept randomization. It is important to appreciate that these methods do not constitute undue persuasion to take part, which would be unethical. Rather, the research has focused for example on the patients' preconceptions about treatments or on the precise way in which practitioners give information, which often conveys the practitioners' own preferences.

Participants' preferences are not a problem when using an NRS, since participants are allowed to chose their preferred treatment. However, allowing participants to choose a preferred treatment can be associated with strong

TABLE 12.3

Challenges Posed by the Patient/Participant

Challenges Posed by the Patient/Participant	Challenge for	
	RCT	NRS
1. A patient/participant may have strong preference	Yes	No
2. Study participants often cannot be blinded to treatment allocation	Yes	Yes

expectations and the risk of biases. Preferences may be associated with other prognostic factors introducing a risk of selection bias and confounding. As in the case of nonblinded practitioners (see the preceding text), awareness of the treatment received can lead participants to adopt health-related behaviors or co-interventions differentially, putting a study at risk of performance bias. For patient-reported outcomes, such as the ability to carry out condition-specific tasks or activities of daily living or health-related quality of life, participants are the outcome assessors; if they are not blinded, these outcomes are at risk of detection bias from participants' prior expectations about the relative effects of alternative interventions (see the preceding text). These challenges are summarized in Table 12.3.

12.3 Commonly Used Nonrandomized Study Designs

I use the term "nonrandomized study" to describe any quantitative study of the effects of an intervention (harm or benefit) that does not use randomization to allocate participants or units of care such as general practices or schools to a control or intervention treatment. I intend this term to include studies where "allocation" occurs in the course of usual treatment decisions or peoples' choices, that is, studies sometimes called "observational."

There are many possible types of NRS, including a cohort study, a case–control study, a controlled before-and-after study (CBA), an interrupted time-series study, and a clinical trial that does not use an appropriate randomization strategy (sometimes called quasi-randomized allocation). The labels, and definitions, often used to describe NRS designs tend to have originated from epidemiology or other study design textbooks [35,36]. However, NRS can be extremely diverse and study design labels are not consistently applied by researchers; this is well illustrated by the varied use of the term case–control study to describe cohort studies and RCTs [37–39]. In the context of appraising primary studies for inclusion in a systematic review, the Cochrane Non-Randomized Studies Methods Group has proposed using a checklist of study design features to describe what researchers actually did [40], rather than relying on study design labels assigned either by the researchers of indexers of medical literature. To avoid ambiguity, researchers should describe their methods carefully, including information about the precise method of allocation of participants (or units of care in the case of interventions delivered at organizational level) to intervention and control groups.

12.3.1 Clustering: Individual versus Cluster Level Allocation and Other Clustering Issues

Before describing particular study designs, the distinction between individual and cluster level allocation, and more general issues of clustering of observations, needs to be highlighted. Clustering of observations by practitioner, or other unit of healthcare provision (Table 12.4), is important because the statistical assumption that all observations are independent may not necessarily hold [41]. Ignoring issues of clustering leads to confidence intervals for treatment effect estimates that are too narrow and *p*-values that are too small.

This distinction between levels of allocation is most clearly seen in RCTs. Individual level randomization occurs when a trial participant is randomized to a control or intervention treatment and is the predominant method of randomization in RCTs of pharmacological interventions. In RCTs of nonpharmacological treatments, randomization is often stratified by practitioner to try to create balance within practitioners with respect to the number of

TABLE 12.4

Types of Clustering

1. Clustering arises in the evaluation of organizational interventions, where treatment is provided to participants defined by cluster membership. Similarity within clusters arises primarily from the cluster characteristics. One or more teams provide the treatment to multiple clusters and clustering by team may also be an important consideration

2. Cluster allocation may also be chosen by researchers to avoid contamination between control and intervention treatments; for example, researchers may worry that aspects of the intervention may transfer to the control group if one practitioner is required to deliver both the intervention and control treatment, or if participants allocated to the intervention and control treatments are likely to meet and discuss their care with one another. The clustering issue is the same as in the case of evaluations of organizational interventions

3. For nonpharmacological treatments provided to individual patients, clustering arises from the similarity of provision within one provider/practitioner, for example, surgical patients operated on by one surgeon. The complexity of nonpharmacological treatments, sometimes involving other healthcare professionals, can give rise to complex clustering situations ("cross"-clustering, i.e., study participants not simply "nested" in one higher level). Information to characterize clustering may not be available in routine datasets

4. Clustering can also be a problem if the measurement of outcome is clustered within observers, giving rise to greater similarity in outcome between participants within an assessor than between participants between assessors [50]

patients allocated to each group and the distribution of important prognostic factors between control and intervention groups.

In contrast, a cluster RCT is an RCT in which clusters of individuals (e.g., clinics, families, geographical areas), rather than individuals themselves, are randomized to different groups. This design is typically used for interventions that naturally occur at the level of clusters, for example interventions to deter children from substance abuse provided to schools [42], or when there is particular concern that there may be contamination between the intervention and control treatments [43].

If randomization is to be effective in creating groups of clusters that are balanced with respect to prognostic factors in a cluster RCT, several clusters have to be randomized; it is strongly recommended that at least eight clusters (and preferably many more) are randomized [41]. If only two clusters are randomized, one to the control treatment and one to the intervention treatment, the comparison between the treatments is completely confounded with the characteristics of the clusters. This makes it impossible to conclude whether any difference in outcome that is observed should be attributed to the difference in treatment or to the differing characteristics of the clusters, irrespective of the number of individual participants included in each cluster.

The issue of cluster level allocation also applies to NRS. For example, outcomes in one geographical region receiving a new intervention (e.g., community-based cardiovascular disease (CVD) prevention program, mammographic screening, intervention to reduce falls among the elderly [44–46]) are often compared with outcomes in another, but without any acknowledgment that the intervention is delivered at the level of a cluster. The need for several control and intervention clusters is equally true for NRS as it is for cluster RCTs.

In cluster RCTs, the need to allocate clusters randomly, and the involvement of a statistician or methodologist in most RCTs, tends to bring clustering to the notice of the research team. Clustering in NRS is often subtle and may go unnoticed but is very common when evaluating nonpharmacological interventions. Clustering may arise by practitioner (as described earlier), or unit of provision (e.g., geographic location, family doctor/general practice, hospital, or ward). Clustering creates complicated statistical issues and it is vital that researchers take relevant methodological and statistical advice to ensure that a project will generate valid results.

In studies that use cluster level allocation, outcomes are often measured before and after an intervention is implemented (see Section 12.3.3.1). When this happens, the numerators and denominators that are typically used to calculate effect sizes warrant closer examination. The study population (denominator) may not be enumerated but described from routine data; this population may change by migration or immigration over the time course of the study or, if the "eligible" population is defined by age, through cohort effects (people too young at the start of the study become eligible as time goes by and some people die). For interventions relating to the provision of particular treatment episodes, for example, surgery, the participants studied before implementing an intervention are likely to be completely different from those studied after (unless some participants require repeat treatment). Numerators for before-and-after periods are often counts of events without considering whether the events arise in the same or different individuals. Numerators alone (event rates per unit time, assuming that the denominator remains constant) may be compared.

Clustering of patients within practitioners is potentially an issue even if practitioners provide both control and intervention treatments, that is, with individual level allocation. This is because outcomes for patients tend to be more similar within than between practitioners (i.e., nonindependent observations), perhaps because of varying expertise between practitioners or because of other aspects of care that vary systematically between practitioners.

12.3.2 Nonrandomized Study Designs with Individual Level Allocation

Studies with individual level allocation use classical epidemiological study designs [35]. Traditionally, such studies have investigated etiological research questions in which exposure to a risk factor or not typically arises from a person's "choice" of health behavior (e.g., smoking, diet) or occupation. The circumstances leading to exposure are, therefore, unlikely to be clustered. As described earlier, this is often not the case with allocation to different treatments.

12.3.2.1 Case Studies/Treatment Registers

Case studies or series can be described as "Observations … made on a series of individuals, usually all receiving the same intervention, before and after an intervention but with no control group" [40]. Such studies are uncontrolled unless the outcome with a specified comparator (the "control" treatment not studied in the case series), for example, palliative care for a terminally ill person, is known. (Even in this example, the outcome is only known for the outcome of survival). Plausible ranges for an outcome frequency when patients are treated with a comparator intervention may be considered in order to estimate effectiveness [47] but cannot estimate effectiveness in the same way as in a controlled study. Case series that are descriptive may be useful for describing the evolution of a technique. Alternatively, a case series may have the objective of describing health outcomes (prognostic study) in a defined population following treatment. For example, a study of total hip replacement sought to describe the distribution of symptom improvement and the risks of comorbidity after surgery [48,49].

12.3.2.2 Cohort Study

A cohort study can be defined as

> A study in which a defined group of people (the cohort) is followed over time, to examine associations between different interventions received and subsequent outcomes. A 'prospective' cohort study recruits participants before any intervention and follows them into the future. A 'retrospective' cohort study identifies subjects from past records describing the interventions received and follows them from the time of those records [40].*

Both retrospective and prospective cohort studies have a "concurrent" control, that is, both intervention and control groups were formed and studied together at the same time (see Figure 12.1).

Ideally, these groups should consist of consecutive eligible and consenting patients classified according to whether or not they received the intervention or control intervention. The researcher "observes" the treatment received (typically, with the same practitioners providing new treatment and comparator). In such studies, there is a substantial risk of confounding by indication, especially if participants are allocated to the intervention or control group substantially on the basis of the practitioners' preferences, and other biases. Other methods can be used to allocate participants in a quasi-experimental manner (e.g., family doctor, therapist, ward, or clinic) but these usually involve clustering of some kind (see Table 12.4).

Some researchers carry out a cohort study but with a "historic" control; this design almost always involves comparing a more recent cohort receiving the intervention of interest with a control cohort receiving care at a more distant (earlier) point in time (see Figure 12.1) [51]. This type of cohort study is at risk of the same kinds of bias as concurrent cohort studies but more so, since confounding can arise from both (a) secular changes in treatment and health outcomes and (b) selection by treatment indication. (In exceptional situations, selection by treatment indication may be avoided if there is a complete switch in treatment practice from control to intervention and no change in the pattern of referring patients for treatment between the early and late period.) The control and intervention groups may also be at differential risk of other biases, for example, detection and attrition biases, unless data collection is carried out strictly in accordance with a protocol during both periods.

Because of the risk of bias, historically controlled cohort studies are not recommended. Nevertheless, this type of study is often favored by surgeons or other practitioners because it allows them to compare patients' outcomes before and after implementing a new intervention. Importantly, it allows

---

* The terms retrospective and prospective are used here to describe studies in which allocation to intervention and control groups arose before (retrospective) or after (prospective) the study was conceived and planned. Note that these terms are not universally adopted and understood and may give rise to confusion.

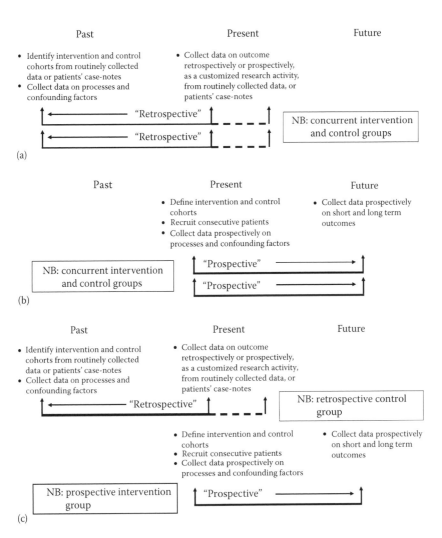

FIGURE 12.1
Diagrams of different types of cohort study. In both retrospective (a) and prospective (b) concurrently controlled cohort studies, both intervention and control groups are formed and studied together at the same time. In a historically controlled cohort study (c), the control and intervention groups are studies over different periods of time; the control group is typically investigated retrospectively and the intervention group prospectively (as shown), although both groups may be studied retrospectively but over different time periods.

them to use the newly learnt technique for all patients for whom they consider the technique suitable, avoiding the problem of strong practitioner preference. (Practitioners who have "invested" in learning a new technique are likely to believe that it provides an important improvement in outcome compared with the older technique.)

It should be noted that some case series or treatment registers in which eligibility is defined by a qualifying disease or receipt of any treatment for a particular condition can provide the opportunity to address comparative research questions about the effectiveness of alternative treatments for the disease of interest [52,53]. In such instances, it is especially important for researchers to specify whether or not decisions to carry out the comparisons were made before or after any preliminary analyses were carried out (see Section 12.4.4).

12.3.2.3 Case–Control Study

A case–control study can be defined as follows: "A study that compares people with a specific outcome of interest ('cases') with people from the same source population but without that outcome ('controls'), to examine the association between the outcome and prior exposure (e.g., having an intervention)" [40].

Case–control studies estimate the association between treatment and outcome by comparing the proportion of cases and controls receiving the intervention of interest. The frequencies of outcome among cases and control are not directly estimated. Such studies are rarely useful for comparing the beneficial effects of alternative treatments except in special circumstances.

However, because of their particular role in studying rare outcomes [35], case–control studies provide a useful tool to investigate serious, rare adverse (or otherwise unexpected) effects of an intervention. In these instances, estimating the magnitude of the effect precisely is often less critical because, if causal, associations between treatment and rare serious adverse effects tend to be strong. (In principle, the same limitations of case–control studies apply to studies of both harmful and beneficial associations with treatment.)

Cohort and case–control designs can be illustrated by the data shown in Table 12.5, which represent favorable or unfavorable health outcome at 6 months among people who had a subarachnoid hemorrhage repaired by surgical or endovascular methods [53]. The study was, in fact, a concurrent cohort study observing the method of treatment used for consecutive patients. A comparison (across columns) could be made between the frequencies of favorable outcome (65% and 66%) among participants who had surgical or endovascular repair, respectively, although this comparison is subject to many limitations [54]. In theory, the same data could have been collected using a case–control design, identifying (ideally, consecutive) people who had an unfavorable outcome at 6 months. A group of controls, that is, people who had an aneurysm repair over the same period and who were known to have a favorable outcome at 6 months would then be identified. Then, the relative proportions of patients who had surgical treatment (57% and 58%, respectively) could be compared among those who did and did not have a favorable outcome (comparison across rows).

TABLE 12.5

Data from a Nonrandomized Study in Which Patients with an Aneurysmal Subarachnoid Hemorrhage Had either a Surgical or an Endovascular Repair of the Aneurysm

(a)	Surgical Repair	Endovascular Repair	Total
Favorable outcome	747 (65.2%)	560 (66.3%)	1307
Unfavorable outcome	399 (34.8%)	285 (33.7%)	684
Total	1146	845	1991

(b)	Surgical Repair	Endovascular Repair	Total
Favorable outcome	747 (57.2%)	560 (42.8%)	1307
Unfavorable outcome	399 (58.3%)	285 (41.7%)	684
Total	1146	845	1991

The study was, in fact, a prospective cohort study, in which the proportion of patients with a favorable outcome at 6 months could be compared by method of repair (Table 12.5a; column percentages; unadjusted risk ratio = 0.97, 95% confidence interval 0.86–1.10). The data could have been collected, at least theoretically, using a case–control design (see text), comparing the relative proportions who had surgical treatment (57% and 58%, respectively) and calculating the odds ratio for the association between method of repair and outcome (Table 12.5b; unadjusted odds ratio = 0.95, 95% confidence interval 0.79–1.15.

12.3.3 Nonrandomized Study Designs with Cluster Level Allocation

There are two commonly used cluster level designs that are considered to be at relatively low risk of bias [36,55].

12.3.3.1 *Controlled before-and-after Study*

A controlled before-and-after study can be defined as follows: "A study in which observations are made before and after the implementation of an intervention, both in a group that receives the intervention and in a concurrent control group that does not" [40]. This definition does not require cluster level allocation but CBAs usually are cluster-allocated, with intervention and control groups being made up of clusters of individuals nested within clusters. Observations can be made at the level of individual members of a cluster, or aggregated across individuals at the level of a cluster.

The design when implemented at a cluster level is shown diagrammatically in Figure 12.2. It involves "before" and "after" periods of observation in intervention and control groups of clusters. In the before period, neither group receives the intervention. In the after period, the intervention is provided in intervention, but not control, clusters. The size of the effect of the intervention compared with the control can be estimated from the interaction of group (intervention or control) and period of observation (before or after). This can be interpreted as either a comparison in outcome between intervention and control clusters in stage 2, "adjusted" for differences in outcome

Study Phase	"Intervention" (Clusters = n_1)	"Control" (Clusters = n_C)
$Time_{before}$	No intervention	No intervention
$Time_{impl}$	Intervention implemented	No intervention
$Time_{after}$	Intervention in place	No intervention

FIGURE 12.2
Diagram showing a controlled before-and-after study using several clusters. The design relies on a comparison between intervention and control clusters over time. An implementation stage is shown in gray since implementation may take considerable time and the intervention may need to become established before the full effect is seen; this stage can also be included in the analysis, if desired, although researchers are likely to have a less clear hypothesis about the observed effect.

at baseline, or the change in outcome over time in the intervention clusters versus the change in outcome over time in the control clusters. It should be noted that the identity of individuals making up a cluster in the two periods of observation is often not clear; individual cluster members may be mainly the same (e.g., if an age cohort is followed over time [46]) or mainly different (e.g., if observations represent episodes of care [56]). The number of clusters per group is equally important when using this design as it is for a cluster randomized controlled trial [41].

12.3.3.2 Interrupted Time Series

An interrupted time series (ITS) can be defined as follows: "A study that uses observations at multiple time points before and after an intervention (the 'interruption'). The design attempts to detect whether the intervention has had an effect significantly greater than any underlying trend over time." The validity of an ITS may be strengthened by having a concurrent control (controlled ITS or CITS) becoming, in effect, a CBA with multiple periods of observation before and after the implementation of an intervention (see Figure 12.3). The number of clusters studied in an ITS is again important; with only one or two clusters, any difference before and after the implementation of the intervention might be explained by a co-intervention or not be applicable to other clusters [41].

12.3.4 Databases

A database is simply an electronic repository for data and should result from any study, irrespective of the design. However, they are considered here as a separate study design category because an increasing number of papers comparing the effects of different interventions for a condition are being written based on secondary analyses of data collected for a different purpose. In this situation, the analyst does not control the method of collection of the data. Databases subjected to secondary analyses may arise from a randomized trial or other study carried out for a different purpose [57], from

Study Phase	"Intervention" (Clusters = 1_i)	"Control" (Clusters = 1_c)
Time 1	No intervention	No intervention
Time 2	No intervention	No intervention
...	No intervention	No intervention
Time 3	No intervention	No intervention
$Time_{impl}$	Intervention implemented	No intervention
Time $n + 1$	Intervention in place	No intervention
Time $n + 2$	Intervention in place	No intervention
...	Intervention in place	No intervention
Time $n + 3$	Intervention in place	No intervention

FIGURE 12.3
Diagram showing the multistage nature of a controlled interrupted time series using several clusters; as with the controlled before-and-after study, the design relies on a comparison between intervention and control clusters over time. (An uncontrolled interrupted time series omits the controls and relies only on a comparison within intervention clusters over time.) An implementation stage is shown in gray since implementation may take considerable time and the intervention may need to become established before the full effect is seen; this stage can also be included in the analysis, if desired, although researchers are likely to have a less clear hypothesis about the observed effect.

routine health care (e.g., custom data collection in one or more sites [58,59], healthcare activity records [56,60], financial or administrative data held by a health insurance company [61]) or purposeful surveillance (e.g., infection control [62], prosthetic joint registry [63]).

Secondary analyses of databases have a number of advantages. If they result from routine data collection or surveillance, they can have large or very large sample sizes; this is a particular advantage if the analyst is interested in rare adverse events. Routine data collection may also allow long-term outcomes to be identified for a defined cohort or to form the "case" group in a case–control study. The database may be made available at no cost providing the opportunity to estimate effects quickly and relatively cheaply.

However, secondary analyses of databases also have a number of disadvantages. The analyst has no control over the data collected; therefore, variation in the application of data definitions across sites or changes in data definitions over time are an intrinsic limitation of the data. Data arising from routine data collection will not have been specified with the intention of addressing the research question of interest and may not have been adequately documented. Unlike a purposefully designed study, the analyst is unlikely to have access to researchers with intimate knowledge of limitations affecting specific variables or certain periods of data collection. Routine data often do not characterize disease severity or comorbidity in sufficient detail to allow satisfactory adjustment for potential confounding factors. The time when a person becomes eligible for inclusion in a database may not be carefully defined and the actual, rather than the intended intervention, is typically recorded [54,64]. Data are often missing because there is no "culture" of complete data collection in routine practice nor validation checks at the time of collection. Data missing at random bias effect estimates toward the null but the big danger is that data are missing differentially, causing informative censoring and systematic bias. Selective reporting is a particular risk for secondary analyses of databases (see Section 12.4.4). The risk of biases from these limitations (which also affect NRS using other study designs) is highly variable and may best be considered as introducing extra uncertainty in the results [65]. This uncertainty acts over and above that accounted for in confidence intervals and, in large studies, may easily be 5–10 times the magnitude of the 95% confidence interval [66]. This additional uncertainty clearly offsets, and can outweigh, the advantage from having a large sample size.

12.4 Role of Nonrandomized Study Designs in Evaluation

12.4.1 Early Stages of Experience of a New Intervention

A model has been proposed to characterize stages of innovation, comprising idea (proof of concept), development, exploration, assessment, and long-term study (IDEAL) [67]. This model was conceived in the context of surgery but may have wider application to nonpharmacological interventions. The model makes explicit that idea and development stages are typically descriptive, not comparative; at these stages, the intervention may continue to evolve (optimizing the intervention) or become standardized for wider adoption. During these stages, the primary aim is to demonstrate that the intervention is feasible and has no obvious safety concern. It is suggested that the exploration stage involves wider adoption and documentation of the wider experience in a prospective research database, shared across practitioners; exploration may also involve a feasibility RCT. However, NRS (prospective databases, registries, or other designs) may be necessary at any stage if randomization is impossible. Such studies need to be designed with features to minimize bias (see Section 12.4.5).

12.4.2 Harms (Unintended Effects) versus Benefits (Intended Effects)

The role of case–control studies to investigate possible harmful effects of an intervention compared with a control treatment has already been described. More generally, RCTs often do not have sufficient statistical power to detect important differences between groups in harmful effects. RCTs may also miss harms that arise a long time after treatment because the harm is unexpected and the investment required to follow a trial cohort is not justified. While RCTs should remain the primary vehicle for studying adverse effects of treatments, there is also increasing recognition that NRS are likely to be required to estimate rare or long-term harmful effects [68]. Although not randomized and, therefore, at risk of confounding, confounding by indication is likely to be less severe because practitioners are unlikely to select patients for intervention or control treatments according to the risk of unexpected harms [69]. However, it should be noted that imbalances between groups and the risk of confounding may still exist, particularly if harmful and beneficial effects are observed in the same organ/physiological systems; one can be more confident that confounding is unlikely to explain an observed association if harmful and beneficial effects are observed in different systems.

12.4.3 Treatments That Are Likely to Be Impossible or Very Difficult to Randomize

There will always be examples of interventions that cannot be randomized. Reasons for not being able to randomize participants or clusters of participants include refusal of patients or practitioners to accept randomization, ethical objections, legal, political or organizational obstacles [70]. When randomization is not possible, naturally occurring variation may allow intervention and control treatments to be compared, providing some answer rather than none. However, these comparisons are inevitably at higher risk of bias than comparisons using RCTs. Perhaps the best consequence of such an NRS is that the findings increase uncertainty rather than resolving it, creating collective equipoise and circumstances in which randomization becomes possible.

Common examples concern interventions conceived with the aim of providing care more efficiently but without any hypothesized health benefit, for example, adoption of a higher threshold for admission to (or a lower threshold for discharge from) intensive care. Other examples concern organizational factors such as the relationship between volume of procedures and outcome [71,72].

12.4.4 Pros and Cons of Choosing to Do a Nonrandomized Study

Advantages of using NRS to estimate the effectiveness of interventions include

- The ability to include practitioners and patients with strong preferences or other reasons for not wanting to randomize patients
- The ability to include a wide range of practitioners and settings, increasing the pragmatism of the study
- The ability to study large numbers (although rarely prospectively; see Section 12.3.4)

Note that cost and the avoidance of regulatory governance or consent should not be considered to be advantages because these come at a price. If a prospective NRS is cheaper than an RCT, this is usually because the quality of data collection is compromised; randomization per se is not expensive. Although some NRS with research objectives are still carried out without seeking consent from individual participants, increasingly research ethics committees require similar protections for patients.

Disadvantages of using NRS to estimate the effectiveness of interventions include

- Allocation/selection bias and the risk of confounding from imbalances in prognostic factors
- Performance bias conditional on knowledge of the treatments that participants actually received (also a risk in RCTs)
- Detection bias from nonblinded assessment of outcomes that are not objectively measured (also a risk in RCTs)
- Complex biases arising from incomplete documentation of the study cohort and unclear time of origin [54,64]
- Poor data quality causing differential or nondifferential misclassification of outcome or confounding factors (assuming key outcomes are documented reliably)
- Loss to follow up causing possible attrition bias/informative censoring
- Poorly defined research question at the outset, for example, poorly defined intervention, comparator or outcomes, or inadequately detailed protocol or pre-specified analysis plan, putting a study at risk of selective reporting and outcome or analysis reporting bias [73]

The risk of selective reporting warrants special mention. Traditionally, the greatest threat to the validity of NRS has been considered to come from confounding. Recent evidence from methodological reviews of RCTs highlights the seriousness of the threat from selective outcome reporting and outcome reporting bias [11,25,73–75], which arise despite the presence of a priori protocols required to obtain research ethics approval and the widespread requirement for trial registration, including a priori specification of key outcomes and other design features. The proportion of NRS that are carried out without protocols is not known but, intuitively, might be suspected to be large, not least because many NRS have historically not had to be submitted in advance for research ethics or other approvals. Without prior specification of key features of a study in a protocol, the opportunity for selective reporting and outcome reporting bias is very much greater. Selective analysis reporting, that is, carrying out many analyses and choosing to report the ones that most favor prior hypotheses, has not been widely discussed. When using an RCT, arguably the risk is relatively small because analyses can often be simple, although choices may still need to be made, for example, about adjustment for baseline covariates. When analyzing data from an NRS, the number of choices that have to be made is very much larger, increasing the opportunity for selective analysis reporting and analysis reporting bias [73].

12.4.5 How to Design a Nonrandomized Study to Estimate the Benefits of an Intervention

Designing a study to estimate unexpected harmful effects of an intervention is a specialized task and is beyond the scope of this chapter [68]. The fundamental principle should be to follow the same steps that would be followed when designing an RCT; researchers should consider the CONSORT items and ensure that those which apply can be adequately reported. A protocol should be written to define the research question; in the case of secondary analyses of databases, this should be done without inspecting/analyzing any data that may be available. Ideally, the protocol should be registered. The notion of *designing* an NRS is very important. Proceeding straightaway, for example, to secondary analyses of an "available" database, is highly likely to provide misleading effect estimates.

- The protocol should define the research question (study population, intervention and comparator, primary and secondary outcomes) and key design features in the same way as would be the case for an RCT.

- Consider how to create groups that are balanced with respect to key prognostic factors (as would be the case in an RCT). This is a key aspect of the design of an NRS and might be achieved by restriction or stratification by possible confounding factors; relevant factors may have been identified in previous case-mix models [76–79]. At the very least, it should involve a careful consideration of factors or variables that may influence (or have influenced) allocation to one or other intervention in the dataset. The study (or secondary analysis) should not be attempted unless data can be collected, or are available, to characterize fully factors underlying allocation since, without adequate characterization of allocation, confounding is likely to arise and effect estimates will be misleading.

- Describe the proposed data source, for example, an existing database that contains the data required (prognostic factors and outcomes of interest), or prospective data collection.

- Describe all patients in the study cohort in a flow diagram, showing in particular the number of patients in the cohort in comparison to the number of patients who data included in the analysis of each outcome.

- The protocol should include an analysis plan, including the methods chosen a priori for model fitting, dealing with missing data, and so on. The analysis should take into account the hierarchical nature of the data, if appropriate.

- An analysis plan for an NRS is likely to be more complicated than for an RCT. Again, the same principle should be followed as for an RCT; if an analysis plan is revised, conditional on knowledge of characteristics of the data such as their distribution but *without* information about allocation to alternative interventions, the revised plan should be dated and registered in some way so that, in the future, someone appraising the findings of the study can compare the analysis plan with the reported analyses.

When designing a prospective NRS, the researchers should in addition

- Choose a primary outcome that is objective or can be assessed by a blinded/independent assessor to prevent detection bias
- Account for all patients approached in a flow diagram, showing in particular the number of patients recruited in comparison with the number of patients with data included in the analysis of each outcome
- Define the time of entry into the study (cf. time of randomization)
- Collect data on the intended intervention as well as actual treatment given
- As far as possible, disguise to participants the comparison of interest to minimize performance bias
- Include RCT-like measures of process so that the "fidelity" with which interventions have been provided can be adequately reported [22]
- Institute high-quality data management and validation, feeding back queries to data collectors in a timely way [80]

These additional considerations highlight the biases that are likely to arise in retrospective cohort studies or secondary analyses of databases, where they cannot be acted on.

12.5 Conclusion

Choosing not to randomize does not avoid most of the challenges of evaluating the effectiveness of nonpharmacological treatments.

If randomization is impossible, consider all the usual sources of bias and institute design features to minimize them (write and ideally register a protocol and analysis plan in advance of carrying out any analysis; document the incipient cohort; blind outcome assessment; institute high-quality data management and validation, etc.).

Nonpharmacological interventions are usually delivered by practitioners, with observations clustered within practitioners or some other unit of healthcare provision. If control and intervention treatments are allocated at the level of clusters of healthcare provision, this also creates a clustering issue. Data hierarchies arising from clustering must be taken into account when estimating the effects of interventions.

References

1. Haynes, R.B., Sackett, D.L., Guyatt, G.H., Tugwell, P. Chapter 1, Forming research questions, in *Clinical Epidemiology: How to Do Clinical Practice Research*, 3rd edn. Philadelphia, PA: Lippincott Williams & Wilkins, 2005, p. 11.
2. UK Medical Research Council Clinical Advisory Board. Report of translational workshop, February 20–21, 2007. http://www.mrc.ac.uk/About/Strategy/Governmentfunding/SingleHealthResearchFund/TranslationWorkshop/index.htm (accessed November 30, 2010).
3. Zwarenstein, M., Treweek, M. What kind of randomized trials do we need? *Can Med Assoc J*, 180, 998, 2009.
4. Molyneux, A.J., Kerr, R.S.C., Bacon, F., Shrimpton, J. Protocol 99PRT18 International Subarachnoid Aneurysm Trial (ISAT). http://www.thelancet.com/protocol-reviews/99PRT-18 (accessed November 30, 2010).
5. Brown, L.C., Epstein, D., Manca, A., Beard, J.D., Powell, J.T., Greenhalgh, R.M. The UK EndoVascular Aneurysm Repair (EVAR) trials: Design, methodology and progress. *Eur J Vasc Endovasc Surg*, 27, 372–381, 2004.
6. Hurley, M.V., Walsh, N.E., Mitchell, H.L., et al. Clinical effectiveness of a rehabilitation program integrating exercise, self-management, and active coping strategies for chronic knee pain: A cluster randomized trial. *Arthritis Rheum*, 57, 1211–1219, 2007.
7. Little, P., Somerville, J., Williamson, I., et al. Randomised controlled trial of self management leaflets and booklets for minor illness provided by post. *BMJ*, 322, 1214, 2001.
8. Jolly, K., Taylor, R., Lip, G.Y.H., et al. The Birmingham Rehabilitation Uptake Maximisation Study (BRUM). Home-based compared with hospital-based cardiac rehabilitation in a multi-ethnic population: Cost-effectiveness and patient adherence. *Health Technol Assess*, 11(35), 2007.
9. Parker, S.G., Oliver, P., Pennington, M., et al. Rehabilitation of older patients: Day hospital compared with rehabilitation at home. A randomised controlled trial. *Health Technol Assess*, 13(39), 2009.
10. Zwarenstein, M., Treweek, S., Gagnier, J.J., et al. CONSORT group; Pragmatic Trials in Healthcare (Practihc) group. Improving the reporting of pragmatic trials: An extension of the CONSORT statement. *BMJ*, 337, a2390, 2008.
11. Hopewell, S., Loudon, K., Clarke, M.J., Oxman, A.D., Dickersin, K. Publication bias in clinical trials due to statistical significance or direction of trial results. *Cochrane Database of Syst Rev*, (1). Art. No.: MR000006, 2009. DOI:10.1002/14651858.MR000006.pub3.
12. Kirkham, J.J., Dwan, K.M., Altman, D.G., Gamble, C., Dodd, S., Smyth, R., Williamson, P.R. The impact of outcome reporting bias in randomised controlled trials on a cohort of systematic reviews. *BMJ*, 340, c365, 2010. DOI: 10.1136/bmj.c365.

13. Core Outcome Measures in Effectiveness Trials. The COMET initiative. http://www.methodologyhubs.mrc.ac.uk/news__events/comet_initiative.aspx (accessed November 30, 2010).

14. Pibouleau, L., Boutron, I., Reeves, B.C., Nizard, R., Ravaud, P. Applicability and generalisability of published results of randomised controlled trials and non-randomised studies evaluating four orthopaedic procedures: Methodological systematic review. *BMJ*, 339, b4538, 2009. DOI: 10.1136/bmj.b4538.

15. Peterson, E.D. Innovation and comparative-effectiveness research in cardiac surgery. *N Engl J Med*, 361, 1897–1899, 2009.

16. Devereaux, P.J., Bhandari, M., Clarke, M., et al. Need for expertise based randomised controlled trials. *BMJ*, 330(7482), 88, 2005.

17. Grobbee, D.E., Hoes, A.W. Confounding and indication for treatment in evaluation of drug treatment for hypertension. *BMJ*, 315, 1151–1154, 1997.

18. Bandolier. Confounding. Available at: http://www.medicine.ox.ac.uk/bandolier/booth/glossary/confound.html (accessed November 30, 2010).

19. Watters, M.P., Ascione, R., Ryder, I.G., Ciulli, F., Pitsis, A.A., Angelini, G.D. Haemodynamic changes during beating heart coronary surgery with the 'Bristol Technique'. *Eur J Cardiothorac Surg*, 19(1), 34–40, 2001.

20. Caputo, M., Reeves, B.C., Rogers, C.A., Ascione, R., Angelini, G.D. Use of control charts for monitoring the performance of residents during training in off-pump coronary surgery. *J Thorac Cardiovasc Surg*, 128, 907–915, 2004.

21. Barkun, J.S., Aronson, J.K., Feldman, L.S., et al. Evaluation and stages of surgical innovations. *Lancet*, 374(9695), 1089–1096, 2009.

22. Boutron, I., Moher, D., Altman, D.G., Schulz, K.F., Ravaud, P. CONSORT group. Extending the CONSORT statement to randomized trials of nonpharmacologic treatment: Explanation and elaboration. *Ann Intern Med*, 148(4), 295–309, 2008.

23. Ergina, P.L., Cook, J.A., Blazeby, J.M., et al.; Balliol Collaboration. Challenges in evaluating surgical innovation. *Lancet*, 374(9695), 1097–1104, 2009.

24. Lopes, R.D., Hafley, G.E., Allen, K.B., et al. Endoscopic versus open vein-graft harvesting in coronary-artery bypass surgery. *N Engl J Med*, 61, 235–244, 2009.

25. Higgins, J.P.T., Altman, D.G. Assessing risk of bias in included studies, in Higgins, J.P.T., Green, S. (eds.), *Cochrane Handbook for Systematic Reviews of Interventions*. Chichester, U.K.: John Wiley, 2008, pp. 187–242, Chapter 8.

26. Hrobjartsson, A. Observer bias in randomised clinical trials. An analysis of trials with both blinded and unblinded outcome observers. Abstracts of the 2010 Joint Colloquium of the Cochrane and Campbell Collaborations, p. 23.

27. Majeed, A.W., Troy, G., Nicholl, J.P., et al. Randomised, prospective, single-blind comparison of laparoscopic versus small-incision cholecystectomy. *Lancet*, 347(9007), 989–994, 1996.

28. TAP study group. Photodynamic therapy of subfoveal choroidal neovascularisation in age-related macular degeneration with verteporfin: One-year results of 2 randomised clinical trials. TAP report 1. *Arch Ophthalmol*, 117, 1239–1245, 1999.

29. Kong, J.C., Lee, M.S., Shin, B.C., Song, Y.S., Ernst, E. Acupuncture for functional recovery after stroke: A systematic review of sham-controlled randomized clinical trials. *CMAJ*, 182(16), 1723–1729, 2010.

30. McCaskie, A.W., Deehan, D.J., Green, T.P., et al. Randomised, prospective study comparing cemented and cementless total knee replacement: Results of press-fit condylar total knee replacement at five years. *J Bone Joint Surg Br*, 80(6), 971–975, 1998.

31. Angelini, G.D., Taylor, F.C., Reeves, B.C., Ascione, R. Beating heart against cardioplegic arrest studies (BHACAS 1 and 2): Clinical outcome in two randomised controlled trials. *Lancet*, 359, 1194–1199, 2002.

32. Torgerson, D., Sibbald, B. Understanding controlled trials: What is a patient preference trial? *BMJ*, 316, 360, 1998.

33. Wade, J., Donovan, J.L., Lane, J.A., Neal, D.E., Hamdy, F.C. It's not just what you say, it's also how you say it: Opening the 'black box' of informed consent appointments in randomised controlled trials. *Soc Sci Med*, 68(11), 2018–2028, 2009.

34. Donovan, J., Lane, J.A., Peters, T.J., et al. Development of a complex intervention improved randomization and consent in an RCT. *J Clin Epidemiol*, 62, 29–36, 2009.

35. Rothman, K.J., Greenland, S. *Modern Epidemiology*, 2nd edn. Philadelphia, PA: Lippincott-Raven, 1998, pp. 16–114.

36. Shadish, W.R., Cook, T.D., Campbell, D.T. *Experimental and Quasi-Experimental Designs for Generalized Causal Inference*. Boston, MA: Houghton Mifflin, 2002, pp. 148–152.

37. Bombardier, C., Jadad, A., Tomlinson, G. What is the study design? 12th Cochrane Colloquium: Bridging the Gaps; October 2–6, 2004. Ottawa, Ontario, Canada. http://onlinelibrary.wiley.com/o/cochrane/clcmr/articles/CMR-6639/frame.html (accessed November 30, 2010).

38. Grimes, D.A. "Case–control" confusion: Mislabeled reports in obstetrics and gynecology journals. *Obstet Gynecol*, 114(6), 1284–1286, 2009.

39. Zhu, X., Chen, J., Han, F., et al. Efficacy and safety of losartan in treatment of hyperuricemia and posttransplantation erythrocytosis: Results of a prospective, open, randomized, case-control study. *Transplant Proc*, 41(9), 3736–3742, 2009.

40. Reeves, B.C., Deeks, J.J., Higgins, J.P.T., Wells, G.A. Chapter 13: Including non-randomized studies, in Higgins, J.P.T., Green, S. (eds.), *Cochrane Handbook for Systematic Reviews of Interventions*, Version 5.0.2 [updated September 2009]. The Cochrane Collaboration, 2009. Available at: http://www.cochrane.org/training/cochrane-handbook (accessed on 15/8/2011).

41. Ukoumunne, O.C., Gulliford, M.C., Chinn, S., Sterne, J.A.C., Burney, P.G.J. Methods for evaluating area-wide and organisation-based interventions in health and health care: A systematic review. *Health Technol Assess*, 3(5), iii–92, 1999.

42. Faggiano, F., Vigna-Taglianti, F., Burkhart, G., et al., EU-Dap Study Group. The effectiveness of a school-based substance abuse prevention program: 18-month follow-up of the EU-Dap cluster randomized controlled trial. *Drug Alcohol Depend*, 108(1–2), 56–64, 2010.

43. Richards, D.A., Lovell, K., Gilbody, S., et al. Collaborative care for depression in UK primary care: A randomized controlled trial. *Psychol Med*, 38, 297–287, 2008.

44. Eaton, C.B., Lapane, K.L., Garber, C.E., Gans, K.M., Lasater, T.M., Carleton, R.A. Effects of a community-based intervention on physical activity: The Pawtucket Heart Health Program. *Am J Public Health*, 89, 1741–1744, 1999.

45. Tabar, L., Fagerberg, G., Chen, H.H., et al. Efficacy of breast cancer screening by age. New results from the Swedish two county trial. *Cancer*, 75, 2507–2517, 1995.

46. Poulstrup, A., Jeune, B. Prevention of fall injuries requiring hospital treatment among community-dwelling elderly. *Eur J Public Health*, 10, 45–50, 2000.

47. Glasziou, P., Chalmers, I., Rawlins, M., McCulloch, P. When are randomised trials unnecessary? Picking signal from noise. *BMJ*, 334(7589), 349–351, 2007.

48. Hajat, S., Fitzpatrick, R., Morris, R.W., et al. Does waiting for total hip replacement matter? Prospective observational study. *J Health Serv Res Policy*, 7, 19–25, 2002.

49. Williams, O., Fitzpatrick, R., Hajat, S., et al. Mortality and morbidity of primary elective total hip replacement. *J Arthroplasty*, 17, 165–171, 2002.
50. Kramer, M.S., Martin, R.M., Sterne, J.A., Shapiro, S., Dahhou, M., Platt, R.W. The double jeopardy of clustered measurement and cluster randomisation. *BMJ*, 339, b2900, 2009.
51. Sacks, H.S., Chalmers, T.C., Smith, H. Randomized versus historical controls for clinical trials. *Am J Med*, 72, 233–239, 1982.
52. Browne, J.P., Hopkins, C., Slack, R., et al. Health-related quality of life after polypectomy with and without additional surgery. *Laryngoscope*, 116, 297–302, 2006.
53. Langham, J., Reeves, B.C., Lindsay, K.W., et al. for the Steering Group for National Study of Subarachnoid Haemorrhage. Variation in outcome after subarachnoid haemorrhage. A study of neurosurgical units in UK and Ireland. *Stroke*, 40(1), 111–118, 2009.
54. Reeves, B.C., Langham, J., Lindsay, K.W., et al. Findings of the international subarachnoid aneurysm trial and the national study of subarachnoid haemorrhage in context. *Br J Neurosurg*, 21, 318–323, 2007.
55. Study designs for EPOC reviews. Effective Practice and Organisation of Care Group. http://epoc.cochrane.org/sites/epoc.cochrane.org/files/uploads/FAQ%20-%20Included%20Studies%20-%20EPOC%20-%202007-May-01.doc. cs unfav2 coil01 (accessed November 30, 2010).
56. Collin, S., Reeves, B.C., Hendy, J., Fulop, N., Hutchings, A., Priedane, E. Computerised physician order entry (CPOE) and picture archiving and communication systems (PACS) implementation in the NHS: A quantitative before-and-after study. *BMJ*, 337, a939, 2008.
57. Lopes, R.D., Hafley, G.E., Allen, K.B., et al. Endoscopic versus open vein-graft harvesting in coronary artery bypass surgery. *N Engl J Med*, 361, 235–244, 2009.
58. Birkhead, J.S., Weston, C.F.M., Chen, R. Determinants and outcomes of coronary angiography after non-ST-segment elevation myocardial infarction. A cohort study of the Myocardial Ischaemia National Audit Project (MINAP). *Heart*, 95, 1593–1599, 2009.
59. Reeves, B.C., Murphy, G.J. Increased mortality, morbidity, and cost associated with red blood cell transfusion after cardiac surgery. *Curr Opin Anaesthesiol*, 21(5), 669–673, 2008. Reprinted: *Curr Opin Cardiol*, 23(5), 607–612, 2008.
60. Jetty, P., Hebert, P., van Walraven, C. Long-term outcomes and resource utilization of endovascular versus open repair of abdominal aortic aneurysms in Ontario. *J Vasc Surg*, 51(3), 577–583, 2010.
61. Ko, D., Chiu, M., Guo, H., et al. Safety and effectiveness of drug-eluting and bare-metal stents for patients with off- and on-label indications. *J Am Coll Cardiol*, 53(19), 1773–1782, 2009.
62. Anonymous. National Nosocomial Infections Surveillance (NNIS) System Report, Data Summary from January 1992–June 2001, issued August 2001. *Am J Infect Control*, 29, 404–421, 2001.
63. Garellick, G., Kärrholm, J., Rogmark, C., Herberts, P. Swedish Hip Arthroplasty Register. Annual Report 2008. http://www.jru.orthop.gu.se/ (accessed November 30, 2010).
64. Byar, D.P. Problems with using observational databases to compare treatments. *Stat Med*, 10, 663–666, 1991.
65. Greenland, S. Interval estimation by simulation as an alternative to and extension of confidence intervals. *Int J Epidemiol*, 33, 1389–1397, 2004.

66. Deeks, J.J., Dinnes, J., D'Amico, R., et al. Evaluating non-randomised intervention studies. *Health Technol Assess*, 7, 27, 2003.

67. McCulloch, P., Altman, D.G., Campbell, W.B., et al. No surgical innovation without evaluation: The IDEAL recommendations. *Lancet*, 374(9695), 1105–1112, 2009.

68. Loke, Y.K., Price, D., Herxheimer, A. Chapter 14: Adverse effects, in Higgins, J.P.T., Green, S. (eds.), *Cochrane Handbook for Systematic Reviews of Interventions*, Version 5.0.1 [updated September 2008]. The Cochrane Collaboration, 2008. Available from Black, N. Why we need observational studies to evaluate the effectiveness of health care. *BMJ*, 312(7040), 1215–1218, 1996.

69. Golder, S., Loke, Y.K., Bland, M. Meta-analyses of adverse effects data derived from randomised controlled trials as compared to observational studies: Methodological overview. *PLoS Med*, 8, e1001026, 2011.

70. Black, N. Why we need observational studies to evaluate the effectiveness of health care. *BMJ*, 312(7040), 1215–1218, 1996.

71. Berman, M.F., Solomon, R.A., Mayer, S.A., Johnston, S.C., Yung, P.P. Impact of hospital-related factors on outcome after treatment of cerebral aneurysms. *Stroke*, 34(9), 2200–2207, 2003.

72. Hannan, E.L., Popp, A.J., Tranmer, B., Fuestel, P., Waldman, J., Shah, D. Relationship between provider volume and mortality for carotid endarterectomies in New York state. *Stroke*, 29(11), 2292–2297, 1998.

73. Norris, S.L., Moher, D., et al. Selective reporting in nonrandomized studies: Impact on systematic reviews and guidance for reviewers. *Res Synth Methods*, in press.

74. Dwan, K., Altman, D.G., Arnaiz, J.A., et al. Systematic review of the empirical evidence of study publication bias and outcome reporting bias. *PLoS ONE*, 3(8), e3081, 2008.

75. Kirkham, J.J., Dwan, K.M., Altman, D.G., et al. The impact of outcome reporting bias in randomised controlled trials on a cohort of systematic reviews. *BMJ*, 340, c365, 2010.

76. Rockall, T.A. Risk scoring in acute upper gastrointestinal haemorrhage. *Dig Liver Dis*, 38, 10–11, 2006.

77. Hutchison, C.A., Crowe, A.V., Stevens, P.E., Harrison, D.A., Lipkin, G.W. Case mix, outcome and activity for patients admitted to intensive care units requiring chronic renal dialysis: A secondary analysis of the ICNARC Case Mix Programme Database. *Crit Care*, 11, R50, 2007.

78. Berman, M., Stamler, A., Sahar, G., et al. Validation of the 2000 Bernstein-Parsonnet score versus the EuroSCORE as a prognostic tool in cardiac surgery. *Ann Thorac Surg*, 81(2), 537–540, 2006.

79. Nashef, S.A., Roques, F., Michel, P., Gauducheau, E., Lemeshow, S., Salamon, R. European system for cardiac operative risk evaluation (EuroSCORE). *Eur J Cardiothorac Surg*, 16(1), 9–13, 1999.

80. Baigent, C., Harrell, F.E., Buyse, M., Emberson, J.R., Altman, D.G. Ensuring trial validity by data quality assurance and diversification of monitoring methods. *Clin Trials*, 5, 49–55, 2008.

13

Methodological and Reporting Considerations for Systematic Reviews of Nonpharmacological Interventions

Alexander Tsertsvadze, Larissa Shamseer, and David Moher
Ottawa Hospital Research Institute

CONTENTS

13.1 Methodology of Systematic Reviews of Nonpharmacological Intervention Trials

The underlying principles and basic methodological steps needed for conducting and reporting systematic reviews of pharmacological or nonpharmacological intervention trials are similar. The basic steps for conducting a systematic review appropriately have been well established and reviewed recently [1]. Regardless of the nature of health intervention under review, the general sequence of broadly defined steps/activities needed for the conduct of systematic reviews includes the following 12 steps:

1. Identification of gaps in knowledge
2. Formulation and refinement of the review question
3. Development of the review protocol by establishing the objectives, study eligibility criteria (design of primary study, study population,

experimental intervention, control intervention, and outcome measures), literature search (search strategy, bibliographic sources, period of time to be covered), methods/tools to assess risk of bias of individual studies, and analytic approaches to describe and synthesize data

4. Assembling identified bibliographic records

5. Selection of the identified bibliographic reports of primary studies (using broad and strict screen)

6. Extraction of relevant data using a pre-specified piloted form

7. Assessment of risk of bias of the included individual studies

8. Analysis and synthesis of data from results of primary study (qualitative and/or quantitative meta-analytic summary)

9. Presentation of the review results (study selection process diagram, tabulate study and population characteristics, study results regarding outcome measures, and risk of bias for each relevant for the review outcome)

10. Interpretation of the review results (risk of bias of individual studies, the strength and applicability of evidence)

11. Dissemination of the review results (e.g., publication, presentation)

12. Updating the review (since evidence continuously accumulates over time, there is a need to maintain the review results up-to-date)

Ideally, a systematic review of nonpharmacological intervention trials should additionally consider unique methodological aspects inherent to such trials. The methodological challenges in conducting nonpharmacological intervention trials have been widely acknowledged [2–9]. Unlike pharmacological interventions, the mechanisms of action of many nonpharmacological interventions are not clear and they are usually administered as single or discrete applications (e.g., acupuncture, massage, manipulation, surgery, diagnostic test, education/counseling) [10]. Since these interventions represent a combination of treatment-specific (i.e., intervention as a procedure) and contextual effects (e.g., intervention fidelity, maturity of intervention, operator's skill set, care provider's age and length of experience, pre- and post-intervention care, study subject's expectation), they are termed as "complex interventions" [8,11]. In this situation, it is often hard to tease out the specific effects of an intervention from those of context.

Therefore, the assessment of risk of bias (or reporting quality) of nonpharmacological intervention trials additionally needs to take into account the sources of contextual effects and information on specific blinding techniques applied to study participants or assessors if applicable, especially when dealing with "soft" (i.e., subjective) outcome measures. Blinding of study participants in trials evaluating nonpharmacological interventions is often very difficult or impossible and this further confounds the trial results. The exploration of comparability with regards to post-interventional care across the study arms of randomized subjects (e.g., in surgical trials) is also important, since the systematic differences in how post-intervention care is delivered may bias the true effects of administered interventions [8].

Some trials evaluating nonpharmacological interventions (e.g., behavioral therapy, psychotherapy, rehabilitation, education) are often cluster randomized. Since subjects in these clusters are not totally independent in their responses, this leads to a loss of power compared with individually randomized trials. Therefore, sample size calculation for the given cluster randomized trial should incorporate the cluster size and intra-cluster correlation coefficient to ensure an adequate study power [12]. Moreover, special statistical techniques need to be applied to adjust for clustering effects arising from the discordance between the units of randomization (cluster level) and analysis (individual level). Ignoring clustered nature of the data may lead to spuriously significant results and conclusions [13]. Systematic reviews of cluster randomized trials should adequately determine the appropriateness of statistical analyses reported in primary publications.

The aforementioned aspects inherent to trials evaluating nonpharmacological interventions, compounded with ethical concerns, tend to further complicate appropriate interpretation of study results. This in turn hinders an investigator's ability to reach valid conclusions in any given systematic review.

A systematic review of trials evaluating nonpharmacological interventions should ascertain and clearly distinguish blinding status and techniques used in a given trial. For example, for each included study, it is important to document the blinding status of study participants, care providers, and outcome assessors in order to objectively assess the risk of bias. In some trials where blinding is impossible, alternative techniques (e.g., blinding to study hypothesis, centralized assessment of the outcome) thought to mitigate the potential for bias should also be sought. Each such review should accurately report the intervention (e.g., type of procedure, care provider, his or her training and experience) and its maturity for each included trial. The change in maturity or age of any given therapeutic device or procedure may result in different profiles of benefit and/or safety. For example, in earlier studies the use of a newly developed technology or device may be shown to be clinically less beneficial and/or more harmful compared with studies conducted and published later in which the device reached its maturity and its use has been refined [2,3,14,15]. For each included in the review trial, funding source/conflict of interest needs to be determined. The pharmaceutical industry often provides funds to assess benefits and harms of using new technologies (e.g., surgical instruments) and investigators are rewarded for recruiting study subjects. In these situations, the level of objectivity of care providers and outcome assessors may be compromised leading to biased study results [4]. In one recent methodological review of 87 surgical trials, Bhandari and colleagues revealed that industry-funded trials compared with government- or foundation-funded trials were more likely to report statistically significant results in favor of the new industry products [16].

In a recent review, Hartling and colleagues discussed specific challenges while conducting systematic reviews of nonpharmacological interventions [14]. The authors reviewed pros and cons of including gray literature and nonrandomized trials in reviews of therapeutic devices and procedures. Furthermore, they explored applicability issues unique to trials evaluating therapeutic devices and procedures and presented several examples indicating that results of trials evaluating nonpharmacological interventions are less readily applicable to broader populations with the condition of interest than those from trials evaluating pharmaceutical drugs. According to the authors' recommendation, the deliberation of applicability of study results should be based not only on the patients' inclusion/exclusion criteria of the review but also on the eligibility criteria for care providers and institutions in which these trials were conducted. The authors also noted that since the use of many invasive procedures (e.g., surgery) indicates less clinical benefit and an increased rate of complications shortly after their administration, it is advisable for a reviewer to pool the study outcomes measured at similar rather than different post-procedure follow-up time points.

13.2 Reporting Issues of Systematic Reviews of Nonpharmacological Interventions

13.2.1 Why Is Reporting Important?

Collecting, aggregating, implementing, and reporting research is a complex process. For example, the pharmaceutical evaluation framework starts with basic science innovations that result in a number of drugs being tested in a variety of animal models. Successful results are then followed by the "first in human" studies. Those drugs showing some benefit in terms of efficacy and safety in a target population are subsequently evaluated in randomized clinical trials. Several trials addressing a similar clinical question (with or without an economic evaluation) may be incorporated in a systematic review. The results of such reviews will be used by clinicians, policy, and decision makers to inform and formulate practice guidelines. This description, naively, assumes that all research is available to interested readers and that what is published, after peer review, is of such a high quality that the descriptions of the methods and findings are clear, accurate, and transparent enabling readers to use the information.

Problems on how research is reported have been well documented for the last two decades. The issues of poor reporting might partially reflect inadequate conduct of systematic reviews. This problem pervades all areas of health across the continuum of research architecture. Mignini and Khan reviewed 30 systematic reviews of animal studies and 30 laboratory bench studies and determined that many of them were poorly reported [17]. Glasziou and colleagues [18] assessed descriptions of treatments in 80 trials and systematic reviews published over a 1 year period (October 2005 to October 2006) in *Evidence-Based Medicine*—a journal aimed at physicians working in primary care and general medicine. The treatment descriptions were inadequate in 41 of the original published articles, which made their use in clinical practice difficult, if not impossible, to replicate. These are just a few examples of a large literature [19–21] indicating the general failure in reporting health research adequately. Despite the large and ever-increasing amount of research being produced worldwide, many publications lack clarity, transparency, and completeness in how the authors actually carried out their research.

This problem has concerned journal editors for some time. Few journals have the resources to carry out the correct implementation of reporting guidelines such that all research reports can uniformly benefit. This has detrimental downstream consequences, namely, misinterpretation by readers and inappropriate application of treatment in clinical settings. While randomized controlled trials (RCTs) have been the focus of particular scrutiny because of their direct impact on health care, their reporting inadequacies can be passed on at the systematic review level, often leaving conclusions of systematic reviews difficult to interpret [22,23]. The use of such primary data may contribute to improper conclusions in any given systematic review.

Poor reporting of primary studies is only one factor that may affect reporting of systematic reviews. Systematic reviews themselves may suffer from poor reporting, thereby propagating reporting inadequacies of primary studies. For instance, interventions are adequately described in only 60% of trial reports, yet if reviewers are willing to put an additional effort into contacting authors, checking references, and carrying out additional searches, this number could be increased to 90% [18]. Since treatment procedures used in nonpharmacological trials are more complex compared with pharmacological trials, the detailed description of these interventions is of paramount importance for adequate reporting of systematic reviews of such trials.

13.2.2 Reporting Guidelines

Reporting guidelines are newer tools that emerged out of the necessity to ameliorate reporting problems in the health research literature. A reporting guideline, developed using a consensus-based process, has been defined as a checklist, flow diagram, and/or explicit text to guide authors in reporting a specific type of research architecture [24].

The first appearance of reporting guidance for randomized trials of healthcare interventions emerged in the early 1990s leading to the development of the Consolidated Standards of Reporting Trials (CONSORT) Statement in 1996 [25]—a guideline aimed at helping authors prepare reports of parallel group randomized trials. Since then, the statement itself has been updated [26,27] and has been shown to be associated with improved reporting quality of trials [28,29]. In addition to its positive impact on trial reporting, CONSORT seems to have spurred the development of a number of other reporting guidelines related to clinical trials. At least eight extensions have been developed, including one for reporting trials using nonpharmacological interventions [30], and others are in development.

While guidance on systematic review conduct is plentiful [1,31–34], guidance on their reporting has been sparse until recently. Relevant reporting guidance first emerged in the form of the Quality of Reporting Meta-analyses (QUORUM) Statement. However, as QUORUM was primarily intended to address reporting of meta-analyses, which may or may not be a part of any given systematic review, its developers revised it to produce the Preferred Reporting Items for Systematic reviews and Meta-Analyses (PRISMA) Statement in 2009 [35,36]. Since its emergence, the PRISMA Statement has been endorsed by at least 175 journals and several influential organizations (www.prisma-statement.org/endorsers).

The PRISMA Statement is a reporting guideline aimed at helping authors present a clear picture to readers of what was planned, done, and found in a systematic review. It was developed by bringing evidence and experts together in a consensus-based process. The statement consists of a set of 27 items authors should describe when preparing a systematic review report for publication (Table 13.1). The checklist is accompanied by a flow diagram (Figure 13.1) to help authors demonstrate the flow of studies through their review.

Authors wishing to report a systematic review should include text, which, at minimum, clearly describes each PRISMA item and includes a flow diagram, following guidance contained in the PRISMA explanation and elaboration (E&E) paper [36]. The E&E should be read thoroughly prior to and during preparation of a systematic review report; it provides authors with context for each item as well as a published example of good reporting for each item. As more journals are moving toward online publishing, journals' word and page limitations for manuscripts are becoming less of an issue. While some journals, namely, PRISMA endorsers, solicit a PRISMA checklist with review submission, others do not. Where permitted, authors may wish to ask a journal to publish a completed PRISMA checklist alongside their review. The completed checklist should list on which journal page the reader can find text describing each PRISMA item. Doing so will improve efficiency and use of systematic reviews for clinical and policy decision making.

Whether or not authors complied with recommended and/or conventional systematic review methods while carrying out the review should not affect the reporting quality of the review. Methodological inadequacies or limitations encountered in the data should be explicitly stated, with reasons why,

TABLE 13.1

The PRISMA Checklist

Section/Topic	#	Checklist Item	Reported on Page #
TITLE			
Title	1	Identify the report as a systematic review, meta-analysis, or both	
ABSTRACT			
Structured summary	2	Provide a structured summary including, as applicable, the background; objectives; data sources; study eligibility criteria, participants, and interventions; study appraisal and synthesis methods; results; limitations; conclusions and implications of key findings; and systematic review registration number	
INTRODUCTION			
Rationale	3	Describe the rationale for the review in the context of what is already known	
Objectives	4	Provide an explicit statement of questions being addressed with reference to participants, interventions, comparisons, outcomes, and study design (PICOS)	
METHODS			
Protocol and registration	5	Indicate if a review protocol exists, if and where it can be accessed (e.g., web address), and, if available, provide registration information including registration number	
Eligibility criteria	6	Specify study characteristics (e.g., PICOS, length of follow-up) and report characteristics (e.g., years considered, language, publication status) used as criteria for eligibility, giving rationale	
Information sources	7	Describe all information sources (e.g., databases with dates of coverage, contact with study authors to identify additional studies) in the search and date last searched	
Search	8	Present full electronic search strategy for at least one database, including any limits used, such that it could be repeated	
Study selection	9	State the process for selecting studies (i.e., screening, eligibility, included in systematic review, and, if applicable, included in the meta-analysis)	
Data collection process	10	Describe the method of data extraction from reports (e.g., piloted forms, independently, in duplicate) and any processes for obtaining and confirming data from investigators	

(continued)

TABLE 13.1 (continued)

The PRISMA Checklist

Section/Topic	#	Checklist Item	Reported on Page #
Data items	11	List and define all variables for which data were sought (e.g., PICOS, funding sources) and any assumptions and simplifications made	
Risk of bias in individual studies	12	Describe methods used for assessing risk of bias of individual studies (including specification of whether this was done at the study or outcome level), and how this information is to be used in any data synthesis	
Summary measures	13	State the principal summary measures (e.g., risk ratio, difference in means)	
Synthesis of results	14	Describe the methods of handling data and combining results of studies, if done, including measures of consistency (e.g., I^2) for each meta-analysis	
Risk of bias across studies	15	Specify any assessment of risk of bias that may affect the cumulative evidence (e.g., publication bias, selective reporting within studies)	
Additional analyses	16	Describe methods of additional analyses (e.g., sensitivity or subgroup analyses, meta-regression), if done, indicating which were pre-specified	
RESULTS			
Study selection	17	Give numbers of studies screened, assessed for eligibility, and included in the review, with reasons for exclusions at each stage, ideally with a flow diagram	
Study characteristics	18	For each study, present characteristics for which data were extracted (e.g., study size, PICOS, follow-up period) and provide citations	
Risk of bias within studies	19	Present data on risk of bias of each study and, if available, any outcome level assessment (see Item 12)	
Results of individual studies	20	For all outcomes considered (benefits or harms), present, for each study: (a) simple summary data for each intervention group and (b) effect estimates and confidence intervals, ideally with a forest plot	
Synthesis of results	21	Present results of each meta-analysis done, including confidence intervals and measures of consistency	
Risk of bias across studies	22	Present results of any assessment of risk of bias across studies (see Item 15)	
Additional analysis	23	Give results of additional analyses, if done (e.g., sensitivity or subgroup analyses, meta-regression [see Item 16])	

TABLE 13.1 (continued)

The PRISMA Checklist

Section/Topic	#	Checklist Item	Reported on Page #
DISCUSSION			
Summary of evidence	24	Summarize the main findings including the strength of evidence for each main outcome; consider their relevance to key groups (e.g., healthcare providers, users, and policy makers)	
Limitations	25	Discuss limitations at study and outcome level (e.g., risk of bias), and at review-level (e.g., incomplete retrieval of identified research, reporting bias)	
Conclusions	26	Provide a general interpretation of the results in the context of other evidence, and implications for future research	
FUNDING			
Funding	27	Describe sources of funding for the systematic review and other support (e.g., supply of data), role of funders for the systematic review	

Source: From Moher, D. et al., *PLoS Med.*, 6(7), e1000097, 2009.

as recommended by PRISMA. For instance, if data were unavailable from primary study reports, but otherwise obtained (i.e., from study authors or elsewhere), review authors should state this. An example of good reporting for this item—PRISMA item 10—is presented in the PRISMA E&E:

> We developed a data extraction sheet (based on the Cochrane Consumers and Communication Review Group's data extraction template), pilot-tested it on ten randomly-selected included studies, and refined it accordingly. One review author extracted the following data from included studies and the second author checked the extracted data … Disagreements were resolved by discussion between the two review authors; if no agreement could be reached, it was planned a third author would decide. We contacted five authors for further information. All responded and one provided numerical data that had only been presented graphically in the published paper [37].

13.2.3 Unique Issues in Reporting Nonpharmacological Systematic Reviews

As of 2009, systematic reviews of nonpharmacological interventions were the most frequently consulted reviews in the Cochrane library [38]. Furthermore, empirical evidence shows that the reporting quality of nonpharmacological trials is unsatisfactory [39–41]. In order to document and overcome the deficiencies of primary studies, clear reporting of systematic reviews of these interventions is essential.

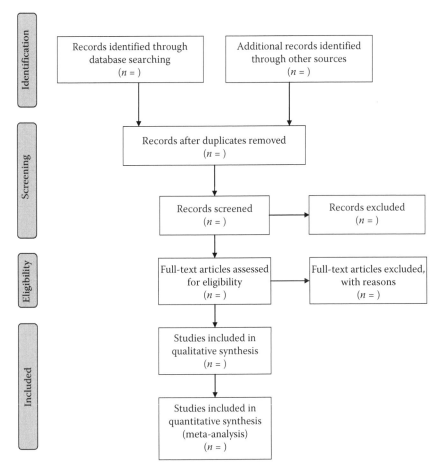

FIGURE 13.1
PRISMA flow diagram. (From Moher, D. et al., *PLoS Med*, 6(7), e1000097, 2009.)

While an extension to the CONSORT Statement has been developed to address the reporting of nonpharmacological randomized trials [30], no specific guidance currently exists to guide reporting of systematic reviews of these interventions. Reporting guidance for primary studies of specific nonpharmacological interventions has also been proposed [42,43]. Building on areas of the PRISMA Statement in which intervention differences may affect reporting and using the CONSORT extension for nonpharmacological trials, PRISMA items that may benefit from additional description are highlighted in the following. These suggestions should be read and used in conjunction with the PRISMA Statement Explanation and Elaboration document [36]. Systematic reviews synthesizing studies of nonpharmacological interventions may require special attention to the following PRISMA items listed below.

Item 6: Eligibility criteria

Since the knowledge of study eligibility criteria is essential for determining the applicability (external validity) of the results of systematic reviews, the criteria used to select studies for inclusion should be explicitly described. Specifically, when describing the intervention under evaluation, there are a number of factors to be reported. Poor description of complex nonpharmacological interventions, such as behavioral, educational, rehabilitation, and surgical interventions, has been a recurring barrier in synthesizing evidence of nonpharmacological studies [44,45].

Information about treatment procedures including care providers (e.g., skill level, experience), concomitant treatments, standardization or uniformity of procedures has been shown to have been better described in pharmacological trials compared with nonpharmacological trials [5]. To minimize or avoid this problem within systematic reviews, an unambiguous description of the non-pharmacological intervention, highlighting the aforementioned components, is needed. Such a description might include what treatment is delivered, the intensity of treatment, frequency of delivery, and if not self-delivered, who delivered it and whether they received specific training to do so [46]. Whether included trials were limited to those describing treatments that were standardized across participants and sites or if those which allowed variations in treatment delivery (and to what extent) were utilized should also be reported. A description of the comparator (i.e., control) treatment of included studies is also important since it may represent a nonpharmacological placebo (i.e., sham or nonactivated device/machine) or standard care, active treatment, or a waiting list. Description of the comparator intervention also helps clinicians place the context of the intervention under evaluation and decide if the described study setting is similar to their own practice setting [30].

Item 7: Information sources

Since nonpharmacological interventions target not only clinical conditions but also a range of behavioral and psychosocial aspects of study populations, information sources (e.g., databases, journals) encompassing nonclinical areas, such as psychology and education, may be considered as part of the review. The exact source and scope of each source should be completely described along with other required PRISMA information.

Item 8: Search

Identifying nonpharmacological trials may require the use of specialized search terminology or, in the case of some nonpharmacological interventions in which randomization may occur in clusters or groups (i.e., practitioners randomized to treatment group), the lack of terminology (i.e., "random-ized"). A description of the complete search strategy for at least one source of information and limits, if used, should be presented in the review report.

Item 12: Risk of bias (quality) in individual studies

Risk of bias assessments in systematic reviews are typically limited by adequacy of reporting of primary studies. This makes assessing nonpharmacological interventions challenging. One empirical study showed that reports of nonpharmacological trials of hip and knee osteoarthritis were of lower quality compared with those of pharmacological trials [5]. For example, nonpharmacological trials reported techniques used to blind study participants, care providers, and outcome assessors less frequently than pharmacological trials. While blinding of participants, personnel, and outcome assessors—one risk of bias domain—is often not possible in non-pharmacological trials, non-standard methods to reduce detection bias are often employed [47]. As such, how review authors deal with blinding during risk of bias assessment of nonpharmacological trials should be stated. For instance, if review authors attributed a low risk of bias to trials adequately describing blinding of outcome assessors but not of participants or personnel, this should be explicitly stated, with rationale for the decision.

Whether or not any changes were considered to the way in which allocation concealment—another domain of risk of bias—was assessed, and what those changes were, should be described. For instance, if some included trials randomized practitioners to treatment groups (typical in nonpharmacological trials), whereas others did not, the decision rules used by authors to judge the study as high, low, or unclear risk of bias, if different from typical pharmacological trials, should be reported. Furthermore, if sensitivity analyses were carried out which were specifically attributable to exploring bias due to the intervention, they should be clearly described.

Evidence also shows that the use of intention-to-treat analysis was better reported for pharmacological versus nonpharmacological trials of hip and knee osteoarthritis [5]. As such, close attention to the reporting of incomplete data in included trials should be paid.

Item 24: Summary of evidence
The external validity of the review findings should be clearly reported, with specific consideration of the applicability of interventions (and comparators) and heterogeneity of treatment regimens of included studies, if present.

13.3 Conclusion

Poor reporting of the medical literature has been recognized as a serious problem for decades [38,48,49]. Systematic reviews are an important conduit of health research; in order for them to be used effectively to make and implement clinical decision, they need to be clearly described. While review authors must do their part to ensure information in their reviews is usable and accessible, the responsibility of adequate reporting also lies on authors of primary studies since systematic reviewers are limited to the information primary authors present and provide.

References

1. Tricco, A.C., Tetzlaff, J., Moher, D. The art and science of knowledge synthesis. *J Clin Epidemiol*, 64(1), 11–20, 2011.
2. van der Linden, W. Pitfalls in randomized surgical trials. *Surgery*, 87(3), 258–262, 1980.
3. McCulloch, P., Taylor, I., Sasako, M., Lovett, B., Griffin, D. Randomised trials in surgery: Problems and possible solutions. *BMJ*, 324(7351), 1448–1451, 2002.
4. McLeod, R.S. Issues in surgical randomized controlled trials. *World J Surg*, 23(12), 1210–1214, 1999.
5. Boutron, I., Tubach, F., Giraudeau, B., Ravaud, P. Methodological differences in clinical trials evaluating nonpharmacological and pharmacological treatments of hip and knee osteoarthritis. *JAMA*, 290(8), 1062–1070, 2003.
6. Brown, M.B. Control groups appropriate for surgical interventions: Ethical and practical issues. *Gastroenterology*, 126(1), S164–S168, 2004.
7. Johnson, A.G. Surgery as a placebo. *Lancet*, 344(8930), 1140–1142, 1994.
8. Ergina, P.L., Cook, J.A., Blazeby, J.M., et al. Challenges in evaluating surgical innovation. *Lancet*, 374(9695), 1097–1104, 2009.
9. Cook, J.A. The challenges faced in the design, conduct and analysis of surgical randomised controlled trials. *Trials*, 10, 9, 2009.
10. Bennet, M.I. Methodological issues in cancer pain: Nonpharmacological trials, in Paice, J.A., Bell, R.F., Kalso, E.A., Soyannwo, O.A. (eds.), *Cancer Pain: From Molecules to Suffering*. Seattle, WA: IASP Press, 2010.
11. Craig, P., Dieppe, P., Macintyre, S., Michie, S., Nazareth, I., Petticrew, M. Developing and evaluating complex interventions: The new Medical Research Council guidance. *BMJ*, 337, a1655, 2008.
12. Campbell, M.K., Mollison, J., Steen, N., Grimshaw, J.M., Eccles, M. Analysis of cluster randomized trials in primary care: A practical approach. *Fam Pract*, 17(2), 192, 2000.
13. Donner, A., Klar, N. Pitfalls of and controversies in cluster randomization trials. *Am J Public Health*, 94(3), 416–422, 2004.
14. Hartling, L., McAlister, F.A., Rowe, B.H., Ezekowitz, J., Friesen, C., Klassen, T.P. Challenges in systematic reviews of therapeutic devices and procedures. *Ann Intern Med*, 142(12), 1100–1111, 2005.
15. Itani, K.M., McCullough, L.B. Randomized clinical surgical trials: Are they really necessary? *Asian J Surg*, 27(3), 163–168, 2004.
16. Bhandari, M., Busse, J.W., Jackowski, D., et al. Association between industry funding and statistically significant pro-industry findings in medical and surgical randomized trials. *Can Med Assoc J*, 170(4), 477, 2004.
17. Mignini, L.E., Khan, K.S. Methodological quality of systematic reviews of animal studies: A survey of reviews of basic research. *BMC Med Res Methodol*, 6, 10, 2006.
18. Glasziou, P., Meats, E., Heneghan, C., Shepperd, S. What is missing from descriptions of treatment in trials and reviews? *Br Med J*, 336(7659), 1472, 2008.
19. Hopewell, S., Dutton, S., Yu, L.M., Chan, A.W., Altman, D.G. The quality of reports of randomised trials in 2000 and 2006: Comparative study of articles indexed in PubMed. *Br Med J*, 340, c723, 2010.

20. Duff, J.M., Leather, H., Walden, E.O., LaPlant, K.D., George, T.J. Adequacy of published oncology randomized controlled trials to provide therapeutic details needed for clinical application. *J Natl Cancer Inst*, 102(10), 702, 2010.

21. de Vries, T.W., van Roon, E.N. Low quality of reporting adverse drug reactions in paediatric randomised controlled trials. *Arch Dis Child*, 95(12), 1023–1026, 2010.

22. Moher, D., Tetzlaff, J., Tricco, A.C., Sampson, M., Altman, D.G. Epidemiology and reporting characteristics of systematic reviews. *PLoS Med*, 4(3), e78, 2007.

23. Kirkham, J.J., Dwan, K.M., Altman, D.G., et al. The impact of outcome reporting bias in randomised controlled trials on a cohort of systematic reviews. *BMJ*, 340, c365, 2010.

24. Moher, D., Schulz, K.F., Simera, I., Altman, D.G. Guidance for developers of health research reporting guidelines. *PLoS Med*, 7(2), e1000217, 2010.

25. Begg, C., Cho, M., Eastwood, S., et al. Improving the quality of reporting of randomized controlled trials. The CONSORT statement. *JAMA*, 276(8), 637–639, 1996.

26. Moher, D., Schulz, K.F., Altman, D.G., CONSORT group. The CONSORT statement: Revised recommendations for improving the quality of reports of parallel group randomized trials. *BMC Med Res Methodol*, 1, 2, 2001.

27. Schulz, K., Altman, D., Moher, D. CONSORT 2010 statement: Updated guidelines for reporting parallel group randomised trials. *BMC Med*, 8(1), 18, 2010.

28. Plint, A.C., Moher, D., Morrison, A., et al. Does the CONSORT checklist improve the quality of reports of randomised controlled trials? A systematic review. *Med J Aust*, 185(5), 263, 2006.

29. The influence of CONSORT on the quality of RCTs: An updated review. 9th Annual Cochrane Canada Symposium, February 16 and 17, 2011.

30. Boutron, I., Moher, D., Altman, D.G., Schulz, K.F., Ravaud, P. Extending the CONSORT statement to randomized trials of nonpharmacologic treatment: Explanation and elaboration. *Ann Intern Med*, 148(4), 295, 2008.

31. Agency for Healthcare Research and Quality. *Methods Reference Guide for Effectiveness and Comparative Effectiveness Reviews*, 1st edn. Rockville, MD: Agency for Healthcare Research and Quality, 2007.

32. Higgins, J., Green, S. *Cochrane Handbook for Systematic Reviews of Interventions.* Chichester, U.K.: Wiley Online Library, 2008.

33. Centre for Reviews and Dissemination. *Systematic Reviews: CRD's Guidance for Undertaking Reviews in Health Care.* York, U.K.: University of York, NHS Centre for Reviews and Dissemination, 2009.

34. Institute of Medicine. *Finding What Works in Health Care: Standards for Systematic Reviews.* Washington, DC: The National Academies Press, 2011.

35. Moher, D., Liberati, A., Tetzlaff, J., Altman, D.G., PRISMA Group. Preferred reporting items for systematic reviews and meta-analyses: The PRISMA statement. *PLoS Med, 6(7), e1000097, 2009.*

36. Liberati, A., Altman, D.G., Tetzlaff, J., et al. The PRISMA statement for reporting systematic reviews and meta-analyses of studies that evaluate health care interventions: Explanation and elaboration. *PLoS Med*, 6(7), e1000100, 2009.

37. Mistiaen, P., Poot, E. Telephone follow-up initiated by a hospital-based health professional for postdischarge problems in patients discharged from hospital to home. *Cochrane Database Syst Rev*, (4), CD004510, 2006.

38. Chalmers, I., Glasziou, P. Avoidable waste in the production and reporting of research evidence. *Obstet Gynecol*, 114(6), 1341, 2009.

39. Agha, R., Cooper, D., Muir, G. The reporting quality of randomised controlled trials in surgery: A systematic review. *Int J Surg*, 5(6), 413–422, 2007.

40. Bhandari, M., Guyatt, G.H., Lochner, H., Sprague, S., Tornetta, P., III. Application of the Consolidated Standards of Reporting Trials (CONSORT) in the fracture care literature. *J Bone Joint Surg Am*, 84-A(3), 485–489, 2002.

41. Hall, J.C., Mills, B., Nguyen, H., Hall, J.L. Methodologic standards in surgical trials. *Surgery*, 119(4), 466–472, 1996.

42. Robb, S.L., Carpenter, J.S., Burns, D.S. Reporting guidelines for music-based interventions. *J Health Psychol*, 16(2), 342, 2011.

43. Des Jarlais, D.C., Lyles, C., Crepaz, N. Improving the reporting quality of non-randomized evaluations of behavioral and public health interventions: The TREND statement. *Am J Public Health*, 94(3), 361, 2004.

44. Michie, S., Fixsen, D., Grimshaw, J.M., Eccles, M.P. Specifying and reporting complex behaviour change interventions: The need for a scientific method. *Implement Sci*, 4(1), 40, 2009.

45. Dijkers, M.P. Ensuring inclusion of research reports in systematic reviews. *Arch Phys Med Rehabil*, 90(11), S60–S69, 2009.

46. O'Connor, D., Green, S., Higgins, J.P.T. Defining the review questions and developing criteria for including studies, in Higgins, J.P.T., Green, S. (eds.), *Cochrane Handbook of Systematic Reviews of Interventions*. Chichester, U.K.: John Wiley & Sons, 2008.

47. Boutron, I., Guittet, L., Estellat, C., Moher, D., Hróbjartsson, A., Ravaud, P. Reporting methods of blinding in randomized trials assessing nonpharmacological treatments. *PLoS Med*, 4(2), e61, 2007.

48. Mahon, W.A., Daniel, E.E. A method for the assessment of reports of drug trials. *Can Med Assoc J*, 90(9), 565, 1964.

49. Gotzsche, P.C. Methodology and overt and hidden bias in reports of 196 double-blind trials of nonsteroidal antiinflammatory drugs in rheumatoid arthritis. *Control Clin Trials*, 10(1), 31–56, 1989.

14

Accounting for the Complexity of the Intervention in Nonpharmacological Systematic Reviews

Sasha Shepperd

University of Oxford

Simon Lewin

Medical Research Council of South Africa

CONTENTS

14.1 Introduction

Healthcare interventions have evolved in response to the complex problems facing health systems around the world. A recent survey of randomized controlled trials, supported by the main noncommercial funders of research in the United Kingdom, reports a diverse set of healthcare interventions evaluated in trials funded by the NHS between 1980 and 2002. In addition to trials of pharmacological and surgical interventions, the interventions evaluated include education and training, service delivery, psychological therapy, and complex care interventions [1]. Other examples of healthcare interventions that might be considered more complex are those acting at the level of the health system. These include interventions directed at health systems governance arrangements, such as the regulation of pharmaceutical sales; interventions directed at health systems financial arrangements, such as mechanisms of financing health care to reduce inequities; interventions directed at health systems delivery arrangements, such as multidisciplinary packages of care to manage chronic diseases; and interventions to promote change in professional practice or care delivery, such as guideline implementation strategies [2].

Evaluating the effectiveness of a complex intervention is methodologically challenging. Researchers face the usual problems with regard to feasibility, recruitment, outcome selection, and funding mechanisms. In addition, there can be difficulties in standardizing the delivery of a complex intervention prior to the evaluation, and ensuring fidelity during the evaluation. As a result, studies are often small and underpowered. It is not surprising, therefore, that an increasing number of systematic reviews of complex interventions are being commissioned to provide the least biased summary of research evidence to guide decision making.

14.2 What Is a Complex Intervention?

While all interventions sit on a spectrum of complexity, the recently revised MRC guidance for single studies of complex interventions helps differentiate between simple and complex interventions. A complex intervention is defined as an intervention that contains several components that may or may not interact with each other. Some interventions may be more complex in relation to one component (e.g., the number of organization levels targeted) and less complex in other ways (e.g., the degree of tailoring allowed). Additional dimensions of complexity include a range of possible outcomes and their variability in the target population, the number of behaviors required by those delivering or receiving the intervention, the number of groups or organizational levels targeted by the intervention, and the degree of flexibility or tailoring of the intervention allowed [3].

Opinion is sharply divided on the potential of systematic reviews of complex interventions to inform healthcare decision making and shape policy [4]. Some argue that systematic reviews provide the least biased summary of evidence to inform those making healthcare decisions to implement, modify, or withdraw an intervention. Others [5] oppose this. While it can be relatively straightforward to systematically review and provide an unbiased estimate of effect of a comparatively simple intervention, for example, drug "A" versus placebo, it is argued that similar methods cannot be applied to more complex interventions. Part of the reasoning behind this revolves around the number of variables that need to be taken into account in a systematic review of a complex intervention, for example, the context of the interventions. It is argued that valuable information about how an intervention works may be lost through the application of the traditional systematic review methodology. Those advocating alternative methods, such as realist synthesis, for reviewing the evidence of complex interventions suggest that the questions concerning how a complex intervention works can best be addressed by including disparate sources of evidence [6]. While this may be so, systematic reviews of randomized controlled trials, including quantitative pooling of outcome data where appropriate and feasible, remain the most robust approach to assessing the average effectiveness of health interventions.

We recognize that for any individual systematic review the reviewer has to assess a range of issues relating to complexity and decide if they are sufficiently important to need addressing. Building on the definition for complex interventions provided by the updated MRC guidance, we describe in more detail the range of issues to be considered when accounting for the complexity of the interventions in nonpharmacological systematic reviews.

14.3 Sources of Complexity in a Systematic Review of Complex Interventions

Complexity in systematic reviews may be due to

1. Differential effects of the intervention across populations and the number of groups or organizational levels targeted by the intervention
2. The definition and characteristics of the intervention, including the number of behaviors required by those delivering or receiving the intervention, the degree of flexibility or tailoring of the intervention allowed, and contextual factors
3. A wide range of outcomes, different measures of the same outcome, similar outcomes but measured at different time points, or proxy outcomes
4. Methodological issues, for example, searching and selecting studies, combining data
5. Other issues such as approaches to implementation and external validity

14.3.1 Differential Effects across Populations and the Number of Groups or Organizational Levels Targeted by the Intervention

Similar interventions may be delivered to a wide range of different populations. For example, audit and feedback interventions may be used to improve the quality of care delivered by nurses, physiotherapists, physicians, or pharmacists [7]. Alternatively, audit and feedback may be directed at a combination of these groups or at teams of professionals, such as a primary care practice. In the context of a systematic review, reviewers have to decide if the findings of all evaluations of an intervention should be combined, or it might be anticipated that the intervention would have different effects across different populations. For example, a systematic review of a hospital at home intervention designed to discharge patients early from hospital assessed, as separate groups, its impact on older people admitted to hospital with a medical condition and those recovering from surgery [8]. Similar interventions may also target different organizational levels. For example, financial incentives aimed at changing behavior may be aimed at primary care physicians [9] or, in the case of conditional cash transfers, at the public [10]. It is likely that financial incentive interventions will have different effects across these groups.

14.3.2 Definition and Characteristics of the Intervention

Simple labels, such as "behavioral intervention," "stroke unit," or "patient education," can be attached to very different interventions. The lack of an agreed explicit definition of the content of complex interventions that carry the same name and the subsequent different, and often inadequate, ways that interventions are described across trials creates inherent difficulties for systematic reviewers [11,12]. This is compounded by poor description of the intervention in trial reports. For example, continuity of care, a concept that is considered to contribute to high-quality care, can be delivered through numerous mechanisms (shared care, telephone follow-up, patient-held records, and case management, to name a few) [13]. Reviewers need to consider whether it makes conceptual "sense" for this range of interventions to promote continuity of care to be included in a single systematic review.

Unlike a fixed dose of a drug, complex interventions may adapt and change. Reviewers need to consider the degree to which interventions are tailored for recipients, the stage at which an intervention is evaluated, the healthcare system and the way an intervention is implemented in practice. The last may result in an intervention becoming a hybrid of the original. For example, a case management intervention to improve continuity of care may be supplemented with additional telephone follow-up, changing the form of the original intervention. Defining an intervention can involve three steps, each of which can be more or less explicitly addressed within studies and thereby be available (or not) to systematic reviewers. The three elements that need to be described are the content (the active ingredients) of an intervention, the methods by which this content is delivered, and the fidelity with which the delivery is achieved. Unfortunately, such clarity is seldom present in published trials. Identifying the core ingredients of even "relatively simple" interventions, including their dose, formulation, and the resulting mechanism of action may not be straightforward for reviewers [14]. Frequently, the comparison intervention is also fairly complex, for example, care being provided on a general medical ward to patients recovering from a stroke is described as not much more than usual or routine care.

Even apparently simple intervention components can be complex. In service delivery, the intervention may appear relatively simple if it can be described in terms of the staff and types of care being delivered, for example, services set up to avoid an acute hospital admission [15]. Although, in such situations, there will be an added behavioral dimension to the intervention if staff are required to perform new behaviors or current behaviors in a new context. The situation becomes even more complicated if the intervention is part of a broader package of care, such as discharge planning combined with enhanced primary care services, that spans more than one healthcare sector [16].

Various approaches have been used within systematic reviews to improve the definition of complex interventions [4]. These include categorizing interventions using typologies, identifying the key characteristics of a complex intervention from data included in the review and seeking supplementary evidence.

1. Typologies have been developed across different areas of health care and can be used to guide the classification of common elements of interventions into homogeneous groups. In some cases, a typology of interventions already exists and can aid the prospective definition of an intervention by being used as originally intended, or modified for a related class of interventions. The typology developed by the Cochrane Effective Practice and Organisation of Care Review Group provides a framework to classify interventions aimed at improving professional practice, as well as organizational, financial, and regulatory interventions (www.epoc.cochrane.org). A similar approach has been used by the Cochrane Consumers and Communication Review Group to classify interventions related to improving communication between healthcare consumers and providers (http://www.latrobe.edu.au/chcp/cochrane/index.html).

2. Identifying the key characteristics of an intervention aided the classification of complexity in a systematic review of audit and feedback, where the aim of the intervention was to change behavior. Two reviewers independently categorized the complexity of the targeted behavior as high, moderate, or low. The categories depended on the number of behaviors required, the extent to which

complex judgments or skills were necessary, whether other factors such as organizational changes were required for the behavior to be improved, and if there was a need for change by a professional (one person), a change to methods of communication or a change in systems. If an intervention was targeted at relatively simple behaviors, but there were a number of different behaviors (e.g., compliance with multiple recommendations for prevention), the complexity was assessed as moderate [7].

3. Supplementary evidence, for example, relevant qualitative or descriptive data, can also shed light on the components of a complex intervention. Although the range of qualitative data available alongside trials of complex health interventions may be limited [17]. The relative contribution of the different types of information will vary by topic area. For example, information on the theoretical basis of interventions such as telephone counseling for smoking cessation [18] may be highly relevant in understanding how an intervention works, whereas for a service innovation, such as hospital at home, the policy context of reducing reliance on inpatient care that drives the development of these types of intervention is a greater priority [19]. The way a policy has been framed can also explain subtle differences in the way interventions have been designed and implemented. For instance, the policy in a number of countries is to provide mental health services for children in the least restrictive setting. This caused some conceptual difficulties in a systematic review of alternatives to mental health care for young people [20]. Varying thresholds of admission between similar services meant that seemingly similar interventions were not always comparable as young people with different levels of mental health need accessed these services.

14.3.3 Outcomes

It is not uncommon for systematic reviews of complex interventions to include a wide range of outcomes, different measures of the same outcomes, similar outcomes but measured at different time points or proxy outcomes. In addition, the measures of outcome may be poorly described and in some cases not validated. For example, in a systematic review of lay health workers, most studies reported multiple measures of effect and many did not specify a primary outcome [21]. Outcome measures included measures of quality of life, psychosocial well-being, and behavior change, all of which can be measured in numerous ways. Calculating a standardized mean difference (SMD), when the same outcomes are being measured in different ways, is one approach to deal with this [22]. This allows comparisons between studies in terms of the magnitude of the effect or strength of the relation, irrespective of whether or not it is statistically significant. However, there are difficulties in interpreting

SMDs because the intervention effect is expressed in standard units and not in the original units of the measurement scale.

The Cochrane Consumers and Communication Review Group has developed a taxonomy of outcomes (see http://www.latrobe.edu.au/chcp/assets/downloads/Outcomes.pdf), which can be used to categorize outcomes reported in included studies. This can guide reviewers in grouping outcomes for later analysis and can also provide a structure for reporting. This has been done in a protocol for an ongoing CJD review [23]. Outcomes have been grouped according to whether they are consumer-oriented outcomes (e.g., understanding of how the risk was acquired), healthcare provider–oriented outcomes (e.g., the level of patient-centered care), or healthcare delivery–oriented outcomes (e.g., the use of care plans or teams).

14.3.4 Methodological Issues

14.3.4.1 Searching for Studies

Identifying all potentially eligible studies is a problem for complex interventions as they generally lack a standard taxonomy and definition. To overcome this difficulty, search strategies will need to be overly sensitive at the expense of being specific and the trade-off is sifting through many 1000s of abstracts in an attempt to identify all possible relevant data. Often data are unpublished and can only be accessed through policy documents, conference proceedings, or book chapters.

Selecting studies for inclusion is often not straightforward. Judgment is required, based on the available information, as to how similar the interventions are and the degree to which local circumstances result in very distinct interventions being implemented under the same policy umbrella. This is a particular problem with multifaceted interventions where the intervention being examined is part of a larger package of care. Shared care, which comes in several forms, illustrates this well. Shared care can be described as a structured intervention involving continuing collaborative clinical care between primary and specialist care physicians in the management of patients with pre-specified chronic diseases. Models include liaison meetings between specialists and primary care team members, shared-care record cards (often patient-held), computer-assisted shared care, and electronic mail (sometimes with a recall mechanism) [24]. Selecting studies that meet the inclusion criteria is even more of a problem when the core purpose of an intervention is to vary according to the characteristics of the participants or the trial setting, or where the intervention concerns a concept such as "family-centered care" [25].

14.3.4.2 Synthesis of Data

To some extent, all systematic reviews will include studies that are heterogeneous to varying degrees. However, it would be naïve to assume that there will always be a common effect across a "class" of complex interventions.

So while one of the advantages of a pooled analysis is to examine the effects of an intervention across larger samples of individuals with more heterogeneous exposures, an intervention with too many ill-defined elements will result in a high degree of heterogeneity. Related to this is the extent to which the review question is broad or narrow.

Identifying and investigating sources of heterogeneity can improve our understanding of how an intervention works, and a failure to do so can distort the results. Guidance for dealing with heterogeneity includes a priori and post hoc specification of effect modifiers, and careful use of fixed or random effects models [26]. However, specific advice on when heterogeneity reaches a level to make meta-analysis invalid is limited [11,27], and it is often left to the reader to explore and interpret heterogeneity. A number of strategies to aid this process are available. Ideally, the variables that might contribute to heterogeneity between interventions should be identified and categorized a priori, for example, the organizational characteristics or therapeutic approaches of an intervention. However, the extent to which this can be done may be limited by the lack of any one plausible or coherent explanation regarding the mechanisms of the intervention, or by insufficient information.

Including different data fields for each component contained within an intervention in the data extraction sheet can also assist reviewers identify the similarities and differences between interventions, thus facilitating an exploration of heterogeneity. For example, in a systematic review of discharge planning with post discharge support, interventions were categorized by the intensity of the post discharge support. This ranged from a single home visit to increased clinic follow-up and frequent telephone contact through to extended home care services [28]. Contacting trialists and obtaining trial protocols may also help to categorize interventions and may be done as part of an individual patient data meta-analysis. In addition, data from process evaluations conducted alongside the trials have the potential to generate hypotheses regarding the mechanism(s) of effects of the intervention being reviewed, although such data are often limited [17].

14.4 Implementation and External Validity

Although several methods for improving the implementation of evidence have been explored [29], a relatively neglected area is the application of evidence from evaluations of complex interventions. An imprecise description of an intervention leaves decision makers (including clinicians, managers, and policy makers) with a wide margin of discretion regarding the form an intervention should take when being implemented. This is a particular

---

BOX 14.1   PRESENTATION OF REVIEW FINDINGS: INFORMATION TO SUPPORT ASSESSMENT OF THE APPLICABILITY OF EVIDENCE OF EFFECTIVENESS

Reviews should provide information on the following:

- Are there important differences or similarities in the structural elements of health systems (i.e., governance, financial, and delivery arrangements) or of health services across the settings of the included studies that might not work in the same way elsewhere?
- Are there important differences in on-the-ground realities and constraints (i.e., governance, financial, and delivery arrangements) between where the research was done and where it could be applied that might substantially alter the potential benefits of the intervention?
- Are there likely to be important differences in the baseline conditions between where the research was done and other settings? If so, this would mean that the intervention could have different absolute effects, even if the relative effectiveness was the same?
- Are there important differences in the perspectives and influences of health system stakeholders (i.e., political challenges) between where the research was done and where it could be implemented that might mean an intervention will not be accepted or taken up in the same way?

Source: From Lewin, S. et al., *Lancet*, 372(9642), 928, 2008.

---

problem with systematic reviews of complex interventions, where a range of configurations or intensity may be possible. A number of attributes that would inform policy makers are seldom present in reviews (see Box 14.1). These include the characteristics of interventions, the healthcare system, the setting and implementation [30–32]. The absence of data on accessibility and risk of adverse events have also been identified as barriers by those attempting to use evidence from systematic reviews for clinical decision making [12]. Systematic reviews do not generally consider the cost-effectiveness or budget impact of interventions, which are also crucial for policy makers. Finally, details are seldom available on the sustainability of interventions over time and what needs to be put in place to ensure sustainability of effect.

A further issue for policy makers who have limited reading time is the presentation of the synthesis of evidence of complex interventions [34].

While attention has been paid to the methodology for completing systematic reviews, relatively little attention has been paid to understanding how evidence from systematic reviews should be presented to optimize the ability of decision makers to understand and apply the evidence appropriately [35]. This is a particular problem for systematic reviews of complex interventions, which frequently rely on lengthy narrative tables to summarize the results. Often the reader is left to navigate their way through pages of summary tables with no clear message to guide the implementation, monitoring, or discontinuation of an intervention. Gruen and colleagues suggest that reviewers can assist readers to generalize the findings of a systematic review by considering the relative importance of the health problem; the relevance of the outcome measures; and the practicality, appropriateness, and cost-effectiveness of the intervention [36]. The development of standardized summary of findings tables for reviews should also aid this process [12].

14.5 Conclusion

Systematic reviews of complex interventions are methodologically challenging for a range of reasons. These include difficulties in defining the intervention, the range of outcome measures in trials, differential effects of the intervention across groups as well the range of groups targeted by the intervention, and methodological issues in identifying and selecting studies. A number of strategies have been developed to facilitate and standardize reviews of complex interventions. However, their application is limited by poor descriptions of interventions in many primary studies. Further work is also needed to refine search strategies to facilitate identification of these studies, to develop validated taxonomies of interventions and outcomes, and to further elaborate methods for managing heterogeneity in the context of a systematic review. Furthermore, methods for presenting the evidence from systematic reviews, such as the summary of findings tables, need to be more widespread. Review evidence also needs to be presented in ways that support assessment of the applicability of evidence across different settings and health systems.

Acknowledgment

We would like to thank Paul Glasziou for his helpful comments.

References

1. Chalmers, I., Rounding, C., Lock, K. Descriptive survey of non-commercial randomised controlled trials in the United Kingdom, 1980–2002. *BMJ*, 27, 1017–1020, 2003.
2. Lavis, J.N., Ross, S.E., Hurley, J.E. et al. Examining the role of health services research in public policy making. *Milbank Q*, 80(1), 125–154, 2002.
3. Craig, P., Dieppe, P., Macintyre, S., Michie, S., Nazareth, I., Petticrew, M. Developing and evaluating complex interventions: The new Medical Research Council guidance. *BMJ*, 337, a1655, 2008. DOI: 10.1136.
4. Shepperd, S., Lewin, S., Straus, S., Clarke, M., Eccles, M., Fitzpatrick, R. Can we systematically review studies that evaluate complex interventions? *PLoS Med*, 6(8), e1000086, 2009.
5. Mays, N., Roberts, E., Popay, J. Synthesising research evidence, in Fulop, N., Allen, P., Clarke, A., Black, N. (eds.), *Studying the Organisation and Delivery of Health Services*. London, U.K.: Routledge, 2001, pp. 188–220.
6. Pawson, R., Greenhalgh, T., Harvey, G., Walshe, K. Realist review—A new method of systematic review designed for complex policy interventions. *J Health Serv Res Policy*, 10, S1:21–S1:33, 2005.
7. Jamtvedt, G., Young, J.M., Kristoffersen, D.T., O'Brien, M.A., Oxman, A.D. Audit and feedback: Effects on professional practice and health care outcomes. *Cochrane Database Syst Rev*, (2), Art. No.: CD000259, 2006. DOI: 10.1002/14651858.
8. Shepperd, S., Iliffe, S. Hospital at home versus in-patient hospital care. *Cochrane Database Syst Rev*, (3), CD000356, 2005. DOI: 10.1002/14651858.
9. Giuffrida, A., Forland, F., Kristiansen, I. et al. Target payments in primary care: Effects on professional practice and health care outcomes. *Cochrane Database Syst Rev*, (4), Art. No.: CD000531, 1999. DOI: 10.1002/14651858.
10. Lagarde, M., Haines, A., Palmer, N. Conditional cash transfers for improving uptake of health interventions in low and middle income countries. *JAMA*, 298(16), 1900–1910, 2007.
11. Davidson, K.W., Goldstein, M., Kaplan, R.M. et al. Evidence based behavioural medicine: What is it and how do we achieve it? *Ann Behav Med*, 26, 161–171, 2003.
12. Glenton, C., Underland, V., Kho, M., Pennick, V., Oxman, A. Summaries of findings, descriptions of interventions, and information about adverse effects would make reviews more informative. *J Clin Epidemiol*, 59, 770–778, 2006.
13. Shepperd, S., Richards, S. Continuity of care—A chameleon concept. *J Health Serv Res Policy*, 7(3), 130–131, 2002.
14. Glasziou, P., Meats, E., Heneghan, C., Shepperd, S. Treatment descriptions in trials and reviews: What is missing? *BMJ*, 336, 1472–1474, 2008.
15. Shepperd, S., Doll, H., Angus, R.M. et al. Avoiding hospital admission through provision of hospital care at home: A systematic review and meta-analysis of individual patient data. *CMAJ*, 180(2), 175–182, 2009. DOI:10.1503/cmaj.081491.
16. Weinberger, M., Oddone, E.Z., Henderson, W.G. Does increased access to primary care reduce hospital admissions? Veterans Affairs Cooperative Study Group on Primary Care and Hospital Readmissions. *N Engl J Med*, 334, 1441–1447, 1996.

17. Lewin, S., Glenton, C., Oxman, A.D. How are qualitative methods being used alongside complex health service RCTs? A systematic review. *BMJ*, 339, b3496, 2009.

18. Stead, L.F., Perera, R., Lancaster, T. Telephone counselling for smoking cessation. *Cochrane Database Syst Rev*, (3), Art. No.: CD002850, 2006. DOI: 10.1002/14651858.

19. Shepperd, S., Doll, H., Broad, J. et al. Early discharge hospital at home. *Cochrane Database Syst Rev*, (1), Art. No.: CD000356, 2009. DOI: 10.1002/14651858.

20. Shepperd, S., Doll, H., Fazel, M. et al. Alternatives to inpatient mental health care for children and young people. *Cochrane Database Syst Rev*, Art. No.: CD006410, 2007. DOI: 10.1002/14651858.

21. Lewin, S., Dick, J., Pond, P. et al. Lay health workers in primary and community health care. *Cochrane Database Syst Rev*, (1), Art. No.: CD004015, 2005.

22. Higgins, J.P.T., Green, S. (eds.). *Cochrane Handbook for Systematic Reviews of Interventions: Version 5.1.0 [updated March 2011]*. The Cochrane Collaboration, 2011. Available from www.cochrane-handbook.org, 2008.

23. Ryan, R., Allen, K., Hill, S., Lowe, D. Notification and support for people exposed to the risk of Creutzfeldt–Jakob disease (CJD) through medical treatment (iatrogenically) (protocol). *Cochrane Database Syst Rev*, (1), Art. No.: CD007578, 2009.

24. Smith, S.M., Allwright, S., O'Dowd, T. Effectiveness of shared care across the interface between primary and specialty care in chronic disease management. *Cochrane Database Syst Rev*, (3), Art. No.: CD004910, 2007. DOI: 10.1002/14651858.

25. Shields, L., Pratt, J., Davis, L., Hunter, J. Family-centred care for children in hospital. *Cochrane Database Syst Rev*, (1), Art. No.: CD004811, 2007.

26. Moher, D., Cook, D.J., Eastwood, S., Olkin, I., Rennie, D., Stroup, D.F. Improving the quality of reports of meta-analyses of randomised controlled trials: The QUOROM statement. Quality of reporting of meta-analyses. *Lancet*, 354(9193), 1896–1900, 1999.

27. Higgins, J., Thompson, S., Deeks, J., Altman, D. Statistical heterogeneity in systematic reviews of clinical trials: A critical appraisal of guidelines and practice. *J Health Serv Res Policy*, 7, 51–61, 2002.

28. Phillips, C.O., Wright, S.M., Kern, D.E., Singa, R.M., Shepperd, S., Rubin, H.R. Comprehensive discharge planning with post discharge support for older patients with congestive heart failure: A meta-analysis. *JAMA*, 291(11), 1358–1367, 2004.

29. Dopson, S., Locock, L., Chambers, D., Gabbay, J. Implementation of evidence-based medicine: Evaluation of the promoting action on clinical effectiveness programme. *J Health Serv Res Policy*, 6, 23–31, 2006.

30. Glasgow, R.E., Green, L.W., Klesges, L.M. et al. External validity: We need to do more. *Ann Behav Med*, 31(2), 105–108, 2006.

31. Green, L.W., Glasgow, R.E. Evaluating the relevance, generalization and applicability of research—Issues in external validation and translation methodology. *Evaluat Health Prof*, 29(1), 126–153, 2006.

32. Rothwell, P.M. External validity of randomised controlled trials: "To whom do the results of this trial apply?" *Lancet*, 365, 82–93, 2005.

33. Lewin, S., Lavis, J.N., Oxman, A.D. et al. Supporting the delivery of cost-effective interventions in primary health-care systems in low-income and middle-income countries: An overview of systematic reviews. *Lancet*, 372(9642), 928–939, 2008.

34. Eccles, M. What is the role of research and evidence in policy making?, in Rawlins, M., Littlejohns, P. (eds.), *Delivering Quality in the NHS*. Oxford, U.K.: Radcliffe Medical Press, 2004.
35. Oxman, A., Schunemann, H.J., Fretheim, A. Improving the use of research evidence in guideline development: 8. Synthesis and presentation of evidence. *BMC Health Res Policy Syst*, 20(4), 2006. DOI: 10.1186/1478-4505-4-20.
36. Gruen, R.L., Morris, P.S., McDonald, E.L., Bailie, R.S. Making systematic reviews more useful for policy-makers. *Bull World Health Organ*, 83(6), 480, 2005.

15

Reporting Guidelines for Nonpharmacological Trials

Isabelle Boutron and Philippe Ravaud

Paris Descartes University and
Public Assistance—Hospitals of Paris

CONTENTS

15.1 Transparency

The results of randomized controlled trials are essential for choosing among different treatment options in clinical practice. The communication of trial results mainly relies on scientific publications that play an important role in disseminating and implementing research results. However, to be useful, published reports must be completely transparent and contain, in addition to a complete description of the benefits and harm of the treatment, a description of all elements necessary to judge the internal and external validity of the trial. The role of transparency is actually an ethical requirement. The Declaration of Helsinki states that "Authors have a duty to make publicly available the results of their research on human subjects and are accountable for the completeness and accuracy of their reports."

Nevertheless, despite this ethical requirement, transparency has long been neglected, and poor reporting is an important barrier to the critical appraisal of trial results. For example, a systematic review of all PubMed-indexed reports of randomized trials published in December 2000 showed that power calculation, primary outcomes, random sequence generation, allocation concealment, and handling of attrition were adequately described in less than half of the publications [1].

15.2 The CONSORT Initiative

The CONSORT initiative was born in the mid-1990s, when researchers faced difficulties in evaluating the methodological quality of published reports because of poor reporting. Lack of transparency in published reports of randomized controlled trials rendered the critical appraisal of the literature difficult and the implementation of the study findings in clinical practice challenging. Further, this inadequate reporting was worrying because some methodological studies highlighted a link between poor reporting and a biased estimation of treatment effect [2].

The CONSORT group grew out of two initiatives aimed at developing reporting guidelines, and the first CONSORT Statement, on improving the quality of reporting of randomized controlled trials, was developed and published in 1996 [3]. The Statement was rapidly endorsed and implemented by several journal editors and is now endorsed by the International Committee of Medical Journal Editors (www.icmje.org). Other prominent editorial groups, the Council of Science Editors and the World Association of Medical Editors, officially supported CONSORT. Meanwhile, research focusing on the transparency of published reports began to flourish, with several articles published in high–impact factor journals [4]. Due to this evolving research field, the CONSORT Statement was regularly updated, first in 2001 [5] and more recently in 2010 [6]. The CONSORT Statement is reported with an Elaboration Explanation manuscript [7] that provides the underlying evidence, data related to the quality of reporting and examples of adequate reporting for each item. Although the level of evidence is limited, the CONSORT Statement seems to improve the quality of reporting [8–10].

To improve the dissemination and implementation of the CONSORT Statement, the CONSORT group developed several extensions addressing the various specificities related to the study design (cluster randomized controlled trials [11], non-inferiority and equivalence trials [12], pragmatic trials [13]), type of treatment (nonpharmacological treatments [14], herbal therapies [15], acupuncture intervention [16]), and data (reporting of harms [17], abstracts [18]).

15.3 The Extension of the CONSORT Statement for Nonpharmacological Trials

The CONSORT group developed an extension for trials of nonpharmacological treatments to consider the specific methodological issues in assessing nonpharmacological treatments (difficulties of blinding, complexity of interventions, influence of care providers, and center expertise) and the evidence of inadequate reporting of trials assessing nonpharmacological treatments. For example, in the field of surgery [19], the setting and the center volume of activity were reported in <10% of a series of articles; selection criteria for care providers were reported in 40%; and the number of care providers performing the intervention was reported in 32%. Reporting the description of the intervention in surgical trials is poor, with another study finding important components such as anesthesia management described in only 35% of articles, preoperative care in 15%, and postoperative care in 49% [19]. In a review of behavioral medicine interventions, insufficient intervention detail was cited as a barrier to reviewing evidence and developing guidelines [20–22]. The reporting of blinding in the surgical literature is also inadequate. A review of clinical trials in the cardiothoracic surgical literature showed that the reporting of blinding, when practical, was low for participants (33%), healthcare providers (54%), and outcome assessors (40%) [23]. In another study, only 11% of the surgical trials described in the fracture care literature reported blinding of outcome assessors [24].

The CONSORT group organized a consensus meeting that resulted in the modification of 11 items and the addition of a new item to assess trials evaluating nonpharmacological treatments. The CONSORT flow diagram was modified to include data on number of care providers and centers in each group, as well as the number of patients treated by each care provider. The statement and the diagram are reported in Table 15.1 and Figure 15.1.

This CONSORT extension highlights important issues when assessing nonpharmacological treatments, particularly the need to correctly take into account the difficulties of blinding, the complexity of the intervention, and the care providers and centers.

Blinding is difficult to achieve in trials assessing nonpharmacological treatments [25]. The traditional terminology used to report blinding in randomized controlled trials relies on the terms "single-blind," "double-blind," and sometimes "triple-blind" [26]. This terminology is particularly ambiguous in trials of nonpharmacological interventions because of the high number of key trial participants. Usually, such trials involve not just one care provider but several people administering interventions and co-interventions. For example, in a surgical trial, blinding of care providers could concern the surgeon, the surgical staff, the anesthetist, and the nurses. Further, blinding usually relies on more complex methods [27] as described in the chapters dedicated to blinding and placebo in this book. Consequently, published reports must clearly indicate who was blinded, whether those administering co-interventions were blinded, and how blinding was achieved.

Nonpharmacological treatments typically involve several components, each of which can potentially influence the estimated treatment effect [28–33]. For example, anesthesia and surgical closure procedures have been shown to influence the risk of harm after carotid endarterectomy [32,34–38]. The CONSORT extension for nonpharmacological treatments recommends

TABLE 15.1

Extension of the CONSORT Statement for Reporting the Results of a Randomized Trial Assessing a Nonpharmacological Treatment[a]

Section/Topic	Item #	CONSORT 2010 Checklist Items	Extension for Trials of Nonpharmacological Treatments
TITLE AND ABSTRACT	1.a	Identification as a randomized trial in the title	In the abstract, description of the experimental treatment, comparator, care providers, centers, and blinding status
	1.b	Structured summary of trial design, methods, results, and conclusions; for specific guidance, see CONSORT for Abstracts (29;42)	
INTRODUCTION			
Background and objectives	2.a	Scientific background and explanation of rationale	
	2.b	Specific objectives or hypotheses	
METHODS			
Trial design	3.a	Description of trial design (e.g., parallel, factorial) including allocation ratio	
	3.b	Important changes to methods after trial commencement (e.g., eligibility criteria), with reasons	
Participants	4.a	Eligibility criteria for participants	When applicable, eligibility criteria for centers and those performing the interventions
	4.b	Settings and locations where the data were collected	
Interventions	5	The interventions for each group with sufficient details to allow replication, including how and when they were actually administered	Precise details of both the experimental treatment and comparator
	5.a		Description of the different components of the interventions and, when applicable, descriptions of the procedure for tailoring the interventions to individual participants

	5.b		Details of how the interventions were standardized
	5.c		Details of how adherence of care providers with the protocol was assessed or enhanced
Outcomes	6.a	Completely defined pre-specified primary and secondary outcome measures, including how and when they were assessed	
	6.b	Any changes to trial outcomes after the trial commenced with reasons	
Sample size	7.a	How sample size was determined	When applicable, details of whether and how the clustering by care providers or centers was addressed
	7.b	When applicable, explanation of any interim analyses and stopping guidelines	
Randomization			
Sequence	8.a	Method used to generate the random allocation sequence	When applicable, how care providers were allocated to each trial group
Generation	8.b	Type of randomization; details of any restriction (e.g., blocking and block size)	
Location concealment mechanism	9	Mechanism used to implement the random allocation sequence (e.g., sequentially numbered containers), describing any steps taken to conceal the sequence until interventions were assigned	
Implementation	10	Who generated the random allocation sequence, who enrolled participants, and who assigned participants to interventions	
Blinding	11.a	If done, who was blinded after assignment to interventions (e.g., participants, care providers, those assessing outcomes) and how	Whether or not those administering co-interventions were blinded to group assignment

(continued)

TABLE 15.1 (continued)

Extension of the CONSORT Statement for Reporting the Results of a Randomized Trial Assessing a Nonpharmacological Treatment[a]

Section/Topic	Item #	CONSORT 2010 Checklist Items	Extension for Trials of Nonpharmacological Treatments
	11.b	If relevant, description of the similarity of interventions	If blinded, method of blinding and description of the similarity of interventions
Statistical methods	12.a	Statistical methods used to compare groups for primary and secondary outcomes	When applicable, details of whether and how the clustering by care providers or centers was addressed
	12.b	Methods for additional analyses, such as subgroup analyses and adjusted analyses	
RESULTS			
Participant flow (a diagram is strongly recommended)	13.a	For each group, the numbers of participants who were randomly assigned, received intended treatment, and were analyzed for the primary outcome	The number of care providers or centers performing the intervention in each group and the number of patients treated by each care provider or in each center
	13.b	For each group, losses and exclusions after randomization, together with reasons	
Implementation of intervention	New item		Details of the experimental treatment and comparator as they were implemented
Recruitment	14.a	Dates defining the periods of recruitment and follow-up	
	14.b	Why the trial ended or was stopped	
Baseline data	15	A table showing baseline demographic and clinical characteristics for each group	When applicable, a description of care providers (case volume, qualification, expertise, etc.) and centers (volume) in each group
Numbers analyzed	16	For each group, number of participants (denominator) included in each analysis and whether the analysis was by original assigned groups	

Outcomes and estimation	17.a	For each primary and secondary outcome, results for each group, and the estimated effect size and its precision (e.g., 95% confidence interval)	
	17.b	For binary outcomes, presentation of both absolute and relative effect sizes is recommended	
Ancillary analyses	18	Results of any other analyses performed, including subgroup analyses and adjusted analyses, distinguishing pre-specified from exploratory	
Harms	19	All important harms or unintended effects in each group; for specific guidance see CONSORT for Harms (30)	
DISCUSSION			
Limitations	20	Trial limitations, addressing sources of potential bias, imprecision, and, if relevant, multiplicity of analyses	
Generalizability	21	Generalizability (external validity, applicability) of the trial findings	Generalizability (external validity) of the trial findings according to the intervention, comparators, patients, and care providers and centers involved in the trial
Interpretation	22	Interpretation consistent with results, balancing benefits and harms, and considering other relevant evidence	In addition, take into account the choice of the comparator, lack of or partial blinding, and unequal expertise of care providers or centers in each group
OTHER INFORMATION			
Registration	23	Registration number and name of trial registry	
Protocol	24	Where the full trial protocol can be accessed, if available	
Funding	25	Sources of funding and other support (e.g., supply of drugs); role of funders	

[a] We strongly recommend reading this Statement in conjunction with the CONSORT explanation and elaboration nonpharmacological treatments [14] and the 2010 CONSORT explanation and elaboration [7].

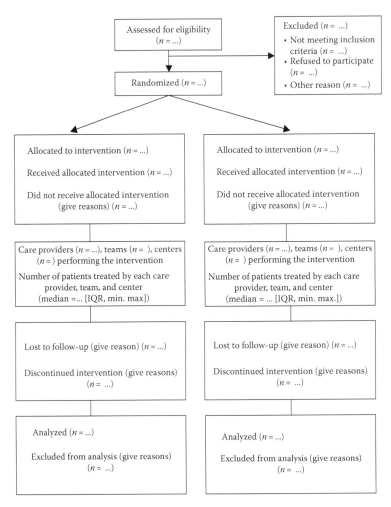

FIGURE 15.1
Example of modified CONSORT flowchart for individual randomized controlled trials of non-pharmacological treatments with an extra box per intervention group relating to care providers. For cluster randomized controlled trials, authors should refer to the extension related to such trials. IQR; interquartile range.

reporting the procedure of standardization, the method used to assess or enhance treatment adherence, and the details of the intervention as it was actually implemented.

In nonpharmacological trials, usually care provider and center expertise will affect the success of the intervention. For example, the Asymptomatic Carotid Artery Surgery (ACAS) trial selected only surgeons with good safety records and therefore excluded 40% of all possible surgeons. In this trial, the postoperative mortality rate was lower than that from a meta-analysis of 46 surgical case series that found operative risks in the 5 years after ACAS

(0.14% vs. 1.11%) [39–41]. Consequently, reporting data related to care providers and centers is essential to adequately appraise the internal validity (i.e., extent to which bias is avoided in trial conduct [42]) and external validity of trials [43] (i.e., extent to which results of trials can be generalized to other circumstances [42]).

Finally, variation in the outcome will be less for patients treated by the same care provider [44]. Consequently, the assumption that the observed outcomes of participants are independent is false, and observations of participants treated by the same care provider may be clustered [45]. As with cluster RCTs, this type of clustering within individually randomized trials of nonpharmacological treatments is likely to affect the effect estimates because it will inflate the standard error and reduce the effective sample size, thus reducing the power of the trial [45].

The CONSORT extension for nonpharmacological trials highlights the need to report all these features.

15.4 Conclusion

The extension of the CONSORT Statement for trials assessing nonpharmacological treatments was developed to improve the quality of reporting the results of such trials. To modify authors' behavior, journal editors must endorse and actively implement these guidelines. For example, instructions to authors for journals could require compliance with this extension and could request that authors clearly indicate in a checklist where each item is reported in the manuscript. Journal editors could systematically verify whether the manuscript complies with the CONSORT Statement extension before sending it for peer review. These active endorsements of the CONSORT Statement by journal editors are probably necessary to improve transparency of nonpharmacological trials reports.

References

1. Chan, A.W., Altman, D.G. Epidemiology and reporting of randomised trials published in PubMed journals. *Lancet*, 365(9465), 1159–1162, 2005.
2. Schulz, K.F., Chalmers, I., Hayes, R.J., Altman, D.G. Empirical evidence of bias. Dimensions of methodological quality associated with estimates of treatment effects in controlled trials. *JAMA*, 273(5), 408–412, 1995.
3. Begg, C., Cho, M., Eastwood, S., et al. Improving the quality of reporting of randomized controlled trials. The CONSORT statement. *JAMA*, 276(8), 637–639, 1996.

4. Egger, M., Juni, P., Bartlett, C. Value of flow diagrams in reports of randomized controlled trials. *JAMA*, 285(15), 1996–1999, 2001.
5. Altman, D.G., Schulz, K.F., Moher, D., et al. The revised CONSORT statement for reporting randomized trials: Explanation and elaboration. *Ann Intern Med*, 134(8), 663–694, 2001.
6. Schulz, K.F., Altman, D.G., Moher, D. CONSORT 2010 statement: Updated guidelines for reporting parallel group randomised trials. *PLoS Med*, 7(3), e1000251, 2010.
7. Moher, D., Hopewell, S., Schulz, K.F., et al. CONSORT 2010 explanation and Elaboration: Updated guidelines for reporting parallel group randomised trials. *BMJ*, 340, c869, 2010.
8. Plint, A.C., Moher, D., Morrison, A., et al. Does the CONSORT checklist improve the quality of reports of randomised controlled trials? A systematic review. *Med J Aust*, 185(5), 263–267, 2006.
9. Hopewell, S., Dutton, S., Yu, L.M., Chan, A.W., Altman, D.G. The quality of reports of randomised trials in 2000 and 2006: Comparative study of articles indexed in PubMed. *BMJ*, 340, c723, 2010.
10. Moher, D., Jones, A., Lepage, L. Use of the CONSORT statement and quality of reports of randomized trials: A comparative before-and-after evaluation. *JAMA*, 285(15), 1992–1995, 2001.
11. Campbell, M.K., Elbourne, D.R., Altman, D.G. CONSORT statement: Extension to cluster randomised trials. *BMJ*, 328(7441), 702–708, 2004.
12. Piaggio, G., Elbourne, D.R., Altman, D.G., Pocock, S.J., Evans, S.J. Reporting of noninferiority and equivalence randomized trials: An extension of the CONSORT statement. *JAMA*, 295(10), 1152–1160, 2006.
13. Zwarenstein, M., Treweek, S., Gagnier, J.J., et al. Improving the reporting of pragmatic trials: An extension of the CONSORT statement. *BMJ*, 337, a2390, 2008.
14. Boutron, I., Moher, D., Altman, D.G., Schulz, K.F., Ravaud, P. Extending the CONSORT statement to randomized trials of nonpharmacologic treatment: Explanation and elaboration. *Ann Intern Med*, 148(4), 295–309, 2008.
15. Gagnier, J.J., Boon, H., Rochon, P., Moher, D., Barnes, J., Bombardier, C. Reporting randomized, controlled trials of herbal interventions: An elaborated CONSORT statement. *Ann Intern Med*, 144(5), 364–367, 2006.
16. MacPherson, H., Altman, D.G., Hammerschlag, R., et al. Revised STandards for Reporting Interventions in Clinical Trials of Acupuncture (STRICTA): Extending the CONSORT statement. *PLoS Med*, 7(6), e1000261, 2006.
17. Ioannidis, J.P., Evans, S.J., Gotzsche, P.C., et al. Better reporting of harms in randomized trials: An extension of the CONSORT statement. *Ann Intern Med*, 141(10), 781–788, 2004.
18. Hopewell, S., Clarke, M., Moher, D., et al. CONSORT for reporting randomized controlled trials in journal and conference abstracts: Explanation and elaboration. *PLoS Med*, 5(1), e20, 2008.
19. Jacquier, I., Boutron, I., Moher, D., Roy, C., Ravaud, P. The reporting of randomized clinical trials using a surgical intervention is in need of immediate improvement: A systematic review. *Ann Surg*, 244(5), 677–683, 2006.
20. Pignone, M.P., Ammerman, A., Fernandez, L., et al. Counseling to promote a healthy diet in adults: A summary of the evidence for the U.S. Preventive Services Task Force. *Am J Prev Med*, 24(1), 75–92, 2003.

21. U.S. Preventive Services Task Force. Behavioral counseling in primary care to promote physical activity: Recommendation and rationale. *Ann Intern Med*, 137(3), 205–207, 2002.

22. Davidson, K.W., Goldstein, M., Kaplan, R.M., et al. Evidence-based behavioral medicine: What is it and how do we achieve it? *Ann Behav Med*, 26(3), 161–171, 2003.

23. Anyanwu, A.C., Treasure, T. Surgical research revisited: Clinical trials in the cardiothoracic surgical literature. *Eur J Cardiothorac Surg*, 25(3), 299–303, 2004.

24. Bhandari, M., Guyatt, G.H., Lochner, H., Sprague, S., Tornetta, P., 3rd. Application of the Consolidated Standards of Reporting Trials (CONSORT) in the fracture care literature. *J Bone Joint Surg Am*, 84-A(3), 485–489, 2002.

25. Boutron, I., Tubach, F., Giraudeau, B., Ravaud, P. Blinding was judged more difficult to achieve and maintain in nonpharmacologic than pharmacologic trials. *J Clin Epidemiol*, 57(6), 543–550, 2004.

26. Devereaux, P.J., Manns, B.J., Ghali, W.A., et al. Physician interpretations and textbook definitions of blinding terminology in randomized controlled trials. *JAMA*, 285(15), 2000–2003, 2001.

27. Boutron, I., Guittet, L., Estellat, C., Moher, D., Hrobjartsson, A., Ravaud, P. Reporting methods of blinding in randomized trials assessing nonpharmacological treatments. *PLoS Med*, 4(2), e61, 2007.

28. Tansella, M., Thornicroft, G., Barbui, C., Cipriani, A., Saraceno, B. Seven criteria for improving effectiveness trials in psychiatry. *Psychol Med*, 36(5), 711–720, 2006.

29. Herbert, R.D., Bo, K. Analysis of quality of interventions in systematic reviews. *BMJ*, 331(7515), 507–509, 2005.

30. Campbell, M., Fitzpatrick, R., Haines, A., et al. Framework for design and evaluation of complex interventions to improve health. *BMJ*, 321(7262), 694–696, 2000.

31. Hawe, P., Shiell, A., Riley, T. Complex interventions: How "out of control" can a randomised controlled trial be? *BMJ*, 328(7455), 1561–1563, 2004.

32. Bond, R., Warlow, C.P., Naylor, A.R., Rothwell, P.M. Variation in surgical and anaesthetic technique and associations with operative risk in the European carotid surgery trial: Implications for trials of ancillary techniques. *Eur J Vasc Endovasc Surg*, 23(2), 117–126, 2002.

33. Kwakkel, G., Wagenaar, R.C., Koelman, T.W., Lankhorst, G.J., Koetsier, J.C. Effects of intensity of rehabilitation after stroke: A research synthesis. *Stroke*, 28(8), 1550–1556, 1997.

34. Halm, E.A., Hannan, E.L., Rojas, M., et al. Clinical and operative predictors of outcomes of carotid endarterectomy. *J Vasc Surg*, 42(3), 420–428, 2005.

35. Rockman, C.B., Halm, E.A., Wang, J.J., et al. Primary closure of the carotid artery is associated with poorer outcomes during carotid endarterectomy. *J Vasc Surg*, 42(5), 870–877, 2005.

36. Bond, R., Rerkasem, K., AbuRahma, A.F., Naylor, A.R., Rothwell, P.M. Patch angioplasty versus primary closure for carotid endarterectomy. *Cochrane Database Syst Rev*, (2), CD000160, 2004.

37. Bond, R., Rerkasem, K., Naylor, R., Rothwell, P.M. Patches of different types for carotid patch angioplasty. *Cochrane Database Syst Rev*, (2), CD000071, 2004.

38. Rerkasem, K., Bond, R., Rothwell, P.M. Local versus general anaesthesia for carotid endarterectomy. *Cochrane Database Syst Rev*, (2), CD000126, 2004.

39. Endarterectomy for asymptomatic carotid artery stenosis. Executive Committee for the Asymptomatic Carotid Atherosclerosis Study. *JAMA*, 273(18), 1421–1428, 1995.

40. Moore, M.J., Bennett, C.L. The learning curve for laparoscopic cholecystectomy. The Southern Surgeons Club. *Am J Surg*, 170(1), 55–59, 1995.

41. Rothwell, P.M., Goldstein, L.B. Carotid endarterectomy for asymptomatic carotid stenosis: Asymptomatic carotid surgery trial. *Stroke*, 35(10), 2425–2427, 2004.

42. Juni, P., Altman, D.G., Egger, M. Systematic reviews in health care: Assessing the quality of controlled clinical trials. *BMJ*, 323(7303), 42–46, 2001.

43. Rothwell, P.M. External validity of randomised controlled trials: "To whom do the results of this trial apply?" *Lancet*, 365(9453), 82–93, 2005.

44. Roberts, C. The implications of variation in outcome between health professionals for the design and analysis of randomized controlled trials. *Stat Med*, 18(19), 2605–2615, 1999.

45. Lee, K.J., Thompson, S.G. Clustering by health professional in individually randomised trials. *BMJ*, 330(7483), 142–144, 2005.

Part II

Assessing Nonpharmacological Treatments: Practical Examples

16

Assessing Cardiothoracic Surgery: Practical Examples

Tom Treasure and Martin Utley

University College London

CONTENTS

16.1 Introduction

In this chapter, we will consider the design, implementation, challenges, and conclusions of two controlled trials. They have been chosen because the issues that arose are common to many trials in surgery (Table 16.1). Both of them are published [1,3]. The first, the Guy's Hospital Suction Trial (GHST), is a single institution study of a very common mechanistic intervention in which the research question was whether to apply or not apply suction to a

TABLE 16.1

Features of the Two Trials, the GHST and the MARS Trial, to Show Key Features and Differences

	GHST [1]	MARS [2]
Settings	Single thoracic department	Multicenter
	Single discipline	Multidisciplinary
N	139	50
Allocation	Minimization computer software	Randomization phone call to center
Concealment	None	None
CONSORT diagram	Yes	Yes
Recruitment randomized/eligible	254/328	50/57/112 (see text for explanation)
Funding	None	Cancer Research, United Kingdom

drain left in the chest cavity after surgery. The second is the Mesothelioma and Radical Surgery (MARS) trial. We will indicate the extent to which they have informed practice and the extent to which they have not.

16.2 The Guy's Hospital Suction Trial

16.2.1 Background and Reasons for Doing the Study

It is standard and universal practice after lung surgery to leave tubes exiting from the chest cavity to permit the escape of any air leaking from the lungs, along with any blood and pleural fluid. The research question in this study was related specifically to air leak.

The lung is of its nature prone to air leak if injured or when operated on. Some basic knowledge is required of the anatomy of the lungs and the physiology of breathing to understand the research question. For a brief summary, see Box 16.1.

For an illustration of the configuration of the chest tubes, water seal, and suction, see Figure 16.1.

The tubes (colloquially known as chest drains or more classically as thoracostomy tubes) are necessary for as long as there is an air leak and a consequent risk of the lung collapsing. Once the leak has stopped, the tubes can be removed. Although under particular circumstances, patients can leave hospital with the tubes still in place, in usual postoperative care, removal of the tubes is a rate-limiting step. Although water seal (or some other form of one-way valve) is universally regarded as essential, suction is an optional addition. When used, it is applied to the sealed space above the water and thus indirectly applying

---

**BOX 16.1   THE ESSENTIAL ANATOMY
AND PHYSIOLOGY OF THE LUNG**

On either side of the chest are cavities of about 3 L capacity which contain the right and left lungs. The lung is in physical continuity with the body only where the blood and air enter and leave at the hilum; otherwise, it sits free within the chest cavities rather like the inner tube of a tire connected only by its valve to the wheel structure, although contained within it (Figure 16.1). The chest wall acts as a bellow generating negative pressure to suck air into the lung through the nose and mouth. The lung itself is a sponge-like structure held in expansion by sub-atmospheric pressure within the chest cavity. If air enters the chest cavity, the lung's elastic tissue recoils and the lung reduces in volume, hence the common term—collapsed lung (technically called pneumothorax). In most cases after surgery, the membrane around the lung is breached and it is inherent in the nature of the lung that it will leak and collapse. Furthermore, as the patient breaths, more air is encouraged out of the lung through the breach. But it is also inherent in the nature of the lung, now a compressed sponge, that the air cannot re-enter it. The air pressure builds up and this is likely to be fatal if not relieved. The remedial action is to leave a flexible tube connected to a rigid pipe of similar bore placed under water so that air can bubble out but cannot re-enter the chest cavity due to the water seal. There are other forms of valve available other than water seal but this basic but sufficient technology is the norm. In the surgical world where variation in practice and "I do it my way" are common features, the virtually universal application of a water seal is noteworthy in its own right. The terms used vary (underwater seal, water trap) but we will use water seal throughout for this one-way valve arrangement.

---

suction to the chest cavity via the chest tube. This accelerates the removal of air from the space but it also increases the rate of air leaking through the fragile lung tissue, potentially interfering with natural sealing and healing.

Two experienced thoracic surgeons at Guy's Hospital, London, had come through different training routes and had received, modified, or derived from their own experience, radically different habits with respect to suction on the drain bottles. To put this in context for the nonmedical reader, mishaps related to chest drains can be fatal in the short term and cause serious complications in the medium to long term, so surgeons tend to take a firm line on this area of clinical practice. When joined by a third experienced surgeon (the first author), it was agreed that it would be rational and indeed safer to have a common policy and so we looked for evidence to guide practice.

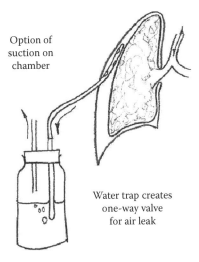

Option of
suction on
chamber

Water trap creates
one-way valve
for air leak

FIGURE 16.1
Diagram of a tube in the pleural space between the lung and the chest wall allowing the non-
return escape of air in to a water-seal chamber.

TABLE 16.2

Publications Concerning Suction versus No Suction on Chest
Drains, prior to Our Study

Publication	Subjects	N	Outcome
Cerfolio et al. [4]	Lung resections	140	Favored no suction
Marshall et al. [5]	Lung resections	78	Favored no suction
Ayed [6]	VATS for pneumothorax	100	Favored no suction
Brunelli et al. [7]	Lobectomy	145	No difference

Source: Alphonso, N. et al., *Eur J Cardiothorac Surg*, 27(3), 391, 2005.

There had been four previous randomized studies (Table 16.2) in all of
which patients had suction applied for the first day after surgery and allocation by randomization was between continuing or discontinuing suction after about 24 h (Figure 16.1). Thus, the question remained unresolved
because theoretically the most critical period for benefit or harm of either
policy was soon after surgery.

It is something of a digression but note that the titles of all four studies
are in fact fundamentally misleading: the comparison suggested is "suction
versus water seal" [4–7]. The methods sections are similarly imprecise. In
all studies, both groups had a water-seal drainage system throughout—as
really had to be the case (see Figure 16.1). The origin of the study design is
also of interest. In the first study, the Institutional Review Board "required
patients to be on suction until the morning of POD #2 [second postoperative
day]." The design in the subsequent three studies and the erroneous titles
appear to have been influenced by the first study.

TABLE 16.3

The PICO for the GHST

Population	All patients undergoing surgery to remove any part but not all of a lung
Intervention	To NOT apply suction to the chamber above the water seal (Figure 16.1)
Comparison	Applying suction to the air in the collection chamber above the fluid
Outcome	Primary: time to cessation of air leak
	Secondary: time to removal of chest tube

In the Guy's Hospital Suction Study, we sought to establish whether the addition of suction to the water seal made any difference to the outcome (Table 16.3) and so it was agreed to conduct a study with one arm having no suction from the outset.

16.2.2 Methodology

16.2.2.1 Concealment/Masking/Blinding

Since suction involves a very visible connection, and to connect to an inactive pump is effectively to obstruct the drain, a dummy connection to conceal the groups was considered unsafe. There appeared to us to be no practical way of concealing the allocation. However, the primary outcome, the observation of continuing air leak, is an objective YES/NO decision.

16.2.2.2 Means of Unbiased Allocation

The first author has used minimization ever since being introduced to the method by Stephen Evans in the early 1990s and has argued the case for it [8]. It has the merit of providing balanced groups by design rather than trusting chance and so has advantages in smaller studies, particularly where the effect of the intervention may be lost among confounding effects. In this instance, the greater the scale of the lung surgery the greater is the likelihood of air leak. Smoking increases both emphysematous change and airway pressure, which are both likely to make leaks persist. Minimization is used to guarantee that two arms are balanced with respect to such factors such that any difference attributable to the intervention under study is not "swamped" by other effects allowing an important difference that might otherwise be lost among confounding factors to be evident.

Factors for minimization were

- Age
- Sex
- Smoker YES/NO

- Operator: consultant/trainee
- Operative approach: video/open
- Type of resection: lobe/wedge/biopsy/pneumothorax
- Prediction of leak likelihood: YES/NO

The second author is among those with reservations about minimization as a reliable means of unbiased allocation. His argument is that those close to the clinical practice, with knowledge of the factor used for minimization, and with knowledge of recent preceding allocations, would in some circumstances be able to predict which way the allocation was likely to go and introduce bias into the allocation process.

Senn is one of the critics of minimization and wrote in BMJ correspondence (some details curtailed):

> There are two constituencies that reject [minimisation]. The first is of those who are interested in randomisation based inference. It remains an open question as to whether this can be validly applied to minimised trials. The second constituency is of those who are interested in optimal design of trials. Minimisation is not, in fact, the best algorithm at improving efficiency. Those who are interested in wide acceptability would be advised to avoid it [9)].

Senn's warning did not appear until after our opinion was published but we were already aware, again through BMJ correspondence, that minimization was not universally accepted [10]. The arguments have to be weighed in each case but for this and similar relatively small surgical trials where an important effect might be swamped by major confounding factors in necessarily heterogeneous clinical groups minimization may allow us to discern an important signal among the noise.

Whatever the means of allocation, it must be made as late as possible in the process. In designing the suction trial, we considered that if a surgeon knows that suction will or will not be applied to the drains, it might well influence the operative decisions, technical details, and the care and time taken in controlling air leaks. For this reason, allocation was deferred until surgery was complete. "The allocation to suction yes/no was made at completion of the operations, after the drain placement, so that it could not influence any part of the operative practice" [11].

16.2.3 Conduct of the Trial

The flowchart of this trial is described in Figure 16.2. Note that 74/328 eligible patients did not consent to enter the trial for reasons not revealed. Compared with most trials, 77% recruitment is high but we know that near 100% would have been possible. Patients appeared very ready to accept random allocation to one or other of this quite technical and to them rather peripheral detail of postoperative management (at least so it appeared to a

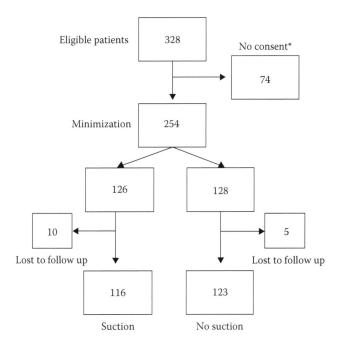

FIGURE 16.2

CONSORT flowchart for the suction trial. Two patients allocated to no suction received suction on the basis of clinical decision but remained in the allocated group for intention to treat analysis. The reason for these patients not consenting was not clear in the analysis of the trial. Whether they refused or were never asked might be important. See text for analysis of this problem, which is common in trials.

senior surgeon) when the trial was put to patients in the course of a clinical discussion concerning their forthcoming treatment. The lapses were more likely than not to be accidental lapses in the busy lives of junior doctors on the evening before surgery. Recruitment to this trial depended on individual junior doctors' awareness of the project, their commitment to research, and their willingness to engage with the tasks involved.

However, on the more general point, there are well-recognized patterns of deliberate non-recruitment. One is when a senior doctor is not fully signed up to handing over a clinical decision to unbiased allocation and repeatedly and perhaps selectively overrides the trial. The same can happen at a shop floor level if, for example, a well meaning nurse believes that particular patients under her care might not do well with on or other of the treatments. This is more likely to have happened with the second study described in this chapter.

In 10 and 5 of 254 randomized patients, data were not collected. There is the possibility of bias given this was a nonblinded study. Of the non-suction patients, 2/123 had suction applied contrary to allocation (crossover). This was defined within the protocol as an acceptable deviation when the

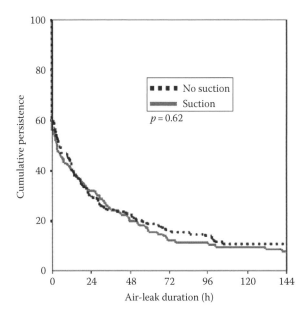

FIGURE 16.3
Kaplan–Meier curves of the cumulative persistence of air leaks in the two arms of the trial.

consultant surgeon determined that it was in the best interest of the patient. Analysis was on intention to treat.

16.2.4 Results

As shown in Figure 16.3, there was a lack of any association between the use of suction and cessation of air leak. It is clear from inspection that there is negligible difference in air leak persistence between the two groups and the log rank test for difference ($p = 0.62$) showed no significant difference between the two curves. The incidence of air leaks persisting beyond six complete days was calculated as 7.8% (95% CI 3.6–14.2) for suction (9/116) and 10.1% (5.7–17.4) for non-suction (13/123). There were two cases of recurrent pneumothorax among those receiving suction and three cases among those not receiving suction.

The study was presented at an international meeting EACTS September 28, 2004; revised by November 29, 2004; accepted December 6, 2004; available online January 13, 2005.

16.2.5 Impact on Practice

Non-suction was implemented immediately at Guy's Hospital on completion of the trial in 2004. A subsequent "best evidence" article reviewed all five randomized studies plus a further non-random retrospective study [11,12]. The authors concluded that of six studies no studies found in favor of continued suction. Four studies favored water-seal drainage without suction. Five of the six studies used a short period of suction in the immediate postoperative period and therefore were designed so that they were limited in their ability to address the question. Nevertheless, the evidence overall is very clearly in favor of less or no suction.

The use of suction requires used of disposable tubing, more equipment, electrical power, takes up nursing time, and provides more opportunities for error. Specifically, the connection of the approximately one in 10 patients who have had a pneumonectomy (removal of a whole lung) to suction can be fatal. Then suction will shift the heart and great vessels toward the now empty space and the distortion of the low pressure veins can obstruct them. In the absence of a benefit, it seems evident that suction need not, indeed should not, be used routinely. That was the policy decision.

However, the three consultants who instigated the study (TT) have now all retired and the practice has since reverted to using suction for the first 24 h with four different protocols (related to timing of drain removal in relationship to air, fluid leakage, and elapsed postoperative time) among the four new surgeons. Evidence-based medicine purists have to come to terms with the fact that beliefs hold sway in spite of evidence and that clinicians by their behavior often express a preference for individuality rather than conformity.

16.3 The Mesothelioma and Radical Surgery Trial

16.3.1 Background

Malignant pleural mesothelioma is a cancer caused by asbestos exposure 30–40 years earlier. It appears insidiously and because it creeps along the pleura, the membrane lining the chest wall and enveloping the lung, it had always appeared to the surgical author inherently unlikely that a surgical procedure would encompass and thus eradicate the cancer. Attempted eradication by radical surgery called extrapleural pneumonectomy (EPP) had been reported in 1976 [13]. The death rate of 30% and invariable recurrence of mesothelioma dampened enthusiasm but a few surgeons kept trying and began to report "better results" in patients who had surgery as part of multimodality therapy. This typically included preoperative chemotherapy and/or postoperative radiotherapy (Table 16.4).

It was a time of increasing incidence (a consequence of asbestos exposure 30–40 years before [14,15] and the number of cases has not yet peaked [14–16]. On the face of it, surgery seemed unlikely to cure this cancer and in the absence of evidence for benefit, it was not performed by the surgeon author nor by the majority of U.K. surgeons. And yet some patients were being referred to whoever would "give them the benefit of the doubt." Probably due to the ease of access to institutional advertising on the Internet, some patients raised the money to travel abroad. It appeared important to resolve

TABLE 16.4

The PICO for the MARS Trials

Population	Malignant pleural mesothelioma judged to be removable with radical surgery. All patients had three cycles of chemotherapy and were reassessed
Intervention	Radical surgery (extrapleural pneumonectomy) followed by radiotherapy
Comparison	Clinical decision as to best nonsurgical treatment often driven by patient pressure
Outcome	Expressly the feasibility of randomization plus survival and quality of life

Source: Treasure, T. et al., *J. Thorac. Oncol.*, 4(10), 1254, 2009.

this issue. If the surgery is effective, it should be available to all patients who might benefit so the case for a trial was argued [17].

A systematic search was made for reports of results of EPP [18]. There were no controlled trials or direct comparisons. Follow-up studies claimed survival better than expected from the known natural history. These publications had all the characteristic flaws of surgical follow-up series. No care was taken about finding a realistic "natural history" against which to make the comparison of better survival or about the degree of selection. All survival data were presented according to completed treatment, introducing a surviving patient bias. We have drawn attention to these flaws [19].

So the MARS trial was proposed and funding was obtained. This has been a high-profile trial throughout with interest from respiratory physicians, oncologists, and surgeons, in Europe, the United States, and Australia. It was confidently predicted by opinion leaders that it would be impossible to randomized patients because no patient would want to forgo the chance of cure. It was also declared to be unethical to randomize by opposing camps: some saw it as unethical to deny patients the chance of cure, whereas others considered this surgery as so predictably "futile" that it was unethical to ever do it. We argued that this polarization, while not quite the usual concept of "equipoise," did indicate uncertainty and the need for a trial.

16.3.2 Methodology

16.3.2.1 Two-Stage Consent

A two-phase consent process was employed for practical reasons. If all patients were to be staged as if for surgery, they would all have to have staging surgery that would not otherwise have been performed. This was to biopsy the mediastinal lymph nodes (mediastinoscopy) as spread of cancer to the lymph nodes would preclude any prospect of removing all the disease with surgery. The initial consent was therefore required by the ethics committee to allow this operation and other investigations to be carried out because they were not useful in management in the 50% randomized to not have surgery and were thus unnecessary outside of the trial. It should be noted that if the trial were to show EPP to be ineffective, the proffered management would then not be what was delivered in the control arm because it would not include mediastinoscopy.

About 3 months elapsed before a second consent to be allocated at random to surgery or not. This interval between entering the study and randomization allowed the trial center to record data and the clinicians to treat all patients according to a standard preoperative protocol. It also had the unplanned advantage that there was a long delay before the difficult question of randomization had to be addressed by which time patients and trialists found it easier to have this conversation.

16.3.2.2 Allocation

Randomization was the means of allocation. This was done by telephone to the trial center after the second consent and review by teleconference of a multidisciplinary team. Figure 16.4 shows the trial design.

16.3.3 Conduct of the Trial

The recruitment rate was slow and the process laborious. Of the patients who came via screening logs, 67/161 were recruited to first consent but it is likely that many patients were not even considered and the true proportion of patients in whom EPP was a realistic option was very low in what is an increasingly elderly population as the asbestos exposed cohort ages.

We obtained our 50 randomizations from 112 registrations into the first phase of the study. In the early part of the study, two patients randomized to no surgery (and that is two out of a small number) insisted on having surgery. Later the patient preference swung the other way, probably under the influence of trial nurses.

16.3.4 Results

The published results show that patients in the radical surgery and radiotherapy arm had shorter median survival and worse quality of life than those not allocated to surgery. [3] It is too soon to judge it this will be seen as conclusive evidence. MARS was a feasibility study to see if randomization would be possible. About half of 57/112 agreed to have their treatment allocated by randomization.

16.3.5 Impact on Practice

The mood has changed in 5 years between 2004 and the time of writing albeit without high-level evidence. Instead, several influential institutions have published data in the form of case series placing emphasis on the

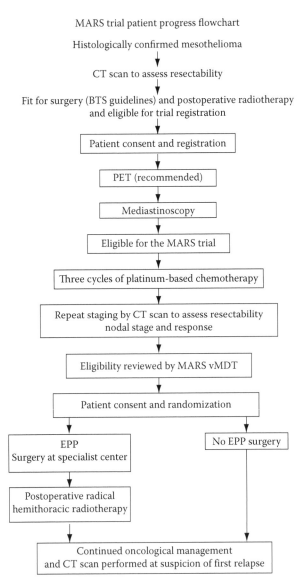

FIGURE 16.4
Protocol flowchart for the MARS trial.

morbidity [20,21] and reporting their survival data which we would regard as universally poor [21–23] and where data were available in non-operated patients, not obviously better [24]. In fact, there is little in the reports, which was not becoming clinically evident 5 years ago but the poor outcomes are being viewed with more realism. The MARS trial may deserve some credit for that: it was known that the question was being addressed and the bullish approach to this surgery has eased off. Some surgeons still want to operate on this disease but most are now advocating a lung sparing operation, which is less devastating for the patient but less likely, if there ever was a prospect of cure, to eradicate the disease. There appears to be an implicit acceptance that EPP will not show a benefit.

16.4 Conclusions

Equipoise is unusual and uncertainty actively discouraged in the character and the training of surgeons. This was illustrated by the disparity of opinions on the use of suction in the first study described. In the exploratory phase for the trial, a discussion was held on the matter at the Thoracic Forum of the Society of Cardiothoracic Surgeons of Great Britain and Ireland. There are certain agreed issues that are summarized as follows:

- There are clear mechanistic benefits in the use of suction in individual patients. For example, an incompletely inflated lung may fully re-expanded after the application of suction or a collection of blood or fluid not draining under gravity may be rapidly aspirated with the use of suction. These are immediate and readily seen mechanistic effects that benefit in short and long terms.

- On the other hand, there are instances in individuals where suction has caused harm by distorting the mediastinum, tearing the lung thus perpetuating air leak, or increasing bleeding from an injured lung.

In a consensus process where participants are asked to grade appropriateness of use of suction from 1 to 9, this is an instance where one might expect a consensus to take the middle ground since there are benefits and harms to balance [25]. In fact, the surgeons' views were polarized, strongly held, and fiercely argued. The surgeons were not engaging in a consensus process.

There is some general truth in this observation. In the middle of an operation, decisions cannot be constantly revisited. Vacillation and uncertainty about what action to take in the midst of an operation can be counterproductive. Similarly, when surgeons have to dictate a policy, having chosen one course of action over another, they need to believe in the rightness of that course of action. Equipoise does not come easily, if at all, and uncertainty is not a trait that is encouraged. This makes the design and conduct of trials in surgery problematic.

One of the most difficult aspects of randomized trials in surgery is the nature of surgery itself. The choice of operation and the expertise with which it is performed are often inextricably linked. This has both technical and emotional components.

- The surgeons may have honed their skills over years in a particular technique. An RCT in which the outcomes are compared between this and another technique previously unpractised would not be reliable. Blind prescription of drug A versus drug B does not have the same problems.
- A personal belief that one or the other operation is in the best interest of the patient may make it unacceptable for a surgeon to perform a long arduous operation in which the surgeon has no faith. That is clearly evident for a massive operation such as in the second example, EPP for mesothelioma.

This unwillingness of surgeons to hand over the decision between two operations to be allocated by some external arbiter, whether is a computer or a coin toss, has led to an acceptance of keeping the surgeon and the surgery as a package—the concept of the expertise based trial [26]. The same two arguments apply in physical and talking therapies [27]. There are situations where there are very good reasons to take this approach, where the required personal commitment means that performing the alternative treatment simply would not be acceptable.

There are grounds for caution, however. It might be that an unusually high level of innate skill is required to get good results in some operations and better results achieved in the hands of surgeons who do the operations with good outcomes might not be generalizable. Then the possibly better treatment cannot be separated from the attributes of the operator. This applies also in the more mundane end of surgical practice. In the suction trial, for example, skill and attention to detail are likely to reduce the degree of air leaks from the lung. Any difference would be falsely attributed to the technique of drainage if the expertise model were pursued.

Given the difficulties discussed earlier, when randomized studies are done in surgery, patient numbers tend to be small. As a result, many surgical trials are underpowered for clinically important end points [28]. There is an ethical dimension to embarking on any trial, that is of insufficient power to be conclusive but there are other considerations. Less rigorous study designs can lead to false inference being drawn [19] and the very process of a trial puts claims of benefit to formal evaluation and under the scrutiny of others. By acting as a focus for discussion and debate, a trial can bring benefits to patient care independent of the results as we believe to be the case with MARS. For these reasons, we subscribe to the view that "some unbiased evidence is better than none" [29].

References

1. Alphonso, N., Tan, C., Utley, M., et al. A prospective randomized controlled trial of suction versus non-suction to the under-water seal drains following lung resection. *Eur J Cardiothorac Surg*, 27(3), 391–394, 2005.
2. Treasure, T., Waller, D., Tan, C., et al. The Mesothelioma and Radical Surgery (MARS) randomised controlled trial: The feasibility study. *J Thorac Oncol*, 4(10), 1254–1258, 2009.
3. Treasure, T., Lang-Lazdunski, L., Waller, D., et al. Extra-pleural pneumonectomy versus no extra-pleural pneumonectomy for patients with malignant pleural mesothelioma: clinical outcomes of the Mesothelioma and Radical Surgery (MARS) randomised feasibility study. *Lancet Oncol*, 12(8): 763–72, 2011.
4. Cerfolio, R.J., Bass, C., Katholi, C.R. Prospective randomized trial compares suction versus water seal for air leaks. *Ann Thorac Surg*, 71(5), 1613–1617, 2001.
5. Marshall, M.B., Deeb, M.E., Bleier, J.I., et al. Suction vs water seal after pulmonary resection: A randomized prospective study. *Chest*, 121(3), 831–835, 2002.
6. Ayed, A.K. Suction versus water seal after thoracoscopy for primary spontaneous pneumothorax: Prospective randomized study. *Ann Thorac Surg*, 75(5), 1593–1596, 2003.
7. Brunelli, A., Monteverde, M., Borri, A., et al. Comparison of water seal and suction after pulmonary lobectomy: A prospective, randomized trial. *Ann Thorac Surg*, 77(6), 1932–1937, 2004.
8. Treasure, T., MacRae, K.D. Minimisation: The platinum standard for trials? *BMJ*, 317(7155), 362–363, 1998.
9. Senn, S. Controversies concerning randomization and additivity in clinical trials. *Stat Med*, 23(24), 3729–3753, 2004.
10. Treasure, T., MacRae, K.D. Minimisation is much better than the randomised block design in certain cases. *BMJ*, 318(7195), 1420, 1999.
11. Sanni, A., Critchley, A., Dunning, J. Should chest drains be put on suction or not following pulmonary lobectomy? *Interact Cardiovasc Thorac Surg*, 5(3), 275–278, 2006.
12. Antanavicius, G., Lamb, J., Papasavas, P., Caushaj, P. Initial chest tube management after pulmonary resection. *Am Surg*, 71(5), 416–419, 2005.
13. Butchart, E.G., Ashcroft, T., Barnsley, W.C., Holden, M.P. Pleuropneumonectomy in the management of diffuse malignant mesothelioma of the pleura. Experience with 29 patients. *Thorax*, 31(1), 15–24, 1976.
14. Peto, J., Hodgson, J.T., Matthews, F.E., Jones, J.R. Continuing increase in mesothelioma mortality in Britain. *Lancet*, 345(8949), 535–539, 1995.
15. Peto, J., Decarli, A., La Vecchia, C., Levi, F., Negri, E. The European mesothelioma epidemic. *Br J Cancer*, 79(3–4), 666–672, 1999.
16. Rake, C., Gilham, C., Hatch, J., Danton, J., Peto, J. Occupational, domestic and environmental mesothelioma risks in the British population: A case control study. *Br J Cancer*, 100(7), 1175–1183, 2009.
17. Treasure, T., Waller, D., Swift, S., Peto, J. Radical surgery for mesothelioma. *BMJ*, 328(7434), 237–238, 2004.
18. Treasure, T., Sedrakyan, A. Pleural mesothelioma: Little evidence, still time to do trials. *Lancet*, 364(9440), 1183–1185, 2004.

19. Treasure, T., Utley, M. Ten traps for the unwary in surgical series: A case study in mesothelioma reports. *J Thorac Cardiovasc Surg*, 133(6), 1414–1418, 2007.

20. Sugarbaker, D.J., Jaklitsch, M.T., Bueno, R., et al. Prevention, early detection, and management of complications after 328 consecutive extrapleural pneumonectomies. *J Thorac Cardiovasc Surg*, 128(1), 138–146, 2004.

21. Schipper, P.H., Nichols, F.C., Thomse, K.M., et al. Malignant pleural mesothelioma: Surgical management in 285 patients. *Ann Thorac Surg*, 85(1), 257–264, 2008.

22. Rice, D.C., Stevens, C.W., Correa, A.M., et al. Outcomes after extrapleural pneumonectomy and intensity-modulated radiation therapy for malignant pleural mesothelioma. *Ann Thorac Surg*, 84(5), 1685–1692, 2007.

23. Flores, R.M., Pass, H.I., Seshan, V.E., et al. Extrapleural pneumonectomy versus pleurectomy/decortication in the surgical management of malignant pleural mesothelioma: Results in 663 patients. *J Thorac Cardiovasc Surg*, 135(3), 620–626, 2008.

24. Flores, R.M., Zakowski, M., Venkatraman, E., et al. Prognostic factors in the treatment of malignant pleural mesothelioma at a large tertiary referral center. *J Thorac Oncol*, 2(10), 957–965, 2007.

25. Tan, C., Treasure, T., Browne, J., Utley, M., Davies, C.W.H., Hemingway, H. Appropriateness of VATS and bedside thoracostomy talc pleurodesis as judged by a panel using the Rand/UCLA appropriateness method (RAM). *Interact Cardiovasc Thorac Surg*, 5, 311–316, 2006.

26. Devereaux, P.J., Bhandari, M., Clarke, M., et al. Need for expertise based randomised controlled trials. *BMJ*, 330(7482), 88, 2005.

27. Boutron, I., Moher, D., Altman, D.G., Schulz, K.F., Ravaud, P. Methods and processes of the CONSORT Group: Example of an extension for trials assessing nonpharmacologic treatments. *Ann Intern Med*, 148(4), W60–W66, 2008.

28. Anyanwu, A.C., Treasure, T. Surgical research revisited: Clinical trials in the cardiothoracic surgical literature. *Eur J Cardiothorac Surg*, 25(3), 299–303, 2004.

29. Lilford, R.J., Thornton, J.G., Braunholtz, D. Clinical trials and rare diseases: A way out of a conundrum. *BMJ*, 311(7020), 1621–1625, 1995.

17

Assessing Obstetrics and Gynecology: Practical Examples

Vincenzo Berghella

Thomas Jefferson University

Jorge E. Tolosa

Oregon Health Science University

CONTENTS

17.1 Introduction

Obstetrics derives from the Latin "ob" and "stare," which mean to "stand by," or "to stand near." It refers to pregnancy and therefore involves the pregnant mother from conception to about 6 weeks postpartum, and the fetus. Gynecology derives from the Greek "gyne" and "logos," meaning "study of women," and refers to all things specific to women's health.

Women, and in particular pregnant women, have historically been underrepresented in randomized controlled trials (RCTs). Pregnancy used to be the condition with the lowest number of level 1 evidence studies (derived

TABLE 17.1

Examples of Nonpharmacological Interventions
in Obstetrics and Gynecology

- Cerclage for prevention of prematurity
- Cesarean delivery (indicated both for maternal or fetal conditions or on maternal request)
- Hysterectomy
- Dilatation and curettage (and/or evacuation)
- Pessary for uterine prolapse or prevention of preterm birth
- Bed rest for several ob-gyn conditions
- Acupuncture for pain
- Moxibustion for breech presentation
- Episiotomy
- Timing of umbilical cord clamping
- Amniocentesis
- Chorionic villus sampling
- Percutaneous umbilical blood sampling with or without transfusion

from RCTs) [1]. This has dramatically changed in the last 20 years. Obstetrics has become one of the medical specialties with the most RCTs [2]. There are in 2009 almost 1000 Cochrane Reviews of the Pregnancy and Childbirth group [3]. Of these 1000s of RCTs and meta-analyses, several concern non-pharmacological interventions. Certainly, RCTs in obstetrics and gynecology (ob-gyn) are feasible and desirable, including for nonpharmacological interventions. Practical examples are shown in Table 17.1, and many others exist. They involve mostly interventions either curative or aimed at decreasing disease for both women and babies.

The aim of this chapter is to improve the design, conduct, and reporting of RCTs in ob-gyn. Cervical cerclage is used as the main example.

17.2 Specific Methodological Issues in Assessing This Treatment

There are several issues in evaluating treatments of nonpharmacological interventions in ob-gyn. Perhaps the most important ethical issue refers to the fact that the same intervention may be beneficial for the mother, but not for the baby. For example, the only cure for preeclampsia (blood pressure >140/90 in a previously normotensive woman and proteinuria >300 mg in 24 h) is delivery. Delivery in cases of preeclampsia is always the treatment of choice for the mother. If severe preeclampsia happens preterm, let us say at 28 weeks (term birth is 40 weeks), delivery can severely affect the fetus (and then neonate), with high incidence of morbidity and about a

10% chance of mortality. Interestingly, the last well-designed study of timing of delivery for severe preeclampsia is from 1994 [4].

Another ethical issue in obstetrics regards the evaluation of interventions in labor by RCTs. This is difficult due to the issues raised by trying to obtain informed consent of a woman who is in pain. The ethics of approaching a woman in labor for research when so many other issues are happening (medical, personal, familial, etc.) has been insufficiently studied.

Blinding is often difficult, if not impossible. Cesarean delivery is the most common major surgery in the United States, with over 1 million performed every year. A sham procedure for cesarean delivery has not been devised, and therefore no such RCT exists. The most common indications for cesarean delivery are prior cesarean, failure of labor progress (cephalopelvic disproportion), and non-reassuring fetal heart status. Given worries regarding harm to the unborn child, and related legal implications, no RCTs exist on these common indications, which represent usually over three-quarters of all cesareans.

Another important issue is the cost of RCTs. An operative intervention part of the RCTs should be paid by the trial itself, not by insurance or other third parties. Often this is not the case, and insurance gets billed, especially in the United States. In these situations, someone randomized to the intervention might get support from the insurance and/or her job, whereas someone randomized to no intervention might not get the same support. For example, a woman receiving a hysterectomy usually gets more nursing attention and medical follow-up, and more time off work, compared with a similar woman who might get no such operation. These differences might affect the outcomes. Moreover, often insurance does not provide reimbursement for a procedure not yet proven to be efficacious.

17.3 Practical Examples

17.3.1 Cerclage

Cervical cerclage (Figure 17.1) involves placing a stitch in the uterine cervix to keep it closed during pregnancy. It is estimated that at least 40,000 of these procedures are performed each year in the United States alone. Cerclage was originally devised and is still used for prevention of preterm birth (birth before 37 weeks' gestation). Preterm birth is the number one cause for perinatal morbidity and mortality, and therefore the main disease in obstetrics. It has been estimated that, for years-of-life lost, preterm birth, affecting over 4 million neonates annually worldwide, may be one of the most important diseases of all, like or even above cancer and cardiac disease.

There are now more than a dozen RCTs on this procedure. It is encouraging that the ratio of RCTs to nonanalytic studies has increased dramatically.

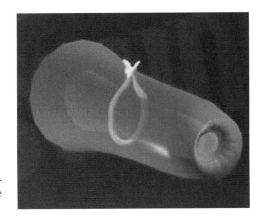

FIGURE 17.1
Schematic representation of uterine cervix, with cerclage (stitch) placed in the middle section.

Cerclage for Women with Prior Preterm Birth Who Develop a Short Cervical Length

Context: A meta-analysis of four small trials revealed that cervical cerclage may be effective in reducing the incidence of preterm birth in women with both a prior preterm birth and a short cervical length in the index pregnancy [5]

Objective: To estimate the efficacy of ultrasound-indicated cerclage in women with a prior preterm birth who develop a short cervical length in the second trimester of the next pregnancy

Primary outcome: Preterm birth <35 weeks' gestation

17.3.1.1 Blinding

None of the RCTs on cerclage had any blinding [6–9]. Blinding would be possible, but difficult and in reality inpractical in several ways. A sham procedure could be devised. This would involve at least taking the pregnant woman to the operating room, placing regional anesthesia, and keeping the woman blind as to the placement or not of the cerclage. Placing a cerclage usually causes bleeding, so no bleeding would point to no cerclage. A small incision could be made in cervix (and repaired) in the control group to cause some bleeding. We believe this would be ethical, as long as the woman was aware and consented to it. Institutional Review Boards may not approve it. Moreover, the woman could palpate her own cervix postoperatively and feel for the presence of a stitch.

Blinding of the physicians involved in the study would be possible if the surgeons placing the cerclage were different than those managing the rest of the pregnancy. Several ultrasounds are done in pregnancy in these women, especially of cervical length. Cerclage is easily identifiable on ultrasound, so physicians would need to be blinded to these post-cerclage ultrasounds. As there is no evidence of effectiveness of these post-cerclage ultrasounds or any interventions based on this information, it would be ethical to blind physicians.

Interestingly, blinding in cerclage RCTs has either not been discussed, or deemed "impossible." We would argue, based on the facts above, that blinding is difficult, but possible. The consequences of nonblinding in these RCTs are unknown. Stress has been associated with a higher incidence of preterm birth. It can be postulated that a woman who knows she is at high risk for preterm birth and is eligible for cerclage through a study but does not get it would have a higher level of stress, and consequently a higher incidence of preterm birth because of stress, not because cerclage is beneficial in the other group.

Moreover, keeping the managing doctors nonblinded may also have consequences. Often, practitioners prescribe bed rest for women at risk of preterm birth. They are more inclined to do so in those without cerclage. While bed rest has not been studied adequately (see below), this and other interventions could differ in the two randomized groups if blinding is not applied. In general, a researcher should follow CONSORT guidelines not just for reporting, but also to help study design.

17.3.1.2 Technical Aspects

Technical aspects of this surgical procedure are often not well defined, making external validity limited. There are several kinds of cerclage. RCTs have used either McDonald or Shirodkar techniques, without comparing the two. Differences in suture material, height of cerclage, depth and spacing of "bites" on the cervix, and other technical aspects may affect results. I believe that standardizing the procedure as much as possible would render the study most externally reproducible and valid. If "cerclage" is the intervention, it is important to know exactly what kind of technique is used to assess results.

Differences in pericerclage interventions such as antibiotics, indomethacin, or other anti-inflammatory agents, progesterone, and so on may affect results. For example, the RCT by Althuisius et al. [6] used indomethacin in the cerclage arm but not in the controls. The benefit found in the cerclage arm could have been due to cerclage, or due to indomethacin, or even due to a combined effect of the two interventions.

There is certainly a learning curve for performing this procedure. Different surgeons, especially if at different level of training and/or with differing expertise, may have differing results. Not only the surgeon's but also the center's expertise may affect results. For example, in obstetrics, chorionic villus sampling has been shown to be safer in high volume units with obstetricians who have done >400 procedures.

As now cerclage has been shown to be effective at decreasing preterm birth in women with a prior preterm and a short cervical length, RCTs should be designed to assess these technical aspects. Cerclage should also be compared with new techniques of stitching the cervix, and to placing a pessary.

17.3.1.3 Ethical Issues

Other ethical issues have been raised regarding RCTs of cerclage. For example, some advocate that cerclage cannot be denied to "certain patients" who need it. Given the fact that there was almost no indication for cerclage supported by level 1 evidence until recently, it could be argued that the ethical practice is to randomize women, as offering the procedure as beneficial would not be based on evidence and could be unethical.

17.3.1.4 Rescue Operations

Most RCTs of cerclage allow "rescue" cerclage in both arms. This means that a woman randomized to receive cerclage may receive another one if the condition worsens. And that a woman randomized to "no cerclage" may receive a cerclage also if her condition worsens. The possibility of these "rescue" cerclages is usually introduced in the study protocol to entice patient recruitment. Up to 10% of controls have received these "rescue" cerclages in some RCTs [7,8,10]. There are also women randomized to cerclage who then decline the procedure. This contamination is real and common.

All these RCTs were analyzed by intention to treat, which is the appropriate analysis. It is interesting that in our RCT [7], patient preferences changed over time, and made recruitment difficult. At the beginning, all wanted a stitch. Later, most wanted expectant management. This could also have been due, in a large percentage of cases, to physician change in counseling. Nonetheless, the effects of cerclage could be diluted if many "controls" receive the same intervention as the "cases," or if "cases" do not receive the assigned intervention. In our opinion, these "rescue" operations should not be allowed. Our IRB approved such a modification in the RCT protocol.

17.3.1.5 Recruitment

Is it ethical to offer a procedure, in this case cerclage, only in the context of an RCT? Absolutely. It is unethical to do otherwise. An RCT is indicated and approved by an Institutional Review Board when its safety and/or efficacy have not been proven yet. When we did an RCT on cerclage for women with a short cervical length [7], its efficacy was not yet proven, with non-RCT studies showing either benefit or no effect.

To improve recruitment, we decided not to offer anymore cerclage for a short cervical length outside of the research trial. Women were so counseled, and they understood. Now that the result of these and other RCTs have been published, in 2009 we can offer cerclage to women who do benefit from the procedure, and advice against it for women who are harmed by it (e.g., multiple gestations such as twins). Adequate and accurate information given to patients is most important.

17.3.2 Cesarean Delivery

There are at least two cesarean deliveries in the United States every minute. There are numerous trials on the technical aspects of this common operation [11]. Unfortunately, there are very few RCTs comparing cesarean to no cesarean. There are no RCTs on cesarean for failure to make progress in labor. It would be fairly easy to design and conduct such important RCTs. There are also no good RCTs on cesarean performed for other indications (Table 17.2).

It is amazing and inexcusable that no adequate RCTs have been done in women with prior cesarean comparing a trial of labor with attempt at vaginal birth after cesarean (VBAC) to repeat cesarean delivery.

Cesarean Delivery on Maternal Request

Context: More and more women request a cesarean delivery and decline a trial of labor and vaginal delivery. There have been no RCTs assessing this common clinical scenario

Objective: To estimate the efficacy of cesarean delivery on maternal request compared with trial of labor and attempt at vaginal delivery

Primary outcome: Maternal and neonatal morbidity and mortality

---

TABLE 17.2

Proposed Improvements to Randomized Trials of Nonpharmacological Interventions in Obstetrics and Gynecology

- Strive for double-blindness (of both women and practitioners); devise sham procedures if necessary
- Technique of procedure must be described and implemented in detail
- Detailed protocols of management for the two arms of the study should be adequately planned, described, and implanted in the conduct of the study
- The possibility of receiving the intervention outside of the RCT should be limited, if not at all banned
- "Rescue" procedures involving the study intervention should be limited, if not at all banned
- Adequately consider insurance and grant issues (e.g., reimbursement, effects of being in intervention versus non-intervention arm, etc.)
- RCTs of surgical procedures are ethical in women and even in pregnant women, and should be encouraged. It is UNETHICAL to do otherwise

A new indication for cesarean delivery is maternal request. These are elective procedures. Before allowing this practice to become common place, it is ethical and "a must" to conduct appropriately designed RCTs. I would say that it is ethical in 2011 to offer cesarean for maternal request in the context of an RCT.

The one large RCT on cesarean was the breech trial [12]. This demonstrated that well-designed RCTs are difficult and complex, but feasible. It was so important that it changed practice worldwide. Unfortunately, it does not address how it would be feasible to blind in RCTs of cesarean delivery.

17.3.2.1 Hysterectomy

Hysterectomy, or removal of the uterus, is a common major surgery. In some countries, up to one-third of women have their uterus removed in their lifetime. Few RCTs have been performed, even for benign indications.

Hysterectomy for Benign Disease (e.g., Fibroid Uterus)

Context: Recently, many alternatives to removal of the uterus (hysterectomy) have been devised for women who may need this procedure

Objective: To estimate the efficacy of hysterectomy for benign disease (e.g., fibroid uterus) in women

Primary outcome: Relief of symptoms (e.g., for fibroids, decrease in bleeding, pain). Even the technical aspects of hysterectomy have been very poorly evaluated, with Few RCTs performed

---

17.3.2.2 Activity Restriction

Activity restriction, including at times even bed rest, is often prescribed to women, especially when pregnant. For example, bed rest has been called "the standard initial therapy for women at risk for preterm birth" [13], even if this statement is not supported by level 1 evidence.

Bed Rest for Obstetrical Complications

Context: It is estimated that about one in four pregnant women is prescribed bed rest. Indications are numerous, and include, among others, vaginal bleeding, threatened preterm labor, preterm premature rupture of membranes, hypertensive disorders, and others

Objective: To estimate the efficacy of bed rest

Primary outcome: Maternal and neonatal morbidity and mortality

---

TABLE 17.3

Important Feasible RCTs to Be Done
in Obstetrics and Gynecology for
Non-pharmacological Interventions

- Cesarean delivery for failure to progress
- Cesarean delivery for non-reassuring fetal testing
- Cesarean delivery for prior cesarean
- Cesarean delivery for other indications
- Hysterectomy versus alternative interventions
- Activity modification (for several ob-gyn indications)

Work conditions, manual labor, working >40 h/week, prolonged standing, and shift work, have been associated with complications of pregnancy, but never properly studied. One problem with RCTs on activity restriction is defining exactly what it is. A second problem is ensuring that the control group maintains "normal" activity. These are both very difficult to define, with infinite questions of "how much" restriction is needed. Nonetheless, use of pedometer, and measuring work duties, exercise, hospital admission, strict bed rest, and other activities, are all possible. Unless "activity restriction" is well defined, there will be difficulties with external validity (Table 17.3).

17.4 Conclusion

RCT are and should be the gold standard for evaluating the safety and effectiveness of non-pharmacological procedures in pregnancy (obstetrics) and in women (gynecology). It is ethical to perform RCTs in women, including (if not especially) pregnant women. To NOT do so is unethical, and a shame of the medical profession, and society in general. We encourage all non-ob-gyn readers to ponder this point. When they are withdrawing treatment (or inclusion in a research study) to a pregnant woman, they are denying health and causing harm.

Blinding in RCTs in ob-gyn is feasible and desired. As it has been done for other procedures (e.g., arthroscopy), blinding for procedures such as cerclage is possible. Rescue "operations" in the control arm should be kept to a minimum, to avoid the potential gap between treatment intended and actually administered. Intention to treat should always be the first and main analysis presented. A strict protocol for each arm of the trial should be devised and adhered to. This may limit external validity to large populations but will make the effects reproducible in clinical practice.

The rate of the outcomes makes obstetric RCTs, especially in the developed world, very difficult as the sample size required to measure an outcome of clinical importance is very large. There has been a methodological need for development of composite outcomes, and the evaluation of long-term outcomes on the neonate for interventions in pregnancy.

Quality RCTs in commons areas, such as mode of delivery, hysterectomy, and activity restriction, are urgently needed and can be, and in fact should be, properly designed and conducted. Based on these, guidelines for improvement in care [14] should be devised and followed to improve the health of women and of their children.

References

1. Cochrane, A.L. 1931–1971: A critical review, with particular reference to the medical profession, in *Medicines for the Year 2000*. London, U.K.: Office of Health Economics, 1979, pp. 1–11.
2. Dickersin, K., Manheimerc, E. The cochrane collaboration: Evaluation of health care and services using systematic reviews of the results of randomized controlled trials. *Clinic Obstet Gynecol*, 41, 315–331, 1998.
3. The Cochrane Collection. http://www.cochrane.org (accessed August 27, 2009).
4. Sibai, B.M., Mercer, B.M., Schiff, E., Friedman, S.A. Aggressive versus expectant management of severe preeclampsia at 28 to 32 weeks' gestation: A randomized controlled trial. *Am J Obstet Gynecol*, 171(3), 818–822, 1994.
5. Berghella, V., Obido, A.O., To, M.S., Rust, O.A., Althiusius, S.M. Cerclage for short cervix on ultrasound: Meta-analysis of trials using individual patient-level data. *Obstet Gynecol*, 106, 181–189, 2005.
6. Althuisius, S.M., Dekker, G.A., Hummel, P., Bekedam, D.J., van Geijn, H.P. Final results of the cervical incompetence prevention randomized cerclage trial (CIPRACT): Therapeutic cerclage with bed rest versus bed rest alone. *Am J Obstet Gynecol*, 185, 1106–1112, 2001.
7. Berghella, V., Odibo, A.O., Tolosa, J.E. Cerclage for prevention of preterm birth in women with a short cervix found on transvaginal ultrasound examination: A randomized trial. *Am J Obstet Gynecol*, 191, 1311–1137, 2004.
8. Rust, O.A., Atlas, R.O., Reed, J., van Gaalen, J., Balducci, J. Revisiting the short cervix detected by transvaginal ultrasound in the second trimester: Why cerclage therapy may not help. *Am J Obstet Gynecol*, 185, 1098–1105, 2001.
9. To, M.S., Alfirevic, Z., Zarko, H., et al. Cervical cerclage for prevention of preterm delivery in women with short cervix: Randomized controlled trial. *Lancet*, 363, 1849–1853, 2004.
10. Owen, J., Hankins, G., Iams, J.D., et al. Multicenter randomized trial of cerclage for preterm birth prevention in women with prior preterm birth and shortened mid-trimester cervical length. *Am J Obstet Gynecol*, 201(4), 375, e1–e8, 2009.
11. Berghella, V., Baxter, J.K., Chauhan, S. Evidenced-based surgery for cesarean delivery. *Am J Obstet Gynecol*, 193, 1607–1617, 2005.

12. Hannah, M.E., Hannah, W.J., Hewson, S.A., Hodnett, E.D., Saigal, S., Willan, A.R., for the term breech trial collaborative Group. Planned caesarean section versus planned vaginal birth for breech presentation at term: A randomised multicentre trial. *Lancet*, 356, 1375–1383, 2000.
13. Sosa, C., Althabe, F., Belizan, J., Bergel, E. Bed rest in singleton pregnancies for preventing preterm birth. *Cochrane Database Syst Rev*, (1), CD003581, 2005.
14. OB Guide: A Guide to Obstetrics. http://www.obguide.org (accessed August 27, 2009).

18

Assessing Lower Limb Arthroplasty: Practical Examples

David Biau

Public Assistance—Hospitals of Paris

Remy Nizard

Public Assistance—Hospitals of Paris and
Paris Diderot University

CONTENTS

18.1 Introduction

Since Sir John Charnley introduced the modern total hip replacement in 1962, arthroplasties of the lower limbs have become one of the most reliable and most successful procedures in orthopedic surgery [1,2]. In 2005, 285,000 hip replacements and 523,000 knee replacements were performed in the United States, and it is expected from projection that the number of procedures performed in 2005 will double by the year 2026 for total hip replacement and by the year 2016 for total knee replacement [3]. Owing to the continuing demand of patients, health authorities, and doctors themselves, these procedures evolved from an intervention that was performed in elderlies only to relieve pain and allow them to resume activities of daily living to an intervention with a wide age range and that aims to relieve pain, allow patients to participate in recreational activities in an accelerated manner, restore high level of function, and last for a life time.

Given the ambitious objectives of today's lower limb arthroplasties (LLA), the evaluation of these procedures encompasses a wide range of areas in the preoperative care of the patients such as education, physiotherapy, and patient optimization [4–6]; in the design of implants and surgical techniques [7–10]; and in the postoperative care such as the rehabilitation [11]. Accordingly, numerous outcomes are being used to measure the effect of these arthroplasties in these different areas. Some outcomes are considered as directly related to the patient such as pain, range of motion, joint function, surgical and medical complications, etc.; some are more related to the implant such as wear of the bearing surfaces, aseptic loosening, revision of the implants, etc.; and some that were considered as irrelevant only 10–20 years ago such as the length of the incision, the duration of hospital stay, costs, etc., have now come into focus. Therefore, clinical research has to cover all these aspects, from the decision of surgery to the end result of the procedure. However considering the objective of LLA, the evaluation of these procedures uses outcomes that usually fall into one of the following three categories, either directly or as "surrogates": pain, function at large, and implant longevity. Accordingly, the methodological issues that have to be considered in the design, conduction, analysis, and reporting of the trial will interfere differently given the intervention considered and the outcome chosen.

The randomized controlled trial (RCT) provides the best methodological framework to assess the effect of an intervention and LLA are not different in that way. It should be mentioned, however, that researchers be particularly careful when planning a RCT in evaluating LLA because of the few trials conducted in this domain and the little experience investigators and other participating staff usually demonstrate. Therefore, problems affecting the internal validity of the trial are more likely to occur and less likely to be detected and corrected. Also, the evaluation of LLA faces unique methodological difficulties such as the remoteness of end points measuring implant longevity, the constant development of surgical techniques and implant designs that cannot be disregarded, and the beliefs and affects that patients and surgeons put in a mechanical intervention.

18.2 Specific Methodological Issues in Assessing This Treatment

The first issue to consider in the design of trials assessing LLA is the willingness of patients and surgeons to participate in the trial. Surgeons are reluctant to participate in a trial for numerous other reasons than those usually observed in trials conducted in the pharmaceutical industry. The first reason is that, due to a permissive regulatory context, joint implants are usually evaluated in a randomized trial after they have been on the market for many years. Consequently surgeons will not be convinced easily that evaluation is still necessary, either because a fair amount of data from noncomparative studies has been reported or because they have gathered some a priori from personal experience. In these cases, the rule of equipoise, or at least uncertainty, can be breached. Next, surgeons may also be reluctant to participate in a trial because it may involve performing procedures they are not comfortable with. Some have proposed that surgeons be part of the treatment group, with some surgeons performing one of the treatment and others performing the other treatment [12]. This design is called the expertise-based RCT. It has, however, significant disadvantages [13,14]. On the other side, patients may also be more reluctant to participate in trials comparing LLA than those comparing drugs. There is usually a mechanical reason, or a reason that can easily be understood, that led to the development of the implant or new surgical technique under evaluation. Consequently, this reason is expected to generate some positive effect and the effect, in turn, yields stronger affect and understanding than the biological effect of some drug. Sometimes patients come to see the surgeon with an idea of the operation they need, such as the demand of some patients to receive hip resurfacing instead of conventional hip replacements, and it proves difficult to convince these patients to enter a trial at the risk of receiving the other treatment. Last, because most implants or surgical techniques evaluated are already in use for some time, investigators should clearly state what treatment will be offered to patients who refuse to enter the trial. For instance, say the aforementioned patient who came for hip resurfacing surgery declines trial participation. Should the patient be offered conventional hip replacement since this is the standard treatment, or should the surgeon be allowed to offer hip resurfacing outside the trial since this procedure has been in use at the center for some years already? This issue should be clearly discussed at the time of designing the trial.

The choice of the primary outcome criteria is a very important step to consider since the conclusion of the trial will be based on this outcome. Also, this choice will have major implications on sample size and trial duration and this should not be disregarded. The investigators must determine the event of interest and the time at which the comparison will be performed. For instance, assume one wishes to compare the survival of a highly reticulated polyethylene acetabular component to a standard polyethylene acetabular component [8]. Revision of any part of the implant for any reason or revision of the acetabular component for aseptic loosening only may both be considered as valid primary outcome criteria. Some would argue that revisions for other than aseptic loosening such as infection or periprosthetic fracture do not depend on reticulation of the polyethylene and that they not should be considered. However, one may argue that first it cannot be proved that these events are independent from reticulation of the polyethylene. For instance, one could perfectly imagine that highly reticulated polyethylene will generate different wear particles leading to more bone resorption or decreased local immunity, and subsequently to increased risk of periprosthetic fracture or infection, respectively. Second, if these events are independent from the event of interest then it should make no difference whether they will be regarded as of interest since both factors are expected to be distributed equally on average under randomization. Also, the time at which the comparison is made must be clearly mentioned in the design of the trial. It is agreed that a minimum follow-up of 5 years is necessary to evaluate joint implants. However, with an estimated 20% revision rate in the first 20 years [15,16], very few failures are expected during the first 5 years and sample size will probably be unreachable. Choosing to compare the event of interest later, say at 15 or 20 years, when more failures are likely to have occurred and the difference, if it exists, likely to have shown up will make the trial very difficult to conduct over time or carry the risk of reporting obsolete results at the time of study end. This is one of the reasons for implementation of joint registries. Alternatively, investigators may consider using surrogate end points such as two-dimensional computerized polyethylene wear analysis [17], implant Roentgen stereophotogrammetry [18] micro-motion, or knee prosthesis coronal alignment in place of implant aseptic loosening. The use of surrogates has caveats. Surrogates were defined, in a rather strict manner, by Prentice [19] 20 years ago as end points that would be correlated with the event of interest and entirely capture the effect of treatment on the event. Such definition renders "surrogates" employed in the design of trials assessing LLA inappropriate because it means that knowing the surrogate would make us apt to determine perfectly the event of interest such as polyethylene wear or implant micro-motion would perfectly predict implant aseptic loosening. This is not the case however. Wear of the polyethylene liner and subsidence of the stem have both been associated with aseptic loosening. Ceramic on ceramic bearings, for instance, introduced some 40 years ago show no wear but the risk for implant revision is not inferior to conventional metal on polyethylene bearings. The Exeter stem is well known for significant subsidence during the first few years but still has demonstrated one of the best implant survivals [16]. Therefore, when using surrogates, it is never certain that the difference observed will be translated, or to what extent, into a difference in the proxy, and when reporting the results of a trial this limit should be made clear. Choosing primary outcome criteria for function or its components is usually easier. It is accepted that maximum function after knee and hip joint replacements is usually obtained and stable after 1 year. Moreover, effective sample size favors the use of continuous outcomes to binary or survival outcomes. Therefore, sample size and duration will render these trials usually feasible without the need for surrogates [20]. Sometimes, the event of interest is rare but has no adequate replacement such as postoperative dislocation and the design of the trial will have to consider an important sample size.

Another issue that is yet problematic in the evaluation of LLA as compared with drugs is that investigators are usually not offered a criterion standard for comparison. There are hundreds of hip or knee implants of which dozens have shown some superiority in terms of implant survival and may be considered as suitable for comparison, but to date there is no best total hip or knee implant, no best surgical approach, no best rehabilitation protocol, etc. And therefore, the results of the comparison of one implant to another are only applicable to this comparison and generalizability may fall short. For instance, in a trial comparing smooth versus rough geometrically identical cemented femoral stems, it cannot be sure that the results are applicable to all femoral stem geometry or alloy [21,22]. Comparisons of class of implants, such as comparing cemented femoral stems to uncemented femoral stems, have better generalizability. However, because of the important heterogeneity in terms of survival in a same class of implants, the comparison may be of little relevance. In the end, investigators should choose the comparator for a precise purpose knowing that it will impede on the generalizability of the results.

LLA are complex interventions where numerous technical issues may interfere with the primary outcome studied and the effect of treatment observed. The type of trial chosen, whether pragmatic or experimental, will usually decide how investigators deal with these issues. If investigators have opted for an experimental trial, they should try to standardize all factors that may affect the primary outcome so that the difference observed may rightfully be attributed to the treatments studied. For instance, in a trial comparing high-flex total knee replacement to standard knee replacement, they should make sure that both femoral implants are implanted with identical tibial implants and that the same brand with similar ancillary device is used. The surgical technique, such as the surgical approach, bone size cuts, removal of osteophytes, release of soft tissues, etc., should be carefully described in the protocol. Surgeons and patients selected for the trials should be sufficiently homogeneous so that only little variation is expected. For instance, because the most important determinant of postoperative knee flexion is preoperative knee flexion, it may make sense not to include patient with very low preoperative knee flexion. On the contrary, those opting for a pragmatic trial will not try to standardize factors affecting the comparison. For instance in a trial comparing the postoperative limb alignment after total knee replacement with or without the use of computer navigation system, it may make sense that surgeons be free to choose the type of implants used, the surgical approach, the navigation system, etc. Also, investigators may wish to have wide selection criteria for patients, surgeons, and centers, so that center, care provider, and volume effect may be looked at. Stratification on important prognostic factors such as center, surgeon, and the presence of preoperative knee deformity will help the adjustment in the analysis so that estimation of treatment effect is more precise, interaction may be sought for, and preplanned subgroup analyses be relevant. Last, it should be remarked that once an explanatory trial has be conducted in very stringent conditions, a pragmatic trial in larger and more heterogeneous settings should confirm the results.

Trials evaluating LLA should take account of relevant clustering such as care provider, center, volume effect, and case-mix. Ignoring clustering effect in the analysis leads to incorrect or inefficient estimation of treatment effect and possibly misleading conclusions [23–25]. It is expected that results be different from one surgeon to another, or from one center to another [26,27]. This heterogeneity should not be disregarded, on the contrary it should be analyzed and reported because it may have strong implications on the applicability of trial results such as interaction between expertise and treatment effect [28]. For instance, a new high-flex total knee replacement may improve the 1-year postoperative knee flexion when implanted by low volume surgeons and not when implanted by high volume surgeons. Similarly, the use of computer navigation system for implantation of total knee replacement may be useful only for patients with important preoperative knee deformity. At the time of planning, randomization should ideally be stratified on clusters so that statistical analysis be more efficient and sample size calculation [29] should account for clustering effects. One specificity of LLA is that they may be performed on the right and left side of the same patient. Some investigators have proposed to randomize the treatment group for first side and to use the other implant on the other side of the patients [30]. The advantage of this design is that the effect of some confusing factors such as patient activity, age, medical conditions, will be the same for both treatment groups. This design presents, however, some significant disadvantages. First, once the first hip or knee replacement is performed, it is not sure that the next one will be possible because the operation may be canceled or postponed to after the end of the study. Second, contamination between groups may occur. For instance, in a trial assessing the effect of preoperative education versus usual care in the prevention of postoperative dislocation, if a patient receives the preoperative education after the first procedure, he or she cannot be considered not to have received the same care for the second procedure. Last, the evaluation of numerous outcomes poses problem of dependency between both treatment groups. For instance, pain felt by the patient after the first procedure will interfere with that felt after the second operation. Similarly, infection of the first joint replacement will interfere with the probability of having infection develop following the second procedure considering antibiotics were given before. Most functional outcomes include item that will be similarly affected by both sides, etc. Therefore preferably, patients should be included for one side only.

The learning curve is a matter to be considered. The learning curve depicts graphically the period during which the performance increases until it reaches a plateau. Usually the learning curve is relevant to surgical techniques more than implants or ancillary devices, but it may still be relevant. For instance, the implantation of a new uncemented femoral stem will require some procedures before surgeons regularly find the stem that gives the best adjustment between the prosthesis and the femur. Until then, oversized implants put femurs at risk of intraoperative fracture and undersized implants put stems at risk for early subsidence. Early experience with hip resurfacing implants is known to be difficult [31]. Including the learning curve in a trial may possibly bias the results in favor of the control treatment since this one is usually known of care providers. The comparison ensuing in such trials is that of the control treatment without the learning curve to the experimental treatment including the learning curve. It does not represent what will occur in practice after the learning curve is finished and it usually does not make sense to include the learning curve in such trials. The learning curve of joint replacement may be regarded as phase 2 trials in drug development: it makes little sense to start a trial before the correct dose for efficacy of the drug has been found. This does not mean that the learning curve should be ignored. On the contrary, the learning curve should be studied thoroughly at the time the implant, surgical technique, or ancillary device are developed. Once trials confirm the benefits of an implant, the learning curve may be studied further in different settings to determine the average number of procedures required to obtain full benefits.

Blinding is an another issue in the evaluation of nonpharmacological treatments and LLA are no different. Almost never will the surgeon be blinded to the implant or technique evaluated. Therefore, adequate countermeasures should be employed to limit the biases ensuing. First of all, whether surgeons are blinded or not, allocation must remain concealed if we want groups to be comparable. Say a surgeon participates to a trial comparing dual mobility to standard acetabular components in the prevention of postoperative dislocation and that allocation is not concealed. If the surgeon sees a patient at high risk of dislocation, because of neuromuscular deficiency for instance and that this patient is to be randomized in the experimental group, he or she may decide not to propose the patient for participation. And little by little, patients in the experimental group will be at lower risk of postoperative dislocation despite an adequate randomization list. Second, although surgeons will rarely be blinded to the treatment group it is often necessary that the best be done so that at least patients and outcome assessors be actually blinded. It is not necessary, because of the time and resources necessary for that purpose, that all participating staff be blinded. For instance, in a trial comparing dual mobility to standard acetabular component, we may imagine masking the implants evaluated from the staff working in theater, such as the scrub nurse. However, it will unlikely bias the result whether or not the scrub nurse knows which implant has been used considering postoperative dislocation as the outcome criteria. It would be paramount, however, that patients be blinded to the group of randomization so that they will not be more careful to avoid movements at risk of dislocation should they be in one group or the other. Similarly, it is most of the time necessary that the outcome assessor be blinded to the treatment group. For instance, in a trial comparing high flexion to standard flexion total knee replacement, radiographic details rendering implants differentiable may be masked to those measuring knee flexion on maximum bend knee lateral radiographs. Last, outcome criteria may be chosen so that although blinding is not feasible it will have little effect on the evaluation of the outcome. For instance, the Harris hip score [32] measured by a blinded outcome assessor may be better than the self-reported modified version of the same score [33] if patients are not blinded. Similarly, the occurrence of revision surgery may be less biased than nonblinded evaluation of radiographic loosening if investigators wish to assess aseptic loosening.

Investigators should be particularly careful when reporting the results of trials evaluating LLA. Careful reporting of the design, conduction, and analysis of the trial is essential to assess the internal and external validity of the trial. The CONSORT statement provides essential guidelines to do so [34]. Reporting the trial with sufficient details is necessary for three reasons. First, it is necessary, before superiority may be chosen, that the trial be reproduced by other investigators and findings confirmed. Second, careful reporting of important details such as the population studied, participating centers, care providers, the surgical technique, and relevant perioperative care will help define the applicability of the results. Last, when trials comparing similar treatments give discordant results, it will be easier to appreciate the reasons of the difference. A specific scale has been described for evaluating the report of nonpharmacological RCT [35].

18.3 Practical Examples

18.3.1 New Highly Reticulated Polyethylene Acetabular Component

---

Evaluation of a New Highly Reticulated Polyethylene Acetabular Component

Context: Polyethylene wear leads to bone resorption and increased risk for aseptic loosening.

Objective: To compare wear of highly reticulated to standard polyethylene in uncemented metal-back press-fit acetabular component.

Primary outcome: Annual two-dimensional computerized polyethylene wear during the second year.

- A double-blind RCT was chosen.
- A conventional polyethylene was chosen for comparison since this is the standard acetabular wear surface.
- Primary outcome criteria is the annual two-dimensional computerized polyethylene wear during the second year. It is measured with a validated method by blinded outcome assessors. Measurements are conducted at 12, 24, 36, and 60 months.
- All surgeons and centers recruited for participation were given specific information with regards to the product. Because implantation

of the component is the same as that of the comparator, no specific learning curve was expected.

- To avoid issues relating to implantation and to maximize accrual rate only high volume, surgeons were selected for this experimental trial.
- Only uncemented metal-back press-fit acetabular components were provided. Wear rates are known to differ between uncemented and cemented implants. Although theoretically stratification and adjustment could account for this confusion factor, as the trial was experimental, it was preferred not to include the cemented version. Possible diameter of femoral heads were 28 and 32 mm. Uncemented acetabular cups were identical and from the same manufacturer in both groups. The femoral stem was uncemented for all operations but no specific brand was imposed. Despite the important variation expected in surgical approaches and postoperative recommendations, no standardization was attempted because of the difficulty to obtain agreement between surgeons and because of the absence of relevant effect of these factors on the primary outcome.
- Due to the exact macroscopic and radiographic resemblance between both implants, patients, surgeons, and outcome assessors were blinded to the treatment group.
- An intention to treat analysis was planned to look for a significant difference; however, in case a significant difference was demonstrated, the treatment effect will be measured on a per protocol analysis.
- Only patients with osteoarthritis, traumatic arthritis, avascular necrosis of the hip proposed a primary hip replacement were selected. No age limit was used. Only one side on a same patient could be included.
- Patients were given the information that the newly tested product may wear less but has reduced fatigue strength properties. Therefore, it is not known whether this decrease in wear will affect the risk for revision positively. However, it is important to demonstrate and quantify the effect on wear first. Improvement in function and relief of pain is expected to be similar and excellent in both groups. Both products are expected to work well for a long time. The patient will not be able to feel the difference between both products because the difference lies at a microscopic level.
- Randomization will be stratified on centers, on cup size, and on head diameter.
- Patient lost to follow-up and patients with missing data will be included in the intention to treat analysis. Missing primary outcome criteria will be estimated from multiple imputation in linear models for longitudinal data.

18.3.2 Evaluation of Computer-Assisted Surgery in Primary Total Knee Replacement

Context: Long-term result and survival of total knee replacement may depend on frontal alignment of the lower limb defined by a line joining the center of the femoral head to the center of the ankle joint. A range of 3° valgus to 3° varus is considered as an optimal technical goal.

Objective: To compare conventional surgery based on mechanical devices to achieve the different cuts necessary to implant a total knee prosthesis to a computer-assisted surgery that allows achievement of the cuts based on data introduced in a computer during surgery.

Primary outcome: Frontal alignment of the lower limb measured on long leg standing radiographs. This angle is measured between two lines one joining the center of the femoral head to the center of the knee defined by a point between the two tibial spines and another joining the center of the knee to the center of the ankle joint.

A RCT appeared the most adapted despite the main following issues:

- The choice of the primary outcome: Frontal alignment of the limb measured on long leg weight-bearing radiographs is the most frequent primary outcome. It has the advantage of being obtained very quickly (6 months or less) and answer in that way to some of the objectives of the companies that finance research and development. Most of the randomized studies considering this primary outcome and the meta-analyses conducted from these randomized studies showed an advantage in using computer-assisted surgery [36]. However, the relationship between this outcome and clinical result or longevity of the prosthesis remains to a great extent uncertain. Some more recent studies tried to focus on clinical result. Unfortunately, the study sample was initially calculated to show differences in coronal alignment and therefore was not adapted to the clinical question that needs much larger sample sizes because of much less important anticipated difference between groups [37,38]. Moreover, the relationship between the measured angles on long leg weight-bearing radiographs and the preoperative measurements obtained with the computer are not strictly related as suggested by Yaffe et al. [39].
- Longevity of the prosthesis represent from the patient point of view a relevant outcome, it is usually measured by survivorship analysis. Most of the published series as well as results from Scandinavian

registers report a survivorship reaching 95% at 10 years [40] with conventional jig-based surgery. In that respect, demonstration of improvement by computer-assisted surgery needs considerable amount of patients followed during a very long period of time, which is practically impossible to do.

- Standardization of the intervention: This crucial problem includes several aspects. For conventional surgery, ancillary devices are specific of one type of prosthesis, they are not used the same way by different surgeons; as an example, some surgeons prefer intramedullary guidance for cuttings, whereas some prefer extramedullary devices. On the other hand, it is still not demonstrated that all computer-assisted systems have the same performance.

- Senior surgeons who are confident in their technique are more reluctant in accepting a standardized technique that is not exactly the way they usually do.

Some technical aspects such as exposure, ligament balance, which is qualitatively assessed, are important factors that may influence the technical and clinical result. These aspects can be hardly standardized.

References

1. Charnley, J. Arthroplasty of the hip. A new operation. *Lancet*, 1, 1129–1132, 1961.
2. Learmonth, I.D., Young, C., Rorabeck, C. The operation of the century: Total hip replacement. *Lancet*, 370, 1508–1519, 2007.
3. Iorio, R., Robb, W.J., Healy, W.L., et al. Orthopaedic surgeon workforce and volume assessment for total hip and knee replacement in the United States: Preparing for an epidemic. *J Bone Joint Surg Am*, 90, 1598–1605, 2008.
4. Clarke, H., Kay, J., Orser, B.A., et al. Gabapentin does not reduce preoperative anxiety when given prior to total hip arthroplasty. *Pain Med*, 11(6), 966–971, 2010.
5. Gill, S.D., McBurney, H., Schulz, D.L. Land-based versus pool-based exercise for people awaiting joint replacement surgery of the hip or knee: Results of a randomized controlled trial. *Arch Phys Med Rehabil*, 90, 388–394, 2009.
6. Kazemi, S.M., Mosaffa, F., Eajazi, A., et al. The effect of tranexamic acid on reducing blood loss in cementless total hip arthroplasty under epidural anesthesia. *Orthopedics*, 33, 17, 2010.
7. Chauhan, S.K., Scott, R.G., Breidahl, W., et al. Computer-assisted knee arthroplasty versus a conventional jig-based technique. A randomised, prospective trial. *J Bone Joint Surg Br*, 86, 372–377, 2004.
8. Garcia-Rey, E., Garcia-Cimbrelo, E., Cruz-Pardos, A., et al. New polyethylenes in total hip replacement: A prospective, comparative clinical study of two types of liner. *J Bone Joint Surg Br*, 90, 149–153, 2008.

9. Gioe, T.J., Glynn, J., Sembrano, J., et al. Mobile and fixed-bearing (all-polyethylene tibial component) total knee arthroplasty designs. A prospective randomized trial. *J Bone Joint Surg Am*, 91, 2104–2112, 2009.

10. Ogonda, L., Wilson, R., Archbold, P., et al. A minimal-incision technique in total hip arthroplasty does not improve early postoperative outcomes. A prospective, randomized, controlled trial. *J Bone Joint Surg Am*, 87, 701–710, 2005.

11. Stockton, K.A., Mengersen, K.A. Effect of multiple physiotherapy sessions on functional outcomes in the initial postoperative period after primary total hip replacement: A randomized controlled trial. *Arch Phys Med Rehabil*, 90, 1652–1657, 2009.

12. Devereaux, P.J., Bhandari, M., Clarke, M., et al. Need for expertise based randomised controlled trials. *BMJ*, 330, 88, 2005.

13. Bednarska, E., Bryant, D., Devereaux, P.J. Orthopaedic surgeons prefer to participate in expertise-based randomized trials. *Clin Orthop Relat Res*, 466, 1734–1744, 2008. Comment in *Clin Orthop Relat Res*, 467, 298–300, 2009; author reply 301–292.

14. Lim, E. Need for expertise based randomised controlled trials: Expertise based design has shortfalls. *BMJ*, 330, 791–792, 2005.

15. Berry, D.J., Harmsen, W.S., Cabanela, M.E., et al. Twenty-five-year survivorship of two thousand consecutive primary Charnley total hip replacements: Factors affecting survivorship of acetabular and femoral components. *J Bone Joint Surg Am*, 84-A, 171–177, 2002.

16. Ling, R.S., Charity, J., Lee, A.J., et al. The long-term results of the original Exeter polished cemented femoral component: A follow-up report. *J Arthroplasty*, 24, 511–517, 2009.

17. Martell, J.M., Berdia, S. Determination of polyethylene wear in total hip replacements with use of digital radiographs. *J Bone Joint Surg Am*, 79, 1635–1641, 1997.

18. Baldursson, H., Hansson, L.I., Olsson, T.H., et al. Migration of the acetabular socket after total hip replacement determined by roentgen stereophotogrammetry. *Acta Orthop Scand*, 51, 535–540, 1980.

19. Prentice, R.L. Surrogate endpoints in clinical trials: Definition and operational criteria. *Stat Med*, 8, 431–440, 1989.

20. Kim, Y.H., Choi, Y., Kim, J.S. Range of motion of standard and high-flexion posterior cruciate-retaining total knee prostheses a prospective randomized study. *J Bone Joint Surg Am*, 91, 1874–1881, 2009.

21. Hinrichs, F., Kuhl, M., Boudriot, U., et al. A comparative clinical outcome evaluation of smooth (10–13 year results) versus rough surface finish (5–8 year results) in an otherwise identically designed cemented titanium alloy stem. *Arch Orthop Trauma Surg*, 123, 268–272, 2003.

22. Rasquinha, V.J., Ranawat, C.S., Dua, V., et al. A prospective, randomized, double-blind study of smooth versus rough stems using cement fixation: Minimum 5-year follow-up. *J Arthroplasty*, 19, 2–9, 2004.

23. Biau, D.J., Halm, J.A., Ahmadieh, H., et al. Provider and center effect in multicenter randomized controlled trials of surgical specialties: An analysis on patient-level data. *Ann Surg*, 247, 892–898, 2008.

24. Goldstein, H., Browne, W., Rasbash, J. Multilevel modelling of medical data. *Stat Med*, 21, 3291–3315, 2002.

25. Lee, K.J., Thompson, S.G. Clustering by health professional in individually randomised trials. *BMJ*, 330, 142–144, 2005.

26. Birkmeyer, J.D., Siewers, A.E., Finlayson, E.V., et al. Hospital volume and surgical mortality in the United States. *N Engl J Med*, 346, 1128–1137, 2002.
27. Birkmeyer, J.D., Stukel, T.A., Siewers, A.E., et al. Surgeon volume and operative mortality in the United States. *N Engl J Med*, 349, 2117–2127, 2003.
28. Biau, D.J., Porcher, R., Boutron, I. The account for provider and center effects in multicenter interventional and surgical randomized controlled trials is in need of improvement: A review. *J Clin Epidemiol*, 61, 435–439, 2008.
29. Vierron, E., Giraudeau, B. Sample size calculation for multicenter randomized trial: Taking the center effect into account. *Contemp Clin Trials*, 28, 451–458, 2007.
30. Kim, Y.H., Kim, D.Y., Kim, J.S. Simultaneous mobile- and fixed-bearing total knee replacement in the same patients. A prospective comparison of mid-term outcomes using a similar design of prosthesis. *J Bone Joint Surg Br*, 89, 904–910, 2007.
31. Nunley, R.M., Zhu, J., Brooks, P.J., et al. The learning curve for adopting hip resurfacing among hip specialists. *Clin Orthop Relat Res*, 468, 382–391, 2010.
32. Harris, W.H. Traumatic arthritis of the hip after dislocation and acetabular fractures: Treatment by mold arthroplasty. An end-result study using a new method of result evaluation. *J Bone Joint Surg Am*, 51, 737–755, 1969.
33. Mahomed, N.N., Arndt, D.C., McGrory, B.J., et al. The Harris hip score: Comparison of patient self-report with surgeon assessment. *J Arthroplasty*, 16, 575–580, 2001.
34. Moher, D., Schulz, K.F., Altman, D.G. The CONSORT statement: Revised recommendations for improving the quality of reports of parallel-group randomised trials. *Lancet*, 357, 1191–1194, 2001.
35. Boutron, I., Moher, D., Tugwell, P., et al. A checklist to evaluate a report of a nonpharmacological trial (CLEAR NPT) was developed using consensus. *J Clin Epidemiol*, 58, 1233–1240, 2005.
36. Mason, J.B., Fehring, T.K., Estok, R., et al. Meta-analysis of alignment outcomes in computer-assisted total knee arthroplasty surgery. *J Arthroplasty*, 22, 1097–1106, 2007.
37. Lutzner, J., Gunther, K.P., Kirschner, S. Functional outcome after computer-assisted versus conventional total knee arthroplasty: A randomized controlled study. *Knee Surg Sports Traumatol Arthrosc*, 18(10), 1339–1344, 2010.
38. Spencer, J.M., Chauhan, S.K., Sloan, K., et al. Computer navigation versus conventional total knee replacement: No difference in functional results at two years. *J Bone Joint Surg Br*, 89, 477–480, 2007.
39. Yaffe, M.A., Koo, S.S., Stulberg, S.D. Radiographic and navigation measurements of TKA limb alignment do not correlate. *Clin Orthop Relat Res*, 466, 2736–2744, 2008.
40. Robertsson, O., Bizjajeva, S., Fenstad, A.M., et al. Knee arthroplasty in Denmark, Norway and Sweden. *Acta Orthop*, 81, 82–89, 2010.

19

Assessing Radiation Therapy: Practical Examples

Hidefumi Aoyama

Niigata University Graduate School of Medical and Dental Sciences

CONTENTS

19.1 Introduction

Radiation therapy (RT) constitutes essential part in cancer management together with surgery and chemotherapy. Modest estimation suggests that at least more than half of cancer sufferers receive RT in the course of whole treatment. RT is used for variety of cancers arising from any kind of organs: central nervous system, head and neck, lung, breast, pancreas, liver, esophagus, stomach, colorectal, prostate, cervical, and so on. RT is also used for different aims: definitive, palliative, or symptom relief, and occasionally combined with chemotherapy or surgery. With regard to the design of RCTs, ~60% deals with different RT techniques or doses of radiation, and the rest are designed to investigate the effect of drugs combined with RT [1].

In the last few decades, a number of new RT modalities, such as stereotactic radiosurgery (SRS), intensity-modulated radiation therapy (IMRT), and proton therapy have emerged. Since those new methods are quite appealing to both physicians and patients, some of them gained popularity before they went through scientifically rigid evaluation processes, randomized control trials (RCTs). SRS is typical to one such example. SRS is a technique to administer precisely directed, high-dose irradiation that tightly conforms to an intracranial target to create a desired radiobiological response, while minimizing radiation dose to surrounding normal tissue. During the last decades, it has been increasingly used for brain malignancies including gliomas and metastatic tumors, arteriovenous malformation, and some other benign tumors including vestibular schwannoma, pituitary adenoma, and meningiomas. The treatment indication is now expanding to extracranial moving tumors such as early-stage lung cancer, liver tumors, and prostate cancer. Compared with this rapid pace in gaining popularity, there are only few RCTs published, including four for brain metastases [2–5], and one for glioblastoma [6] (Table 19.1). Same can be said to IMRT and proton therapy. IMRT is a most advanced form of photon-based RT, which is capable of normal tissue sparing while maintaining or even increasing radiation dose to the irregularly shaped target. Surprisingly, there are only four RCTs that evaluated the benefit of IMRT over conventional RT [7–10,15]. There are four RCTs for proton therapy as well although all of those four trials compared only radiation doses but they were not designed to compare proton with photon RT including IMRT [11–14,16]. In this chapter, issues around nonpharmacological RCTs involving RT will be discussed, with a special focus on the causes that make it difficult to perform RCTs between new RT techniques and conventional ones.

19.2 Specific Methodological Issues in Assessing This Treatment

19.2.1 Difficulty of Blinding

The methodological guidelines in conducting clinical trials have been developed mainly for pharmacological studies. Partly due to this, some of their methodologies do not fit well into RT trials. Difficulty of blinding is one of the characteristics of nonpharmacological RCTs involving RT. It is considered that a trial is "double-blinded" when all of the patients, care providers, and outcome assessors are unaware of the assigned treatment

TABLE 19.1

Randomized Control Trials Involving New Radiation Therapies, Including Stereotactic Radiosurgery (SRS), Intensity-Modulated Radiation Therapy (IMRT), and Proton Therapy

Modality	Disease	Treatment	N Trials	Publication Year	N	Primary End Point	Blinding	Ref.
SRS	Brain metastasis	SRS+WBRT vs. SRS	5	2006	132	OS	No	[2]
		SRS+WBRT vs. WBRT		2004	333	OS	No	[3]
				1999	27	TC	No	[4]
		Surgery+WBRT vs. SRS		2008	62	OS	No	[5]
	Glioblastoma	SRS+RT vs. RT		2004	203	OS	No	[6]
IMRT	Nasopharyngeal cancer	IMRT vs. conventional RT	4	2007	60	Late AE	No	[7]
				2006	51	Late AE	No	[8]
	Breast cancer	IMRT vs. conventional RT		2008	358	Acute AE	Yes	[9]
				2007	306	Late AE	No	[10]
Proton	Ocular melanoma	High vs. low dose	4	2000	188	Late AE	Yes	[11]
	Skull base tumors	High vs. low dose		1998	96	Late AE	No	[12]
	Prostate cancer	High vs. low dose		2005	393	PFS	No	[14]
		High vs. low dose		1995	202	OS, PFS	No	[13]

RT, radiation therapy; SRS, stereotactic radiosurgery; IMRT, intensity-modulated radiation therapy; OS, overall survival; TC, tumor control; AE, adverse effect; PFS, progression-free survival.

throughout the trial [17]. If this rather strict definition is applied to 22 nonpharmacological RCTs involving RT published in three leading medical journals (*New England Journal of Medicine*, *JAMA*, and *Lancet*) between 2000 and January 2009 [2,3,14,18–36], none of them successfully blinded care providers, radiation oncologists in this situation, to the assigned treatment, and therefore it could be said that they did not achieve "double-blinding" (Table 19.2). The use of placebos is considered the established method to achieve "double-blinding" in pharmacological RCTs. In those 22 RCTs, however, "placebo RT" was used only in one trial. This was a study designed to investigate the role of RT for symptom relief of Graves' orbitopathy, in which "sham RT" was used and the assigned treatment was blinded to patients and ophthalmologist who assessed the treatment outcome [33]. Among 13 RCTs involving SRS, IMRT, or proton therapy (Table 19.1), none of them achieved "double-blinding" [9,11] according to that strict definition; however, two of them blinded the patients and the outcome assessors to the assigned treatment. One is the RCT of IMRT vs. conventional RT for breast cancer conducted by Pignol and colleagues [9], and the other is high-dose vs. low-dose proton therapy for choroidal melanoma [11]. The patients and outcome assessors of primary end points, which were skin reaction in the former one [9] or visual function in the latter one [11], were blinded to the type of RT which the patients received. It is worth mentioning that primary end points of these trials had little to do with the survival or the tumor control, and more to do with radiation-induced adverse effects or symptom relief. In my view, the blinding technique could be applied to the nonpharmacological RCTs involving RT, only when the end points were not life-threatening events. In addition, blinding the care providers (radiation oncologists) would be quite difficult, presumably because of the safety concerns.

19.2.2 Selection of End Point

RT is used for aiming not only at extending overall survival (OS) or progression-free survival (PFS), but also at the palliative effects such as relied from the pain of bone metastases. Therefore, selection of end points should reflect the aim of the treatment. In a review of 22 publications (Table 19.2), there was a trend that the end points related to survival (OS or cause-specific survival) were chosen in RCTs dealing with diseases in which only short survival could be expected (brain metastases, glioblastoma, pancreas caner, or other advanced cancers). On the other hand, end points related to tumor control (disease-free survival, local tumor control, metastases-free survival, etc.) were preferably used for the diseases in which longer survival can be expected (breast or prostate cancer). Among 13 trials involving new RT techniques (Table 19.1), treatment-related morbidities were also used for the end points. For instance, all four RCTs of IMRT employed toxicity-related outcomes as primary end points [15], the acute skin reaction [9] or the cosmetic outcome [10] in breast cancer trials, and the salivary gland function in both nasopharyngeal cancer trials [7,8]. In reporting treatment-related morbidity in cancer

TABLE 19.2

Nonpharmacological RCTs Involving RT Published in *NEJM*, *JAMA*, and *Lancet* between January 2000 and January 2009

First Author	Year	Journal	Ref.	Disease	Aim of RT	Survival	Primary End Points				Comparison	Use of Placebo	Blinding		
							Survival	Disease Control	Adverse Events	Symptom			Patients	Provider	Assessor
Andrews	2004	*Lancet*	[3]	Brain metastasis	Palliative	Median, 6 months	○				RT regimen	×	×	×	×
Aoyama	2006	*JAMA*	[2]	Brain metastasis	Palliative	Median, 8 months	○				RT regimen	×	×	×	×
Slotman	2007	*NEJM*	[18]	ED-SCLC, Brain metastasis	Prophylactic	Median, 6 months		○		○	RT vs. no RT	×	×	×	×
Keime-Guibert	2007	*NEJM*	[19]	GBM, elderly	Palliative	Median, 6 months	○				RT vs. no RT	×	×	×	×
Neoptolemos	2004	*NEJM*	[20]	Pancreas	Definitive	Median, 17 months	○				RT vs. no RT	×	×	×	×
van den Bent	2005	*Lancet*	[21]	LGG	Definitive	Median, 7 months	○	○			RT vs. no RT	×	×	×	×
Julien	2000	*Lancet*	[22]	Breast, DCIS	Definitive	95% at 10 years		○			RT vs. no RT	×	×	×	×
Bartelink	2001	*NEJM*	[23]	Breast, early	Definitive	80% at 10 years		○			RT dose	×	×	×	×
Fisher	2002	*NEJM*	[24]	Breast, early	Definitive	47% at 20 years		○			RT vs. no RT	×	×	×	×
Veronesi	2002	*NEJM*	[25]	Breast, early	Definitive	58% at 20 years		○			RT vs. no RT	×	×	×	×
Fyles	2004	*NEJM*	[26]	Breast, early	Definitive	93% at 5 years		○			RT vs. no RT	×	×	×	×

(continued)

TABLE 19.2 (continued)

Nonpharmacological RCTs Involving RT Published in *NEJM*, *JAMA*, and *Lancet* between January 2000 and January 2009

First Author	Year	Journal	Ref.	Disease	Aim of RT	Survival	Sur-vival	Disease Control	Adverse Events	Symptom	Comparison	Use of Placebo	Patients	Provider	Assessor
								Primary End Points					**Blinding**		
Hughes	2004	*NEJM*	[27]	Breast, early, elderly (>70)	Definitive	87% at 5 years	○	○			RT vs. no RT	×	×	×	×
Bentzen	2008	*Lancet*	[28]	Breast, early	Definitive	90% at 5 years		○			RT regimen	×	×	×	Partial
Zietman	2005	*NEJM*	[14]	Prostate	Definitive	96% at last years	○	○	○		RT dose	×	×	×	×
Bolla	2005	*Lancet*	[29]	Prostate	Definitive	92% at 5 years	○	○			RT vs. no RT	×	×	×	×
Thompson, Jr.	2006	*JAMA*	[30]	Prostate, locally advanced	Definitive	MS, 13.8–14.7 years	○	○			RT vs. no RT	×	×	×	×
Widmark	2009	*Lancet*	[31]	Prostate, locally advanced	Definitive	29%–39% at 10 years	○	○			RT vs. no RT	×	×	×	×
Creutzberg	2000	*Lancet*	[32]	Endome trial	Definitive	81%–85% at 5 years	○	○			RT vs. no RT	×	×	×	×
Mourits	2000	*Lancet*	[33]	Graves' orbitopathy	Definitive	Not applicable				○	RT vs. no RT	○	○	×	○
Overgaard	2003	*Lancet*	[34]	HN	Definitive	50% at 5 years	○	○			RT regimen	×	×	×	×
Sauer	2004	*NEJM*	[35]	Rectal	Definitive	75% at 5 years	○				RT timing	×	×	×	×
Macdonald	2001	*NEJM*	[36]	Stomach, resected	Definitive	Median, 27–36 weeks	○				RT vs. no RT	×	×	×	×

management, standardized grading system of the Common Terminology Criteria for Adverse Events (CTCAE) version 3.0 is clinically used, which defines terms and grades for late effects of cancer therapy. However, those criteria are sometimes not sensitive enough to describe some differences among different RT methods. In such cases, other biomarkers should be used depending on what the investigators trying to prove. One of the examples is neurocognitive function, which is increasingly used for RCTs of CNS tumors including brain metastasis [37] and low-grade gliomas [38]. The other example of the end points is the patient-rated health-related quality of life (HRQOL), which is also increasingly used as an end point in RCTs of RT [18,19].

19.2.3 Lack of Equipoise

In general, to design and include patients in RCTs require that investigators are uncertain about the effect [1,39]. Obstacle that prevents us from conducting new vs. conventional RT techniques could be summarized that "The problem with many potential proton vs. photon trials is that all foreseeable benefits—and no specific risks to counterbalance these—are associated with the proton arm" as pointed out by Bentzen [40]. This idea can be easily expanded to SRS or IMRT vs. conventional RT trials. Since all of those new modalities are proven to generate better physical dose distribution compared with conventional ones, most of physicians, including radiation oncologists, tend to believe that these better dose distributions could be directly translated into better treatment outcomes in an actual clinical situation. This belief could be one of the obstacles of conducting the head-to-head comparison, such as proton vs. photon or SRS/IMRT vs. conventional RT.

19.2.4 Recruitment of Patients and Care Providers

Recruitment of patients and care providers is sometimes difficult. One of the typical examples is proton or other heavy particle therapy. It is because radiation oncologists see patients who are referred to their institutions aiming at receiving proton therapy; therefore, it would be quite difficult to persuade patients to receive conventional photon RT instead of proton therapy. Moreover, some of care providers working in proton center are not willing to perform conventional photon RT. We have to take the cost of the treatment into consideration as well. As an example, the cost of particle therapy including proton is ~10 times higher than conventional RT and four times higher than IMRT in Japanese health insurance system. In addition, the cost of particle therapies is sometimes not subsidized by national health insurance for some cancers. Therefore, the issue of the medical cost could not be neglected in conducting RCTs involving different RT techniques. Nevertheless, I think we should conduct RCTs between high-tech RT vs. low-tech RT especially IMRT or proton, because there is no strong evidence that dose escalation by using high-tech RT could really improve the tumor control. In addition, none of us are aware if the dose reduction to the normal tissue obtained from proton therapy could be directly translated into the significant difference of developing normal tissue complications.

19.3 Practical Examples

19.3.1 Whole Brain Radiation Therapy

Stereotactic Radiosurgery (SRS) with/without Whole Brain Radiation Therapy (WBRT) for 1–4 Brain Metastatic Patients [2]

Context: In patients with brain metastases, it is unclear whether adding up-front WBRT to SRS has beneficial effects on mortality or neurological function compared with SRS alone.

Objective: To determine if WBRT combined with SRS results in improvements in survival, brain tumor control, functional preservation rate, frequency of neurological death, and neurocognitive function.

Primary outcome: Overall survival.

What are the specific issues and possible solutions in planning a trial of radiation oncology?

19.3.1.1 Choice of the Design, Comparator

When this trial was designed, there were two "standard" treatments for patients with brain metastases that are limited in number. One was scientifically supported treatment (WBRT+SRS) and the other was community-supported treatment (SRS alone). There had been numerous retrospective data comparing both approaches, which showed conflicting outcome regarding some aspects, including OS. Therefore, we considered that RCT would be the only solution for this situation.

19.3.1.2 Standardization of the Intervention

It was not easy to decide which radiation dose should be regarded as standard. In term of radiation dose of WBRT, 30 Gy in 10 fractions had been a standard for a decade; therefore, it was not difficult to decide to employ this regimen. The problem was the dose of SRS. The SRS dose is usually prescribed to the tumor margin. In the 1990s, the RTOG initiated a phase I dose escalation trial of SRS for brain metastasis [41]. This trial was stopped early before reaching the maximum tolerance dose (MTD). However, they stated size-dependent dose recommendations for SRS; metastases with a maximum diameter of up to 2 cm should be treated with 22–25 and 18–20 Gy for those larger than 2 cm. Therefore, there was no well-known or scientifically recommended dose of SRS combined with WBRT. Even though, we decided to employ 30% reduction dose when SRS was combined with WBRT because of the safety concern based on our experience.

19.3.1.3 Recruitment of Patients and Care Providers

Recruitment of care provider was the first obstacle in designing this RCT because most of the hospitals that have SRS-dedicated machines do not have conventional radiotherapy machine. Physicians in those hospitals had strong opinion against the use of WBRT to brain metastatic patients. Obviously, their opinion was not supported by any scientific evidence, though we gave up having them in this trial, and only the hospitals—most of them were academic ones—involved in this trial.

19.3.1.4 Blinding Patients, Care Providers, or Assessors

Blinding of patients, care providers, as well as outcome assessors was not applied in this study because it was expected that more than half of patients would receive salvage treatment for brain tumor recurrence, and accurate information of RT was indispensable to consider the possible options for the salvage treatment. Blinding assessor could have been used for the assessment of adverse effect to the brain on MRI or the assessment of neurocognitive function. The reason why we did not do this was that those end points were secondary important in this study. However, after this trial was published, neurocognitive function has been frequently used as an end point in the trials dealing with brain metastasis. If neurocognitive function was used as a primary end point, blinding the outcome assessors would be possible.

19.3.2 Breast Intensity-Modulated Radiation Therapy

IMRT versus Conventional RT for Early Breast Cancer Patients [9]

Context: Dermatitis is a frequent adverse effect of adjuvant breast radiotherapy. It is more likely in full-breasted women and when the radiation is distributed non-homogeneously in the breast. Breast IMRT is a technique that ensures a more homogeneous dose distribution.

Objective: To investigate if breast IMRT would reduce the rate of acute skin reaction (notably moist desquamation), decrease pain, and improve quality of life compared with standard radiotherapy using wedges.

Primary outcome: The intensity of acute skin reaction or pain using the National Cancer Institute Common Toxicity Criteria (NCI CTC) version 2.0 scale 22 and the occurrence of moist desquamation.

This was one of few RCTs comparing high-tech and traditional RT techniques, in which IMRT was compared with conventional wedged-pair technique. In this trial, both patients and outcome assessor were blinded to the treatment arm. In addition, the outcome assessment was performed in a separate clinic. This would be one of the solutions for the difficulty of blinding in nonpharmacological RCTs involving RT. However, this could be possible because the primary end point of this trial was non-threading event and the role of early-stage breast cancer had been well-established with satisfactory long-term survival.

19.4 Conclusion

If I summarize the characteristics of nonpharmacological RCT involving RTs, one is the difficulty of applying the blinding techniques. However, a careful choice of end point might possibly the use of blinding technique, at least to the patients and the outcome assessors. A "myth" that high-tech RTs are better than low-tech ones is another obstacle to conduct RCTs. Significant difference in the treatment cost of proton, IMRT, SRS, and conventional RT is also of negligible factors. Nevertheless, we need to cope with those difficulties in order to get rid of the "myth" that something new is always better.

References

1. Soares, H.P., Kumar, A., Daniels, S., et al. Evaluation of new treatments in radiation oncology: Are they better than standard treatments? *JAMA*, 293(8), 970–978, 2005.
2. Aoyama, H., Shirato, H., Tago, M., et al. Stereotactic radiosurgery plus whole-brain radiation therapy vs stereotactic radiosurgery alone for treatment of brain metastases: A randomized controlled trial. *JAMA*, 295(21), 2483–2491, 2006.
3. Andrews, D.W., Scott, C.B., Sperduto, P.W., et al. Whole brain radiation therapy with or without stereotactic radiosurgery boost for patients with one to three brain metastases: Phase III results of the RTOG 9508 randomised trial. *Lancet*, 363(9422), 1665–1672, 2004.
4. Kondziolka, D., Patel, A., Lunsford, L.D., Kassam, A., and Flickinger J.C., et al. Stereotactic radiosurgery plus whole brain radiotherapy versus radiotherapy alone for patients with multiple brain metastases. *Int J Radiat Oncol Biol Phys*, 45(2), 427–434, 1999.
5. Muacevic, A., Wowra, B., Siefert, A., et al. Microsurgery plus whole brain irradiation versus Gamma Knife surgery alone for treatment of single metastases to the brain: A randomized controlled multicentre phase III trial. *J Neurooncol*, 87(3), 299–307, 2008.

6. Souhami, L., Seiferheld, W., Brachman, D., et al. Randomized comparison of stereotactic radiosurgery followed by conventional radiotherapy with carmustine to conventional radiotherapy with carmustine for patients with glioblastoma multiforme: Report of Radiation Therapy Oncology Group 93-05 protocol. *Int J Radiat Oncol Biol Phys*, 60(3), 853–860, 2004.

7. Kam, M.K., Leung, S.F., Zee, B., et al. Prospective randomized study of intensity-modulated radiotherapy on salivary gland function in early-stage nasopharyngeal carcinoma patients. *J Clin Oncol*, 25(31), 4873–4879, 2007.

8. Pow, E.H., Kwong, D.L., McMillan, A.S., et al. Xerostomia and quality of life after intensity-modulated radiotherapy vs. conventional radiotherapy for early-stage nasopharyngeal carcinoma: Initial report on a randomized controlled clinical trial. *Int J Radiat Oncol Biol Phys*, 66(4), 981–991, 2006.

9. Pignol, J.P., Olivotto, I., Rakovitch, E., et al. A multicenter randomized trial of breast intensity-modulated radiation therapy to reduce acute radiation dermatitis. *J Clin Oncol*, 26(13), 2085–2092, 2008.

10. Donovan, E., Bleakley, N., Denholm, E., et al. Randomised trial of standard 2D radiotherapy (RT) versus intensity modulated radiotherapy (IMRT) in patients prescribed breast radiotherapy. *Radiother Oncol*, 82(3), 254–264, 2007.

11. Gragoudas, E.S., Lane, A.M., Regan, S., et al. A randomized controlled trial of varying radiation doses in the treatment of choroidal melanoma. *Arch Ophthalmol*, 118(6), 773–778, 2000.

12. Santoni, R., Liebsch, N., Finkelstein, D.M., et al. Temporal lobe (TL) damage following surgery and high-dose photon and proton irradiation in 96 patients affected by chordomas and chondrosarcomas of the base of the skull. *Int J Radiat Oncol Biol Phys*, 41(1), 59–68, 1998.

13. Shipley, W.U., Verhey, L.J., Munzenrider, J.E., et al. Advanced prostate cancer: The results of a randomized comparative trial of high dose irradiation boosting with conformal protons compared with conventional dose irradiation using photons alone. *Int J Radiat Oncol Biol Phys*, 32(1), 3–12, 1995.

14. Zietman, A.L., DeSilvio, M.L., Slater, J.D., et al. Comparison of conventional-dose vs high-dose conformal radiation therapy in clinically localized adenocarcinoma of the prostate: A randomized controlled trial. *JAMA*, 294(10), 1233–1239, 2005.

15. Veldeman, L., Madani, I., Hulstaert, F., et al. Evidence behind use of intensity-modulated radiotherapy: A systematic review of comparative clinical studies. *Lancet Oncol*, 9(4), 367–375, 2008.

16. Olsen, D.R., Bruland, O.S., Frykholm, G., and Norderhaug, I.N., et al. Proton therapy—A systematic review of clinical effectiveness. *Radiother Oncol*, 83(2), 123–132, 2007.

17. Fergusson, D., Glass, K.C., Waring, D., and Shapiro, S., et al. Turning a blind eye: The success of blinding reported in a random sample of randomised, placebo controlled trials. *BMJ*, 328(7437), 432, 2004.

18. Slotman, B., Faivre-Finn, C., Kramer, G., et al. Prophylactic cranial irradiation in extensive small-cell lung cancer. *N Engl J Med*, 357(7), 664–672, 2007.

19. Keime-Guibert, F., Chinot, O., Taillandier, L., et al. Radiotherapy for glioblastoma in the elderly. *N Engl J Med*, 356(15), 1527–1535, 2007.

20. Neoptolemos, J.P., Stocken, D.D., Friess, H., et al. A randomized trial of chemoradiotherapy and chemotherapy after resection of pancreatic cancer. *N Engl J Med*, 350(12), 1200–1210, 2004.

21. van den Bent, M.J., Afra, D., de Witte, O., et al. Long-term efficacy of early versus delayed radiotherapy for low-grade astrocytoma and oligodendroglioma in adults: The EORTC 22845 randomised trial. *Lancet*, 366(9490), 985–990, 2005.

22. Julien, J.P., Bijker, N., Fentiman, I.S., et al. Radiotherapy in breast-conserving treatment for ductal carcinoma in situ: First results of the EORTC randomised phase III trial 10853. EORTC Breast Cancer Cooperative Group and EORTC Radiotherapy Group. *Lancet*, 355(9203), 528–533, 2000.

23. Bartelink, H., Horiot, J.C., Poortmans, P., et al. Recurrence rates after treatment of breast cancer with standard radiotherapy with or without additional radiation. *N Engl J Med*, 345(19), 1378–1387, 2001.

24. Fisher, B., Anderson, S., Bryant, J., et al. Twenty-year follow-up of a randomized trial comparing total mastectomy, lumpectomy, and lumpectomy plus irradiation for the treatment of invasive breast cancer. *N Engl J Med*, 347(16), 1233–1241, 2002.

25. Veronesi, U., Cascinelli, N., Mariani, L., et al. Twenty-year follow-up of a randomized study comparing breast-conserving surgery with radical mastectomy for early breast cancer. *N Engl J Med*, 347(16), 1227–1232, 2002.

26. Fyles, A.W., McCready, D.R., Manchul, L.A., et al. Tamoxifen with or without breast irradiation in women 50 years of age or older with early breast cancer. *N Engl J Med*, 351(10), 963–970, 2004.

27. Hughes, K.S., Schnaper, L.A., Berry, D., et al. Lumpectomy plus tamoxifen with or without irradiation in women 70 years of age or older with early breast cancer. *N Engl J Med*, 351(10), 971–977, 2004.

28. Bentzen, S.M., Agrawal, R.K., Aird, E.G., et al. The UK Standardisation of Breast Radiotherapy (START) Trial B of radiotherapy hypofractionation for treatment of early breast cancer: A randomised trial. *Lancet*, 371(9618), 1098–1107, 2008.

29. Bolla, M., van Poppel, H., Collette, L., et al. Postoperative radiotherapy after radical prostatectomy: A randomised controlled trial (EORTC trial 22911). *Lancet*, 366(9485), 572–578, 2005.

30. Thompson, I.M., Jr., Tangen, C.M., Paradelo, J., et al. Adjuvant radiotherapy for pathologically advanced prostate cancer: A randomized clinical trial. *JAMA*, 296(19), 2329–2335, 2006.

31. Widmark, A., Klepp, O., Solberg, A., et al. Endocrine treatment, with or without radiotherapy, in locally advanced prostate cancer (SPCG-7/SFUO-3): An open randomised phase III trial. *Lancet*, 373(9660), 301–308, 2009.

32. Creutzberg, C.L., van Putten, W.L., Koper, P.C., et al. Surgery and postoperative radiotherapy versus surgery alone for patients with stage-1 endometrial carcinoma: Multicentre randomised trial. PORTEC Study Group. Post operative radiation therapy in endometrial carcinoma. *Lancet*, 355(9213), 1404–1411, 2000.

33. Mourits, M.P., van Kempen-Harteveld, M.L., García, M.B., et al. Radiotherapy for Graves' orbitopathy: Randomised placebo-controlled study. *Lancet*, 355(9214), 1505–1509, 2000.

34. Overgaard, J., Hansen, H.S., Specht, L., et al. Five compared with six fractions per week of conventional radiotherapy of squamous-cell carcinoma of head and neck: DAHANCA 6 and 7 randomised controlled trial. *Lancet*, 362(9388), 933–940, 2003.

35. Sauer, R., Becker, H., Hohenberger, W., et al. Preoperative versus postoperative chemoradiotherapy for rectal cancer. *N Engl J Med*, 351(17), 1731–1740, 2004.

36. Macdonald, J.S., Smalley, S.R., Benedetti, J., et al. Chemoradiotherapy after surgery compared with surgery alone for adenocarcinoma of the stomach or gastroesophageal junction. *N Engl J Med*, 345(10), 725–730, 2001.
37. Meyers, C.A., Brown, P.D. Role and relevance of neurocognitive assessment in clinical trials of patients with CNS tumors. *J Clin Oncol*, 24(8), 1305–1309, 2006.
38. Brown, P.D., Buckner, J.C., O'Fallon, J.R., et al. Effects of radiotherapy on cognitive function in patients with low-grade glioma measured by the folstein minimental state examination. *J Clin Oncol*, 21(13), 2519–2524, 2003.
39. Glimelius, B., Montelius, A. Proton beam therapy—Do we need the randomised trials and can we do them? *Radiother Oncol*, 83(2), 105–109, 2007.
40. Bentzen, S.M. Randomized controlled trials in health technology assessment: Overkill or overdue? *Radiother Oncol*, 86(2), 142–147, 2008.
41. Shaw, E., Scott, C., Souhami, L., et al. Radiosurgery for the treatment of previously irradiated recurrent primary brain tumors and brain metastases: Initial report of radiation therapy oncology group protocol (90–05). *Int J Radiat Oncol Biol Phys*, 34(3), 647–654, 1996.

20

Assessing Electroconvulsive Therapy: Practical Examples

Keith G. Rasmussen

Mayo Clinic

CONTENTS

20.1 Introduction

Electroconvulsive therapy (ECT) has been in use for psychiatric disorders since 1938 and is used virtually around the world [1]. Arguably, it may be considered the single most important intervention, on a global basis, for severe states of mental illness because of the lack of readily available psychotropic medications in impoverished areas. ECT is an important treatment in modern psychiatry. It is estimated that ~100,000 patients a year in the United States are given ECT [1]. The majority of these patients suffer from major depression, which according to WHO estimates will be the second leading cause of disability by 2020 [2]. Research publications emanate from all populated continents. Herewith, I shall briefly describe ECT to familiarize the reader with the indications, apparatus, and treatment techniques. Then I will discuss the various considerations in designing scientifically rigorous trials.

ECT involves the application of an electrical stimulus to a patient's head to induce a generalized seizure. ECT was developed based on a theory of biologic incompatibility between seizures and psychosis [1]. Thus, it was theorized that psychotic patients, if given seizures, would no longer be psychotic.

Though this was a flawed argument, the procedure turned out to be dramatically effective for a variety of patients with severe psychiatric syndromes.

ECT is given as a series of treatments, typically three times per week until clinical remission occurs. It is most commonly used for patients with severe depression but also for patients with catatonia, mania, or schizophrenia [3]. The procedure is performed under general anesthesia. Thus, the first step for the patient is to be anesthetized. Most commonly, this is accomplished with an intravenous anesthetic agent such as a barbiturate (thiopental and methohexital are the most common), etomidate, propofol, or ketamine. Occasionally, adjunctive use of high-potency opioids may be used to lessen the dosage needed of the anesthetic agent. Also, inhalational anesthesia may be used but is uncommon.

After the patient is anesthetized, ventilation is needed and this is accomplished with bag and mask. Patients are not routinely intubated during ECT. Monitoring includes continuous electrocardiogram (ECG), pulse oximetry, and blood pressure. Since the purpose of ECT is to induce a generalized seizure, it is desirable to block the convulsive movements that may predispose to fractures, so muscular paralysis is undertaken. This is accomplished with succinylcholine in the vast majority of cases. Once the anesthetist confirms the presence of anesthesia and then paralysis, the psychiatrist commences with seizure induction. Oxygenation via positive pressure ventilation continues until resumption of spontaneous breathing after the end of the seizure.

Seizure induction occurs by passing an electrical current through the brain. This is accomplished by placing two electrodes on the patient's head, which are connected via a thin cable to one of several commercially available ECT machines. When the button on the machine is pressed, an electrical current is passed through the electrodes into the brain. Seizure activity is confirmed by inspection of a continuous output electroencephalogram (EEG) record, which is available on the ECT machine. Typically, an ECT seizure lasts from 20 to 60 s and stops spontaneously. Resumption of breathing occurs within the next few minutes with consciousness following shortly thereafter. Treatments can be administered on an outpatient basis as long as the patient is accompanied by a responsible adult.

As indicated earlier, this process is repeated usually two to three times per week until remission of the clinical syndrome occurs. This course of treatments is termed an *index* course of treatments as it treats the index episode of illness. As most psychiatric conditions tend to be recurrent, some sort of prophylactic treatment is instituted after remission in order to prevent relapse. One of these options is a technique known as *maintenance* ECT, in which the treatments are given at spaced intervals (usually once a week to once a month) after completion of the index course of treatments in order to prevent a new illness episode. Maintenance ECT may last from a few weeks to many years, the latter for patients with otherwise highly chronic, recurring illnesses.

20.2 Specific Methodological Issues in Assessing This Treatment

In designing controlled ECT studies, we of course wish to make conclusions about the presumed true inherent biologically mediated therapeutic action of the procedure separate from the natural history of the condition being studied, alternate therapeutic interventions that may be operative without the investigators' knowledge, or expectations of the patient (i.e., placebo effects). Regarding ECT, particular challenges include the blinding of patients, the blinding of outcome assessors and others involved in the patient care process, the choice of placebo (herein referred to as sham) control methods, and controlling for other treatment modalities. It is these issues that we will broach with specific examples from the published literature.

20.3 Practical Examples

20.3.1 Establishing the Efficacy of Electroconvulsive Therapy vis-à-vis Sham Treatments

Context: Using the classical placebo (or sham treatment)—controlled design to establish the inherent biologically mediated therapeutic efficacy of ECT is quite difficult due to the lack of a credible sham procedure.

Objective: To randomly allocate depressed ECT patients to one of two levels of intensity of treatment in order to maintain blindness.

Primary outcome: Depression interview scales conducted at baseline, during, and after treatment.

Of course, one of the bulwarks of randomized controlled trials is blinding of patients and caregivers. If patients know to which treatment they are assigned, then pre-existing expectations about response may influence outcome, that is, the placebo mechanism may become paramount. For medications of course, it is easy to develop placebo pills that appear identical to the active pill and preserve blindness. For procedures, it is usually almost impossible to develop placebo or "sham" procedures that involve no inherent biological therapeutic effects but otherwise are experienced in the same way to patients as the active procedure. In the ECT situation, blinding of patients is perhaps the biggest challenge. The treatments involve being admitted to a room with intravenous access, general anesthetic administration, induction of a generalized tonic clonic seizure, and recovery. Side effects are frequent and are impossible to miss: headaches, muscle aches, nausea/vomiting, disorientation, and persisting memory impairment. In order to make scientifically valid inferences on the true inherent biologically mediated efficacy of ECT, it is imperative that patients in sham-controlled trials do not know to which group they have been assigned.

There are three main aspects of a good sham control for ECT: (1) it is believable to patients (in other words, the patients cannot guess to which treatment they have been assigned); (2) it does not inadvertently mimic the therapeutic action of ECT itself (i.e., it cannot be so close to ECT that it is therapeutically active); and (3) it must not involve biological effects that are themselves therapeutic for the psychiatric disorder being studied.

In order to design a sham control that patients do not know is a sham, it would be important first to know what patients' pre-existing knowledge is about ECT. For example, if patients know absolutely nothing about the ECT procedure (e.g., the use of general anesthesia, recovery room disorientation, other side effects), then it might be easy to design a "believable" sham group. However, patients in modern times relatively sophisticated in their knowledge not only about treatments but also about ethical principles as well as those of good research conduct demand that researchers disclose known aspects about proposed treatments including what the treatments consist of as well as anticipated side effects. Thus, in designing a "believable" sham for an ECT trial, we must assume that patients will know about the various aspects described earlier. Thus, a question is what type of sham procedure will mimic ECT except for the seizure? The core therapeutic aspect of ECT is believed to be the generalized seizure, though the role of the electrical stimulus that induces the seizure is not precisely known. As indicated earlier, another challenge of designing a sham ECT control is not to do something which itself has biologically mediated therapeutic action, or else we are not truly comparing ECT to a sham but to another active treatment.

If one conducts a sham ECT study, how would one assess the adequacy of the sham treatment to maintain double blindness as the study progresses? Strategies may include simply asking patients if they can guess to which treatment they were assigned, and after the study is done see if the guesses are correct beyond what chance would dictate. Another strategy would include a side effect questionnaire. If the real ECT group is associated with better outcomes but also more side effects, then the possibility that as patients experienced side effects they guessed they were receiving real ECT and expected to get better, in other words, a placebo effect, may be the real mechanism of improvement.

There are several sham-controlled ECT studies available from the literature [4–9]. Unfortunately, none of them assessed for adequacy of the sham but attempts were made to mimic the ECT experience without duplicating the therapeutic action. In other words, patients were anesthetized but were not given electrically induced seizures. These were a series of studies conducted in Great Britain in the 1970s and early 1980s. All involved the question of the efficacy of ECT for patients with severe depressive illness. The sham, or placebo, form of ECT chosen consisted of the administration of anesthesia alone without the presentation of the electrical stimulus that normally would cause a seizure. The patients had baseline assessments to determine the presence of depressive illness and rating scales to quantify the severity of the symptoms before treatment. After a series of either real or sham sessions, the rating scales were repeated. In most of the studies, the patients given real ECT fared better than those receiving sham treatments. However, the question is, did the patients know to which treatment they were randomized? If patients were able to guess correctly, as mentioned earlier, then expectationally mediated effects (i.e., placebo) may have been responsible for the outcomes. In none of these studies were attempts made to assess adequacy of the blinding, so we do not know for sure. It is unlikely, though, that the blinding was totally effective, as patients awakening from a few minutes of general anesthesia alone have little lingering effects, whereas patients awakening from ECT treatments have substantial disorientation and other side effects.

Fortunately, there has been another experimental design used in modern ECT studies that allows for better blinding of patients and establishes the biological efficacy of the treatment versus placebo. The methodology is based on the fact that ECT treatment technique is not homogeneous. Rather, there are several different ways the treatments can be delivered. For example, there are different types of electrode placement. The reader may recall that two electrodes are placed on the patient's head and connected to the ECT machine. There are different locations on the scalp where these electrodes can be placed. The earliest such electrode placement is termed bitemporal, in which electrodes are placed in the right and left temporal fossa (i.e., 1 in. above a point halfway between the outer canthus of the eye and the tragus of the ear). Another electrode placement is termed unilateral, in which one electrode is placed in a temporal fossa and the other is placed just next to the vertex of the scalp on the same side as the temporal electrode. Since this technique is almost always done on the right side, this is termed right unilateral placement. In a study in the mid-1980s, patients were randomly assigned to bitemporal or right unilateral electrode placement [10]. The basis for the study was the belief that the unilateral placement would cause less memory impairment, which is the major side effect of ECT. Concerns had been raised about the efficacy of unilateral placements, so the study was designed to compare efficacy and cognitive side effects. Patients in the bitemporal group had twice the remission rates as those in the unilateral group. There was no "sham" or placebo group in this study. However, if one assumes that the unilaterally treated group would have fared no worse than a sham group would have, then by inference we can say that this study shows that at least one form of ECT is better than sham and thus has true inherent therapeutic actions separate from any placebo effects or the natural history of depressive episodes. An analogy for this technique in pharmacological research is to randomize patients to high versus low doses of medications. Again, if one assumes that a true placebo group would fare no better than the least effective medication dose, then finding a clinical outcome difference between the two dosages constitutes proof of a true inherent therapeutic effect of the medication (for at least one dosage strength). One can easily see how this strategy could be used for other neurostimulatory procedures in which different intensities of stimulation can be compared in a study.

20.3.2 Blinding of Outcome Assessors and Other Caregivers When Patient Blinding Is Impossible

Context: When comparing a procedure like ECT to another psychiatric treatment, such as psychotropic medications, a method must be instituted to remove bias from nonblind outcome evaluations.

Objective: To develop a method of videotaping outcome assessment interviews and sending them to a blind evaluator who is blind to treatment status and phase of treatment.

Primary outcome: Videotaped depression interviews conducted serially throughout treatment.

In the earlier discussed example, patients were randomly assigned to two versions of a treatment type (i.e., ECT), thus preserving blinding of patients. In some designs, such as those comparing a nonpharmacological treatment to another treatment, it is impossible to maintain blinding. For example, the comparative efficacy of ECT versus psychotropic medication is an important question in psychiatry. However, if we randomly assign patients to receive either ECT or medications, the patients will not be blind. One possibility is to randomize patients to real ECT/placebo pills versus sham ECT/active medication. Such trials were attempted many decades ago using the anesthesia-only sham method [11,12] but sample sizes were much too small. In modern times, it would not be feasible from an ethical or cost standpoint to conduct such trials.

If patients are not to be blind in an ECT medication comparative trial, then blinding of outcome assessors would be all the more crucial in order to try to preserve some scientific basis. In the case of medication trials, it is easy to make all parties involved with the patient blind to intervention, but this is not the case with ECT. First of all, when patients are receiving treatments, all caregivers in the room know exactly that the patients are receiving ECT, even in a sham ECT paradigm. Thus, it is important for these personnel not to discuss the treatment with anybody involved in outcome assessment.

Depending on the control group, it is relatively easier to blind outcome assessors. If, for example, the study involves comparing different forms of ECT with one another, then outcome assessor blinding is easy. If on the other hand the control group is receiving medications, and the patients then know to which group they are randomized, then it is challenging to blind the outcome assessors. One would have to instruct patients not to discuss their treatment with the outcome assessors. The author was involved in a maintenance ECT study in which patients were followed with ECT treatments at spaced intervals or medications over a 6 month period [13,14]. Outcome assessors could not be blinded. However, a unique design feature was developed to at least partially offset this.

In the trial, patients who were depressed received an index course of ECT treatments. Those who remitted and stayed remitted for 1 week after the end of the thrice weekly course of treatments were randomly allocated to receive 6 months of continued ECT treatments at spaced intervals versus 6 months of aggressive antidepressant medication therapy. Patients in both groups were seen by study personnel for ongoing outcome assessment at the same frequency. Neither patients nor the outcome assessors were blind to treatment. However, the outcome assessment interviews, which consisted of a standard depression interview scale [15], were videotaped. Some of the videotapes were sent to a central investigator who did not participate in any patient care activities of the study and who was responsible for teaching and supervising the conduct of the depression interviews. This person was known as the independent blind rater. Due to the high volume of patients in the study, it was not feasible to have every taped interview sent to her, so only a subset were sent for her evaluation and scoring. As it was desirable to have her blind not only to which treatment the patient was receiving (i.e., medications or ECT) but also to phase of treatment (i.e., early on or near the end), we devised a method of sending her 40% of each cohort of tapes generated. For each cohort of tapes, 60% were sent to the next cohort, from which another 40% were sent to her and 60% on to the next cohort, as schematized. Thus, in an ongoing fashion, the tapes that she received were scrambled in terms of which phase of the study the patients were evaluated at. In other words, toward the end of the study, tapes she received may have been from the early part of the study. She was able to watch the depression interviews and score the severity of depression of each patient and compare this score with that assigned by the on-site study personnel. This provided confidence that even though the on-site personnel were not blind, because their scores closely matched those of the independent blind rater, there was no evidence of bias in our outcome assessment. This technique is highly applicable to psychiatric research, in which the clinical outcomes are virtually always based on subjective information obtained from patient interviews.

Another challenge for blindness in ECT research is the scenario whereby ECT is compared with another neuropsychiatric procedure. For example, transcranial magnetic stimulation (TMS) is an emerging treatment, just recently approved in the United States by the United States Food and Drug Administration. In TMS, a magnetic stimulus is generated by a coil with an electric current flowing through it. The magnetic force passes unimpeded through the skull into the brain, where it generates an electric current. As currently applied, TMS does not cause a seizure. It is hypothesized that it does cause biological changes in the brain, which treat depression. It is given as a series of treatments just like ECT. Since it involves no anesthesia, loss of consciousness, or cognitive impairment, there is intense interest in whether its therapeutic effects match those of ECT. Thus, comparative studies are logical. In designing such a study, one of course needs to pay attention to blinding issues because receiving a TMS session and an ECT session are so different from the patient's perspective. One option would be to randomize patients to sham TMS/real ECT versus real TMS/sham ECT. However, the frequency and intensity of sessions would be prohibitive for such a study, as would cost. Thus, such a study is unlikely ever to be accomplished. An alternative strategy is to design single blind studies whereby the patients and treaters are not blind but outcome assessors are blind. Such studies have been conducted [16]. In none of these was an attempt made to ascertain whether the outcome assessors really were blind; in other words, performing depression interviews may seem different to the interviewers depending on which treatment the patient is receiving. An example of this is the memory impairment of ECT. If an outcome assessor is performing a depression interview on a patient who clearly is having trouble remembering recent symptoms (such as sleep or appetite patterns), then over time they may become very good at guessing patient's treatment assignments, thus militating against the blindness of their evaluations.

20.4 Conclusion

ECT is an important neuropsychiatric treatment modality that is used virtually around the world for serious mental illness. The technique of ECT is radically different from the prescription of psychotropic agents. Thus, the design of controlled trials entails special considerations not inherent in medication trials. In particular, the main goal of establishing the true inherent biologic therapeutic activity of the treatment leads to the challenge of designing a good "placebo" or sham technique that is not detectable as a sham by the patients receiving it. This is a challenge inherent in controlled research of many nonpharmacological procedures. In this chapter, the author has outlined a method of dealing with this challenge, namely, the administration of different types of ECT to test its efficacy and preserve blindness. Additionally, the challenge of comparing the efficacy of ECT versus other psychiatric treatments, such as psychotropic medications and the new procedure known as TMS, meanwhile preserving blindness has also been discussed.

References

1. Abrams, R. *Electroconvulsive Therapy*, 4th edn. New York: Oxford University Press, 2002.
2. Michaud, C.M., Murray, C.J.L., Bloom, R.R. Burden of disease: Implications for future research. *JAMA*, 285(5), 535–539, 2001.
3. American Psychiatric Association Committee on Electroconvulsive Therapy. *Electroconvulsive Therapy: Recommendations for Treatment, Training, and Privileging*, 2nd edn. Washington, DC: American Psychiatric Association, 2001.
4. Brandon, S., Cowley, P., McDonald, C., et al. Electroconvulsive therapy: Results in depressive illness from the Leicestershire trial. *BMJ*, 288, 22–25, 1984.
5. Freeman, C.P.L., Basson, J.V., Crighton, A. Double-blind controlled trial of electroconvulsive therapy (E.C.T.) and simulated E.C.T. in depressive illness. *Lancet*, 1, 738–740, 1978.
6. Gregory, S., Shawcross, C.R., Gill, D. The Nottingham ECT study. A double-blind comparison of bilateral, unilateral, and simulated ECT in depressive illness. *Br J Psychiatry*, 146, 520–524, 1985.
7. Johnstone, E.C., Deakin, J.F.W., Lawler, P., et al. The Northwick Park electroconvulsive therapy trial. *Lancet*, 2, 1317–1320, 1980.
8. Lambourn, J., Gill, D. A controlled comparison of simulated and real ECT. *Br J Psychiatry*, 133, 514–519, 1978.
9. West, E.D. Electric convulsion therapy in depression: A double-blind controlled trial. *BMJ*, 282, 355–357, 1981.
10. Sackeim, H.A., Decina, P., Kanzler, M., Kerr, B., Malitz, S. Effects of electrode placement on the efficacy of titrated, low dosage ECT. *Am J Psychiatry*, 144, 1449–1455, 1987.
11. Harris, J.A., Robin, A.A. A controlled trial of phenelzine in depressive reactions. *J Ment Sci*, 106, 1432–1437, 1960.
12. Robin, A.A., Harris, J.A. A controlled comparison of imipramine and electroplexy. *J Ment Sci*, 108, 217–219, 1962.
13. Kellner, C.H., Knapp, R.G., Petrides, G., et al. Continuation electroconvulsive therapy versus pharmacotherapy for relapse prevention in major depression: A multi-site study from the Consortium for Research in Electroconvulsive Therapy (CORE). *Arch Gen Psychiatry*, 63(12), 1337–1344, 2006.
14. Rasmussen, K.G., Knapp, R.G., Biggs, M.M., et al. Data management and design issues in an unmasked randomized trial of electroconvulsive therapy for relapse prevention of severe depression: The CORE Trial. *J ECT*, 23(4), 244–250, 2007.
15. Williams, J.B.W. Standardizing the Hamilton depression rating scale: Past, present, and future. *Eur Arch Psychiatry Clin Neurosci*, 251(Suppl 2), II/6–II/12, 2001.
16. Rasmussen, K.G. ECT versus TMS for major depression: A review with recommendations for future research. *Acta Neuropsychiatr*, 20, 1–4, 2008.

21

Assessing Acupuncture: Practical Examples

Hugh MacPherson

University of York

CONTENTS

21.1 Introduction

Acupuncture has been a continuously practiced medical tradition for over 2000 years, which is quite remarkable. Its longevity is largely a result of the way it has evolved and adapted over the centuries in reaction to changing social and political circumstances [1]. In the last century, acupuncture and its underlying theories were shaped as a result of contact with what was termed "Western medicine," and the need for its proponents in China to provide a justification for traditional Chinese medicine in a way that was more consistent with the biomedicine that had arrived from the West [2].

Further developments in the last 50 years have led both to an increased standardization of Chinese medicine and a concurrent transmission to the West [3]. Over this period, the use of acupuncture in the West has grown substantially and, with this progressively more widespread provision, there have been innovative developments in the way acupuncture is practiced [4]. New forms of acupuncture have appeared, for example, ear acupuncture, trigger point acupuncture [5], as well as adaptations of traditional therapies, such as the Five Element school of acupuncture in England and the United States [6]. This diverse history is characterized by a dynamic and evolving development of both theoretical frameworks and practice-based techniques [7]. For acupuncturists, this innovation and the potential for further creativity is one of the exciting aspects of being involved in the field. For researchers, this heterogeneity adds considerable challenges, not only in defining an adequate intervention, but also in generalizing the results of trials.

Along with the heterogeneity of practice, another feature that continues to challenge researchers is the complexity of treatment [8]. Acupuncture is not a simple intervention. There are multiple components to treatment, with a variety of subtle techniques used and there are components of treatment beyond needling that are driven by acupuncture theory. These need to be considered integral (and specific) to acupuncture treatment [9]. For example, the number and location of acupuncture points that are needled are commonly customized to the patient, and are likely to be altered over time in order to follow the patient's trajectory of evolving symptoms. In addition, each point is needled in ways that might vary depending on the current symptomatology. Such variations could include depth of needling, the sensation that is to be elicited, and stimulation techniques that might be employed. Beyond needling, there are a number of components to treatment that acupuncturists might provide including, for example, explanation of the patient's condition from the perspective of acupuncture theory, and lifestyle advice that is determined from the acupuncture diagnosis. These complexities add to the challenge of conducting clinical evaluations of acupuncture.

Researching acupuncture in a way that optimizes generalizability requires consideration of who commonly consults acupuncturists for treatment, who provides this treatment, and in what medical care setting. We know that patients commonly consult with chronic pain conditions such as low back pain, neck pain, shoulder pain, and headache [10]. There are also variations geographically and historically. For example, in the United Kingdom, we have evidence that between 1988 and 2002 there was a marked reduction in patients consulting traditional acupuncturists for musculoskeletal conditions, indicating a wider case mix [11]. The characteristics of the providers of acupuncture vary across countries and within countries. For example, in Germany the provision of acupuncture is dominated by ~30,000 physicians who practice acupuncture. Meanwhile, in the United Kingdom the provision is almost equally divided across physicians, physiotherapists, and traditional acupuncturists [12]. The settings within which acupuncture is provided also vary considerably so that, for example, in the United Kingdom, physicians provide acupuncture largely within primary care, physiotherapists within hospital outpatient departments, and traditional acupuncturists at independent clinics.

The challenge of evaluating the clinical effects of acupuncture is one that both motivates and daunts the budding researcher in equal measure. A major task in any research activity is defining the research question. What the right question might be depends on the context one wishes to focus on, and especially the priorities that one assesses to be of most importance. Within pharmacological trials, there has been an accepted route for clinical evaluation which posits a series of phases that start with the identification of potentially beneficial substances and their mechanism of action, and proceeds through animal studies and testing on humans, to efficacy and effectiveness studies. This prioritization, which makes sense in protecting the public from potentially harmful substances, may not be the appropriate way to prioritize when confronted with a medical intervention that is already in widespread use. Fonnebo and colleagues [13] have set out an alternative route map, which posits that the sequencing should proceed from research into current levels of utilization, through questions of safety in routine care, to effectiveness. When overall effectiveness, and if necessary cost-effectiveness, is established, for example, in pragmatic or comparative trials, the active components of the (complex) intervention can be identified, as well as their contribution to the overall effect. For this purpose, some form of sham acupuncture is necessary as a control. When the specific components have been identified, then there is a clear opportunity to explore in a focused way the underlying mechanism(s). While this ordering of priorities may have a simple elegance, the reality is that research into acupuncture needs to proceed at many levels.

21.2 Specific Methodological Issues in Assessing This Treatment

There are number of generic methodological issues that pertain to the field of acupuncture research. While these will be discussed in the context of published trials discussed below, it is perhaps useful to flag up some of these issues as part of setting the context. First, the complexity of routine practice needs to be considered and this is linked to the choice of research design that may be most useful when conducting a clinical evaluation. An example of a relevant methodological approach is the Medical Research Council's framework for the clinical evaluation of complex interventions [14] Another relevant example is "the whole system" approach that is designed to incorporate a number of traditional Chinese medicine components within a single clinical evaluation [15]. With these approaches, a key concern is to define the parameters of the intervention. This is not easy given the multiple components of acupuncture treatment, which are likely to include components beyond needling [9]. For pragmatic trials, there is a relatively simple option of requesting practitioners to deliver the intervention as they normally would. However for explanatory trials, the challenge is more daunting. The dimensions that might be considered include the history of how the condition has been treated in China, how the process of transmission has impacted on the dimensions of treatment, what existing research evidence there might be, expert opinion, and possibly some sort of consensus process with experts. However, there is unlikely to be definitive data on precisely what intervention is optimal as there is usually insufficient research evidence to define a treatment protocol with confidence.

Related to defining the active components of the intervention is the requirement to define an appropriate comparator(s) or control(s). For pragmatic trials, an active comparator such as standard care or no additional treatment is commonly used. However for explanatory trials, the choice of an adequate sham remains problematic. There are two aspects to the problem that are worth setting out at this point. First, we do not know what the active ingredients of acupuncture are and so it is theoretically impossible to select an intervention that mimics acupuncture without entraining any of acupuncture's putative yet unknown active ingredients. Second, the evidence from the literature indicates that commonly used forms of sham acupuncture have effect sizes far greater than the standardized mean differences of around 0.28 to 0.3 that one would expect from an inert placebo [16].

These methodological challenges will be explored in the context of the four trials discussed below: the first two are acupuncture for hypertension trials with explanatory trial designs and the second two are pragmatic trials of acupuncture for chronic pain, namely headache and low back pain. It is intended that the contrasting trial designs, explanatory and pragmatic, will provide a useful backdrop for exploring the issues associated with the clinical evaluation of acupuncture.

21.3 Practical Examples

21.3.1 Two Explanatory Trials of Acupuncture for Hypertension

Two fairly similar explanatory trials of acupuncture for hypertension have been selected to illustrate the particular challenges associated with research questions that seek to establish whether there is a specific effect of acupuncture per se, that is, an effect beyond a placebo. What follows are brief descriptions of the two trials, one from the United States and one from Germany, followed by a discussion of some of the key methodological challenges associated with them.

**BOX 21.1   STOP HYPERTENSION WITH THE
ACUPUNCTURE RESEARCH PROGRAM (SHARP) [17]**

Context: Case studies and small trials suggested that acupuncture may effectively treat hypertension, but no large randomized trials had been reported.

Objective: To evaluate the clinical efficacy of acupuncture for hypertension.

Primary outcome: The mean blood pressure (BP) from baseline to 10 weeks.

In the Macklin et al. [17] trial (see Box 21.1), which was described as a pilot, 192 participants with untreated blood pressure (BP) were enrolled. Participants were weaned off antihypertensives before enrolment and were then randomly assigned to three treatments: individualized traditional Chinese acupuncture, standardized acupuncture at preselected points, or invasive sham acupuncture. Participants received 12 acupuncture treatments over 6 to 8 weeks. The mean BP decrease from baseline to 10 weeks did not differ significantly between participants randomly assigned to active (individualized and standardized) versus sham acupuncture. The authors concluded that active acupuncture provided no greater benefit than invasive sham acupuncture in reducing systolic or diastolic BP.

In the Flachskampf et al.'s [18] trial (see Box 21.2), 160 outpatients with uncomplicated arterial hypertension were randomized in a single-blind fashion to a 6 week course of active acupuncture or sham acupuncture. One hundred and forty patients finished the treatment course. There was a significant ($p < 0.001$) difference in posttreatment blood pressures, adjusted for baseline values, between the active and sham acupuncture groups at 6 weeks, the end of treatment. At 3 and 6 months, mean systolic and diastolic blood

---

**BOX 21.2   RANDOMIZED TRIAL OF ACUPUNCTURE
TO LOWER BLOOD PRESSURE [18]**

Context: Arterial hypertension is a prime cause of morbidity and mortality in the general population. Pharmacological treatment has limitations resulting from drug side effects, costs, and patient compliance.

Objective: To investigate whether traditional Chinese medicine acupuncture is able to lower blood pressure.

Primary outcome: Mean 24 h ambulatory blood pressure levels after the treatment course at 6 weeks, and at 3 and 6 months post-randomization.

---

pressures returned to pretreatment levels in the active treatment group. The authors concluded that acupuncture according to traditional Chinese medicine but not sham acupuncture, after 6 weeks of treatment, significantly lowered mean 24h ambulatory blood pressures; the effect disappeared at 3 months.

Hypertension is a considerable healthcare burden with the potentially serious consequences of myocardial infarction and stroke if untreated. Pharmacological interventions are costly and have side effects and problems with compliance. In addition, many people with this condition are undiagnosed and/or untreated. It is thought worthwhile therefore to investigate whether acupuncture might be a useful modality to treat hypertension. The researchers conducting these two hypertension trials took the view that acupuncture has to function either as an alternative to medication [17] or as an adjunct [18]. In both cases, an explanatory trial design was selected; Flachskampf et al.'s trial compared acupuncture with sham acupuncture and Macklin et al.'s trial compared individualized acupuncture with standardized acupuncture and sham acupuncture. Neither trial had a usual care arm nor were cost-effectiveness analyses possible. Unique to acupuncture trial research are two key challenges for researchers: the design and implementation of an appropriate treatment protocol and an adequate sham control. As Macklin et al. acknowledge, "What constitutes a valid acupuncture treatment and an appropriate control is a question of much debate in the field" [17]. The next two sections look at the ways these hypertension trials addressed these challenges.

21.3.1.1 Acupuncture Treatment Protocols for Explanatory Trials Evaluating the Treatment of Hypertension

The rationale for conducting trials of acupuncture for hypertension is exemplified by Flachskampf et al.'s statement that acupuncture is "an ancient technique anchored in traditional Chinese medicine (which) has been reported to have potential for treating cardiovascular disease, including arterial hypertension" [18]. In a similar vein, Macklin and colleagues claim that acupuncture "has been used in traditional Chinese medicine (TCM) to treat symptoms of hypertension for 2,500 years" [17]. This raises one of the first challenges in conducting trials of acupuncture, the legitimacy of the claims regarding the authenticity of the tradition. Hypertension is a relatively new condition in the West, and even more recent in China. Clearly, it was not possible for this condition to have been understood before the technology was in place to measure blood pressure. Historical analysis shows that the diagnostic categories currently promoted for the treatment of hypertension are relatively recent and have not had the centuries of empirical testing that support diagnostic and treatment approaches for many other conditions.

It is always challenging to delineate an optimal acupuncture treatment protocol, especially in an explanatory trial where parameters of treatment are required to be prespecified. In this context, Macklin and colleagues state that "The design of the trial combined rigorous methodology and adherence to principles of traditional Chinese medicine" [17]. A challenge here is how to use the principles of Chinese medicine, which require a diagnosis based on the patient's signs and symptoms, when hypertension can often be symptom-free? In an extended discussion of diagnostic criteria in the Macklin et al. trial, it is stated that "not all the signs and symptoms (for each diagnostic category) need be present" [19]; however, there is no discussion of diagnostic processes when signs and symptoms were completely absent.

Furthermore acupuncture, when practiced either according to traditional principles or according to the trigger point method as developed in the West more recently [5], is normally provided to patients on a customized basis. Patients presenting with the same conventional medical diagnosis will usually have different treatments related to the underlying patterns of disharmony. Over a course of treatment, the details of the intervention will evolve and follow the trajectory of symptom change over time. Therefore, there is an additional challenge in explanatory trials, such as the ones described in this section, that of accommodating the flexibility that is integral to authentic practice. The two trials described in this section have developed compromises to the authenticity of the acupuncture, such that the protocols delineate treatment that probably lies somewhere between fully authentic practice and totally constrained treatment. For example, Macklin and colleagues state that "Diagnoses and treatment followed TCM principles" [17] yet then their protocol limited the acupuncturist to a pool of possible points, thereby reducing flexibility.

21.3.1.2 Adequacy of the Sham Control for Acupuncture in Explanatory Trials

There has been much written about sham acupuncture as a control for acupuncture. The key problem is that we do not know the specific mechanism of acupuncture; therefore, designing a control or sham that captures all the nonspecific effects of acupuncture but none of the specific effects is theoretically impossible. Despite this inherent problem, various sham methods have been designed to provide an adequate control for explanatory trials [20]. As an example, in the Flachskampf et al. trial, acupuncture points were selected for needling on the basis that they were, "without relevance for blood pressure lowering according to traditional Chinese medicine concepts" [18]. The Macklin et al. trial had two controls: a standardized acupuncture treatment (with fixed acupuncture points for all patients) and sham acupuncture, the latter designed on the basis that the control points used were "not active according to TCM theory." The conundrum of a medical tradition built on empirical evidence of what is known to work, but not on evidence of what is known to not work, was not made explicit by the trialists. The problem is that we have no hard evidence on causal mechanisms and consequently we have no clear parameters defining what is "active" and what is not. To conclude, we cannot be sure that these control points have no antihypertensive properties.

21.3.1.3 Interpretation of Results from the Two Acupuncture for Hypertension Trials

The longer-term effects of acupuncture in these two trials were similar, both reporting no benefit favoring acupuncture at 3 months. The main difference in outcomes was that the Flachskampf et al. trial showed temporary short-term benefits favoring acupuncture immediately posttreatment, at 6 weeks. How does one interpret this difference? Perhaps, it was because the Flachskampf et al. trial provided their patients in the acupuncture arm with 22 sessions over 6 weeks, whereas the Macklin et al. trial only provided up to 12 sessions over 6 to 8 weeks. Another explanation might lie in the fact that the patients in the Macklin et al. trial may have had a posttreatment benefit, but this was not a follow-up point, so we do not know what their posttreatment blood pressures were and cannot check congruence with the Flachskampf et al. trial. Nevertheless, the results of both trials lead one to conclude that acupuncture is not more effective than sham for hypertension over the longer term. To some acupuncturists this may be a surprise, but to many others, it could be an expected result. In terms of current provision, it should be noted that few patients seem to consult acupuncturists for this condition [11], perhaps because antihypertensives are relatively effective in controlling blood pressure. There is also a concern among acupuncturists that removal of antihypertensives is something that should only happen in controlled circumstances, because of the risk of stroke or myocardial infarction if high blood pressure is untreated. For these reasons, acupuncture is unlikely to be seen as a referral option in everyday practice, even if it did outperform sham acupuncture.

21.3.2 Two Pragmatic Trials of Acupuncture for Chronic Pain

Chronic pain is a major healthcare burden in primary care and because of perceived limitations in conventional treatments [21], patients are seeking acupuncture in increasing numbers. The evidence base is steadily growing and, now in 2009 there is some evidence that acupuncture is more beneficial than sham acupuncture for headache [22] and low back pain [23]. The question was less clear when the two trials discussed below were commissioned by the funders, whose interest was in the practical question of whether acupuncture should be offered routinely in primary care for chronic pain. To answer such questions, pragmatic designs focused on comparative effectiveness were deemed appropriate. With the comparator being usual care, it was possible for questions regarding cost-effectiveness to be addressed in both trials. What these two trials do not evaluate are answers to research questions that seek to determine the relative contribution to the overall effect from the components of treatment that are specific to acupuncture. In this section, we discuss this issue and other challenges associated with pragmatic trials, including the implementation of the acupuncture intervention and the possibility that patients will seek care from interventions that are different from the one to which they are allocated.

Two pragmatic trials have been selected to illustrate the particular challenges associated with research into whether there is an overall effect of acupuncture on chronic pain conditions when it is offered as a referral option in a primary care context. These two trials of chronic pain are useful exemplars of the challenges of conducting trials in which acupuncture is provided to everyday patients in primary care by practitioners who are not constrained in terms of the treatment that they provide. What follows are brief descriptions of the two trials, both conducted in the United Kingdom, followed by a discussion of the key methodological challenges associated with them.

In Vickers et al.'s [24] randomized controlled trial (see Box 21.3), 401 patients with chronic headache, predominantly migraine, were recruited. Patients were randomly allocated to receive up to 12 acupuncture treatments over 3 months or to a control intervention offering usual care. Physiotherapists provided treatments to those allocated to the acupuncture arm. Headache scores at 12 months were found to be lower in the acupuncture group than in controls. Patients in the acupuncture group also experienced fewer days of headache per year and used less medication, made fewer visits to general practitioners, and took fewer days off work. The authors concluded that acupuncture is associated with persistent, clinically relevant benefits for primary care patients with chronic headache, particularly migraine. In a separate publication, they reported that acupuncture for chronic headache was cost-effective [25].

BOX 21.3   ACUPUNCTURE FOR CHRONIC HEADACHE IN PRIMARY CARE: LARGE, PRAGMATIC, RANDOMIZED TRIAL [24]

Context: A Cochrane review concluded that although the existing evidence supported the value of acupuncture, the quality and amount of evidence was not fully convincing. There was a need for a pragmatic trial to estimate the effects of acupuncture in practice.

Objective: To determine the effects of a policy of "use acupuncture" on headache, health status, days off sick, and use of resources in patients with chronic headache, compared with a policy of "avoid acupuncture."

Primary outcome: Headache score at 12 months.

**BOX 21.4   RANDOMIZED CONTROLLED TRIAL OF A
SHORT COURSE OF TRADITIONAL ACUPUNCTURE
FOR PERSISTENT NONSPECIFIC LOW BACK
PAIN COMPARED WITH USUAL CARE [26]**

Context: While there was emerging evidence of short-term relief of pain
using acupuncture compared with no treatment or sham therapy, the
evidence for long-term effectiveness was sparse and for cost-effectiveness was nonexistent.

Objective: To determine whether a short course of traditional acupuncture improved longer-term outcomes for patients with persistent nonspecific low back pain in primary care.

Primary outcome: The primary outcome was SF-36 bodily pain, measured at 12 and 24 months.

In Thomas et al.'s [26] randomized controlled trial (see Box 21.4), 241
patients with chronic low back pain were recruited from general practices
in York, UK. Patients were randomly allocated to receive up to 10 individualized acupuncture treatments plus usual care over 3 months or to usual care
alone. Professional acupuncturists with 3 years training in acupuncture or
equivalent provided the treatments to those allocated to the acupuncture
arm. Weak evidence was found of an effect of acupuncture on persistent
nonspecific low back pain at 12 months, but stronger evidence of a small
benefit at 24 months. The authors also reported that referral to a qualified traditional acupuncturist for a short course of treatment seems safe and acceptable to patients with low back pain. In a separate work, the trialists reported
that acupuncture for low back pain was highly cost-effective [27].

21.3.2.1 Interventions in the Two Pragmatic Trials
of Acupuncture for Chronic Pain

In both these trials, acupuncturists were instructed to treat as they normally would. Vickers et al. reported that "The acupuncture point prescrip-
tions were individualised to each patient" [24] and Thomas et al. reported
that "Acupuncture care comprised up to 10 individualised treatment sessions
over three months. Acupuncturists determined the content and the number
of treatments according to patients' needs" [26]. While one of the strengths of
pragmatic trials is their generalizability or applicability, one of the challenges
by contrast is their replicability. In the two trials under discussion, some level
of standardization was achieved by ensuring that the practitioners were in
clearly identifiable professional groupings. In the former trial, they were physiotherapists who were members of the Acupuncture Association of Chartered

Physiotherapists and in the latter they were traditional acupuncturists who were members of the British Acupuncture Council. Another challenge is reporting the details of the interventions, given the degree of variability that exists between the treatments provided for patients, and within patients. Given the limitations often imposed by journals, one solution is to publish the details of the interventions in a separate work, as was done for the acupuncture for low back pain trial [28]. Of assistance in providing a template for reporting acupuncture interventions are the STRICTA recommendations [29].

An additional requirement in pragmatic trials is that the intervention parameters in the control arm be defined. Both trials had a "usual care" control, where usual care was left open with regard to advice and referral from general practitioners and decisions that patients might make were based to some extent on patient choice. This meant that there was considerable variation in what constituted usual care, and there was a clear need to report all treatments received, primarily so that a check could be made to determine if there were problematic differentials. In addition, Vickers et al. [24] specified the avoidance of acupuncture as a requirement for their usual care control arm.

21.3.2.2 Potential Limitations in Pragmatic Trials of Acupuncture for Chronic Pain

One of the key limitations to these pragmatic trials was that, because there was no sham or placebo-like control group, it could be argued that the effects in the acupuncture group resulted from the placebo effect. Two examples of immediate responses to Thomas et al.'s [26] work were as follows:

> pragmatic trials such as this of Thomas et al. may seriously mislead healthcare policy, and even the most rigorous cost-analysis may only demonstrate the cost-effectiveness of placebo for a self-limiting condition. To put it bluntly, hugging a tree may even be more cost-effective (and safer) than acupuncture [30].

And,

> Any claimed benefit (which is open to statistical challenge) may be the result of the complex interaction between patient and acupuncturist rather than the needle therapy [31].

The challenge remains for those who conduct pragmatic trials to continue to argue that the methodology should follow the research question. Neither of the two trials described in this section was designed to separate out the proportion of overall effect that could be ascribed to the placebo. There are a number of arguments to support the importance of evaluating practical and pragmatic research questions [32]. As Vickers et al. put it, "Such an argument (that all trials must evaluate placebo effects) is not relevant since in everyday practice, patients benefit from placebo effects" [24].

Patient blinding in sham-controlled trials is aimed at minimizing differences in patient experience and expectation resulting from practitioner contact. While Vickers et al. [24] were concerned to minimize practitioner-related bias in their open trial, they argued that the differences they found between their two arms were far larger (odds ratio for response of 2.5) than empirical estimates from a failure to blind (odds ratio of 1.2). Another potential source of bias in pragmatic trials, in which patients are encouraged to do what they would normally do for reasons of ecological validity, is related to possible differentials between arms in terms of "off-study" interventions. There is a clear need in such trials to record and report off-study interventions in order that judgments can be made as to the extent that they might act as confounders. Vickers et al. reported that "the use of medication and other therapies (such as chiropractic) was lower in patients assigned to the acupuncture, indicating that the superior results in this group were not due to confounding by off-study interventions" [24].

Pragmatic trials are also vulnerable to other sorts of bias. For example, patients' preferences, expectations, and beliefs could contribute to bias, especially if they were confounding variables, that is, if there were differentials between groups at randomization and also such variables were effect modifiers. There is some evidence, for example, that expectations of improvement prior to acupuncture are associated with better outcomes [33]. However, there is some contrasting evidence from Thomas et al.'s trial, whose authors observed that patients with "neutral" beliefs that acupuncture would help their back pain gained relatively more benefit from acupuncture care than those with a "positive" belief [26]. Additional discussions regarding the generic challenges of pragmatic trials have been set out elsewhere [32].

21.3.2.3 Interpretation of Results from These Two Pragmatic Trials of Acupuncture for Chronic Pain

Reading these two trials, one is struck by the remarkable fact that both trials show improvements long after the 3 month courses of treatment were concluded. For example, in Vickers et al.'s trial, the gap between arms in the weekly headache scores that favored the acupuncture at 3 months, with a score of 3.9 (95% CI 1.6–6.3), increased over time so that at 12 months there was a score of 4.6 (95% CI 2.2–7.0). In a similar fashion, Thomas et al. reported that the gap between acupuncture and usual care grew from 5.6 (95% CI 0.2–11.4) at 12 months to a score of 8.0 (95% CI 1.8–16.2) at 24 months. Typically, trials of interventions for chronic low back pain are susceptible to effects arising from regression to the mean, yet what was observed in these two trials was an ongoing and sustaining effect from acupuncture. Two questions of relevance and interest here are as follows: What is the reason for these unexpectedly sustained outcomes over time? and how long do these significant differences last? The answer to the second question has in part been answered by a follow-up study conducted by Prady et al. [34], which showed that 7 years after randomization, there were no sustained differences in patients' outcomes in the low back pain trial.

More difficult to answer is the question about why the benefits of acupuncture lasted so much longer than the 3 month treatment period in these two trials. It is possible to speculate that it may be because practitioners were allowed to treat patients as they normally would. Paterson and Britten have shown that patients in explanatory trials tend to "play their part," and though being treated "as a whole person" was associated with traditional acupuncture, this did not occur within the setting of a research trial [35]. Paterson and colleagues reported in a separate work based on in-depth interviews of patients who had participated in an explanatory acupuncture trial, that "The resulting changes to their (the patients) normal expectations and behavior influenced how the intervention was delivered and experienced" [36]. In contrast, interviews that were conducted with acupuncturists on the Thomas et al. low back pain trial found evidence that they specifically engaged in discussions about acupuncture-related lifestyle advice aimed to elicit long-term change [9,37]. Further research is called for to better understand the key factors that are associated with changes over the longer term.

Acknowledgment

Thanks are due to Sally Brabyn who read and commented on an earlier version of this chapter.

References

1. Unschuld, P. *Medicine in China: A History of Ideas*. Berkeley, CA: University of California Press, 1985.
2. Taylor, K. Divergent interests and cultivated misunderstandings; the influence of the West on modern Chinese medicine. *Soc Hist Med*, 17(1), 93–111, 2004.
3. Taylor, K. *Chinese Medicine in Early Communist China (1945–1963): Medicine in Revolution*. London, U.K.: Routledge, 2004.
4. MacPherson, H., Katchuk, T.J. (eds.). *Acupuncture in Practice: Case Histories from the West*. Edinburgh, U.K.: Churchill Livingstone, 1997.
5. Baldry, P.E. *Acupuncture, Trigger Points and Musculoskeletal Pain*, 2nd edn. Edinburgh, U.K.: Churchill Livingstone, 1998.
6. Hicks, A., Hicks, J., Mole, P. *Five Element Constitutional Acupuncture*. London, U.K.: Churchill Livingstone, 2007.
7. Hsu, E. (ed.). *Innovation in Chinese Medicine. Needham Research Institute Studies (No. 3)*. Cambridge, U.K.: Cambridge University Press, 2001.

8. Paterson, C., Dieppe, P. Characteristic and incidental (placebo) effects in complex interventions such as acupuncture. *BMJ*, 330(7501), 1202–1205, 2005.
9. MacPherson, H., Thorpe, L., Thomas, K. Beyond needling—Therapeutic processes in acupuncture care: A qualitative study nested within a low-back pain trial. *J Altern Complement Med*, 12(9), 873–880, 2006.
10. Sherman, K.J., Cherkin, D.C., Eisenberg, D.M., Erro, J., Hrbek, A., Deyo, R.A. The practice of acupuncture: Who are the providers and what do they do? *Ann Fam Med*, 3(2), 151–158, 2005.
11. MacPherson, H., Sinclair-Lian, N., Thomas, K. Patients seeking care from acupuncture practitioners in the UK: A national survey. *Complement Ther Med*, 14(1), 20–30, 2006.
12. House of Lords. *Report of the Select Committee on Science and Technology: Complementary and Alternative Medicine*. London, U.K.: The Stationary Office, 2000.
13. Fonnebo, V., Grimsgaard, S., Walach, H., et al. Researching complementary and alternative treatments—The gatekeepers are not at home. *BMC Med Res Methodol*, 7, 7, 2007.
14. Medical Research Council. *A Framework for Development and Evaluation of RCTs for Complex Interventions to Improve Health*. London, U.K.: Medical Research Council, 2000.
15. Ritenbaugh, C., Verhoef, M., Fleishman, S., Boon, H., Leis, A. Whole systems research: A discipline for studying complementary and alternative medicine. *Altern Ther Health Med*, 9(4), 32–36, 2003.
16. Hrobjartsson, A., Gotzsche, P.C. Is the placebo powerless? Update of a systematic review with 52 new randomized trials comparing placebo with no treatment. *J Intern Med*, 256(2), 91–100, 2004.
17. Macklin, E.A., Wayne, P.M., Kalish, L.A., et al. Stop hypertension with the Acupuncture Research Program (SHARP): Results of a randomized, controlled clinical trial. *Hypertension*, 48(5), 838–845, 2006.
18. Flachskampf, F.A., Gallasch, J., Gefeller, O., et al. Randomized trial of acupuncture to lower blood pressure. *Circulation*, 115(24), 3121–3129, 2007.
19. Kalish, L.A., Buczynski, B., Connell, P., et al. Stop Hypertension with the Acupuncture Research Program (SHARP): Clinical trial design and screening results. *Control Clin Trials*, 25(1), 76–103, 2004.
20. Dincer, F., Linde, K. Sham interventions in randomized clinical trials of acupuncture—A review. *Complement Ther Med*, 11(4), 235–242, 2003.
21. Fisher, P., Van Haselen, R., Hardy, K., Berkovitz, S., McCarney, R. Effectiveness gaps: A new concept for evaluating health service and research needs applied to complementary and alternative medicine. *J Altern Complement Med*, 10(4), 627–632, 2004.
22. Linde, K., Allais, G., Brinkhaus, B., Manheimer, E., Vickers, A., White, A.R. Acupuncture for tension-type headache. *Cochrane Database of Syst Rev*, (1), CD007587, 2009.
23. Manheimer, E., White, A., Berman, B., Forys, K., Ernst, E. Meta-analysis: Acupuncture for low back pain. *Ann Intern Med*, 142(8), 651–663, 2005.
24. Vickers, A.J., Rees, R.W., Zollman, C.E., et al. Acupuncture for chronic headache in primary care: Large, pragmatic, randomised trial. *BMJ*, 328(7442), 744, 2004.
25. Wonderling, D., Vickers, A.J., Grieve, R., McCarney, R. Cost effectiveness analysis of a randomised trial of acupuncture for chronic headache in primary care. *BMJ*, 328(7442), 747, 2004.

26. Thomas, K.J., MacPherson, H., Thorpe, L., et al. Randomised controlled trial of a short course of traditional acupuncture compared with usual care for persistent non-specific low back pain. *BMJ*, 333(7569), 623–626, 2006.

27. Ratcliffe, J., Thomas, K.J., MacPherson, H., Brazier, J. A randomised controlled trial of acupuncture care for persistent low back pain: Cost effectiveness analysis. *BMJ*, 333(7569), 626–628, 2006.

28. MacPherson, H., Thorpe, L., Thomas, K., Campbell, M. Acupuncture for low back pain: Traditional diagnosis and treatment of 148 patients in a clinical trial. *Complement Ther Med*, 12(1), 38–44, 2004.

29. MacPherson, H., White, A., Cummings, M., Jobst, K., Rose, K., Niemtzow, R. Standards for reporting interventions in controlled trials of acupuncture: The STRICTA recommendations. *Complement Ther Med*, 9(4), 246–249, 2001.

30. Ernst, E. Acupuncture or tree-hugging? http://www.bmj.com/cgi/eletters/333/7569/623#142850, March 14, 2009.

31. Grevitt, M. Acupuncture: Exact treatment effect yet to be determined. http://www.bmj.com/cgi/eletters/333/7569/623#142850, March 14, 2009.

32. MacPherson, H. Pragmatic clinical trials. *Complement Ther Med*, 12(2–3), 136–140, 2004.

33. Linde, K., Witt, C.M., Streng, A., et al. The impact of patient expectations on outcomes in four randomized controlled trials of acupuncture in patients with chronic pain. *Pain*, 128(3), 264–271, 2007.

34. Prady, S.L., Thomas, K., Esmonde, L., Crouch, S., MacPherson, H. The natural history of back pain after a randomised controlled trial of acupuncture vs usual care—Long term outcomes. *Acupunct Med*, 25(4), 121–129, 2007.

35. Paterson, C., Britten, N. The patient's experience of holistic care: Insights from acupuncture research. *Chronic Illn*, 4(4), 264–277, 2008.

36. Paterson, C., Zheng, Z., Xue, C., Wang, Y. "Playing their parts": The experiences of participants in a randomized sham-controlled acupuncture trial. *J Altern Complement Med*, 14(2), 199–208, 2008.

37. MacPherson, H., Thomas, K. Self-help advice as a process integral to traditional acupuncture care: Implications for trial design. *Complement Ther Med*, 16(2), 101–106, 2008.

22

Assessing Orthosis: Practical Examples

Katherine Sanchez, Claire Jourdan,
François Rannou, and Serge Poiraudeau

*Public Assistance—Hospitals of Paris and
Paris Descartes University*

CONTENTS

22.1 Introduction

Orthoses and braces are medical devices applied to the external surface of the body to achieve one or more of the following: relieve pain, immobilize musculoskeletal segments, reduce axial load, prevent or correct deformity, and improve function [1,2].

Various materials are available for the fabrication of orthoses. The choice of material determines to some extent the fabrication technique used. The selection of material and design for an orthosis is influenced by patient needs, the specific objectives for the orthosis, and the mechanical principles [2].

When prescribing the orthosis, the physician should educate the patient about its use and purpose. The patient needs to be competent in wearing and positioning the orthosis. The goals of the device should be discussed with patients so that they have a clear understanding of the role of the orthosis [3].

Orthosis therapy is one of the nonpharmacological treatment strategies for diverse pathologies and corresponds to the rehabilitation option. The types of limb and spine orthoses are varied (Table 22.1).

TABLE 22.1

Orthoses by Body Segments and Pathologies

Body Segments	Pathologies	Orthosis[a]
Upper limb	*Shoulder*	
	Subacromial impingement	Functional brace (Coopercare-Lastrap shoulder brace) [4]
	Acute bursitis/tendinitis	Sling [2]
	Elbow	
	Osteoarthritis	Posterior elbow splint [3]
	Rheumatoid arthritis	Dynamic spring splint, static resting splint [5]
	Tennis elbow	Brace [6], dynamic extensor brace [7]
	Wrist	
	Osteoarthritis/ rheumatoid arthritis	Working wrist splint [8], resting splint [5], cock-up wrist splint [3]
	Carpal tunnel syndrome	Cock-up wrist splint (20° extension), nocturnal splinting [9]
	Postoperative	Resting splint [10]
	Hemiplegics	Thermoplastic resting splint [11]
	Hand	
	Osteoarthritis/ rheumatoid arthritis	Oval eight splints [3], functional thumb orthosis [12]
		Silver ring splint [3], gutter splint, metacarpophalangeal (joint) splint [3]
	Basal joint osteoarthritis	CMC splint or C splint [3], neoprene thumb splint [13], thermoplastic CMC support [14]
	Rheumatoid arthritis	Static work ulnar deviation splint, resting volar splint, static resting splint [5]
	Tendon rupture, metacarpophalangeal (joint) implant	Dynamic outrigger splint with extension [15]
	DeQuervain's tendinitis	Thumb post-splint with wrist strap [2]
	Scleroderma	Extension splint, rolling splint, static orthosis [2]
Lower limb	*Hip*	
	Osteoarthritis	Canes, hip joint moment reduction brace, S-form hip brace [3]
	Knee	
	Osteoarthritis	Knee braces [16]
		Lateral wedge orthotics [17], medial wedge insole [18]
		Knee sleeve [19]
	Acute anterior cruciate ligament tear	Knee brace (SofTec Genu) [20]
	Postoperative anterior cruciate ligament reconstruction	DonJoy Iron brace [21]

TABLE 22.1 (continued)

Orthoses by Body Segments and Pathologies

Body Segments	Pathologies	Orthosis[a]
	Patellofemoral pain syndrome	Patellar brace, foot orthosis, flat insoles [22]
	Posterior tibial tendon dysfunction	Foot orthosis [23]
	Ankle	
	Osteoarthritis	Lace-up ankle brace (neoprene sleeve), semirigid foot orthosis, ankle-foot orthosis (solid thermoplastic) [3]
	Ankle sprain	Aircast ankle brace, below-knee cast, Bledsoe boot, Tabigrip, tubular bandage [24]
	Low-risk ankle fractures in children	Aircast (ankle brace), below-knee fiberglass walking cast [25]
	Foot	
	Juvenile rheumatoid arthritis, rheumatoid arthritis	PTB, shoe modifications, hindfoot orthosis [26]
	Achilles tendinopathy	AirHeel brace [27]
	Plantar fasciitis	Customized and prefabricated foot orthosis [28]
	Chronic stroke patients	Ankle foot orthosis, cane, slider shoe [29]
	Excessively pronated feet	Rigid foot orthosis (rear-foot wedge) [30]
	Pes cavus	Custom-made foot orthosis [31]
	Elderly woman	Custom-made insoles [32]
Spinal	*Cervical*	
	Single-level anterior cervical fusion with plate	Cervical collar [33]
	Whiplash injury	Collar [34,35]
	Benign paroxysmal positional vertigo	Soft cervical collar [36]
	Dorsal and lumbar	
	Adolescent idiopathic scoliosis	Rigid spinal orthosis (Boston brace) [37], flexible spinal orthosis (SpineCor) [38]
	Traumatic thoracolumbar spine fracture	Jewett hyperextension orthosis [39]
	Secondary prevention of low back pain	Lumbar corset (LumboTrain, LumboTrainLady, LumboLoc, LordoLoc) [40]
	Postoperative lumbar spinal arthrodesis for degenerative conditions, diskogenic low back pain, lumbar disk herniation	Lumbar corset [41]

[a] The orthoses listed have been assessed in at least one randomized clinical trial.

22.2 Specific Methodological Issues in Assessing This Treatment

Evaluating the effectiveness of an orthosis device raises specific methodological issues, and many randomized controlled trials (RCTs) in this field are considered to be of low quality.

To characterize the difficulties and specific issues in RCTs assessing orthoses, we searched MEDLINE and the Cochrane Central Register of Controlled Trials for reports of trials by using the search terms "orthosis" OR "ortheses" OR "splint," with a limitation to RCTs published in the five last years. Of 339 articles identified, 99 were selected, of which 43 referred to the lower extremities, 41 to the upper extremities, and 15 to the spine (Table 22.2).

The pathologies related to upper extremities were impacted proximal humeral fracture, traumatic anterior dislocation of the shoulder, subacromial impingement syndrome of the shoulder, displaced midshaft clavicular fractures, tennis elbow tendinitis, wrist contraction after stroke, rheumatoid arthritis, extensor tendon repair, carpal tunnel syndrome, wrist buckle fracture in children, torus fracture of the distal forearm, and hand osteoarthritis. Pathologies related to the lower extremities were patellofemoral pain syndrome, knee osteoarthritis, dislocation of the patella, anterior cruciate ligament tear and reconstruction, gait biomechanical alterations, ankle sprains, ankle fractures, plantar fasciitis, walking disability in chronic stroke patients, juvenile idiopathic arthritis, postural stability, pronated feet, prevention of falls in older people, quality of life in elderly women, tibial tendon dysfunction, flexible excess pronation of the feet, and tendinopathy of the Achilles tendon. Spine pathologies were whiplash injury, benign paroxysmal positional vertigo, idiopathic scoliosis, traumatic thoracolumbar spine fractures, prevention of recurrent low back pain among home care workers, lumbar spinal arthrodesis, diskogenic low back pain, and lumbar disk herniation.

To evaluate methodological differences in clinical trials of nonpharmacological and pharmacological treatments of hip and knee osteoarthritis, we used the list elaborated by Boutron et al. [42] to evaluate methodological differences in clinical trials of nonpharmacological and pharmacological treatments of hip and knee osteoarthritis. The methodological characteristics of the RCTs of orthoses classified by body segments are in Table 22.2. RCTs assessing orthoses for the lower limb were of higher quality than those evaluating the spine and upper limb. Some pitfalls, such as absence of reporting of allocation concealment, the use of per-protocol analysis, and absence of sample size justification, were similar to those reported for other clinical studies assessing pharmacological treatment [42].

The specific pitfalls of RCTs of orthoses most often identified were a low rate of reporting care provider experience (18%), issues about the choice of the control intervention (often reported to be an active control treatment

TABLE 22.2

Characteristics of Selected Trials Assessing Orthoses[a]

Characteristics	Upper Extremity (*n* = 41)	Lower Extremity (*n* = 43)	Spine (*n* = 15)	All Trials (*n* = 99)
Randomization				
Generation of allocation sequence				
Adequate	19 (46.3)	25 (58.1)	8 (53.3)	52 (52.5)
Inadequate	3 (7.3)	5 (11.6)	0	8 (8.1)
Not reported	19 (46.3)	13 (30.2)	7 (46.7)	39 (39.4)
Allocation concealment				
Adequate	7 (17.1)	16 (37.2)	3 (20.0)	26 (26.3)
Inadequate	2 (4.9)	1 (2.3)	1 (6.7)	4 (4.0)
Not reported	33 (80.5)	26 (60.5)	11 (73.3)	70 (70.7)
Intervention reproducible	37 (90.2)	39 (90.7)	14 (93.3)	90 (90.9)
Individualization	35 (85.4)	40 (93.0)	15 (100.0)	90 (90.9)
Influence of care provider skill	6 (14.6)	7 (16.3)	5 (33.3)	18 (18.2)
Control intervention				
Placebo	11 (26.8)	16 (37.2)	4 (26.7)	31 (31.3)
Active control treatment	25 (61.0)	26 (60.5)	9 (60.0)	61 (61.6)
Usual care or waiting list	5 (12.2)	2 (4.7)	3 (20.0)	10 (10.1)
Potential similar placebo effect	28 (68.3)	30 (69.8)	11 (73.3)	69 (69.7)
Blinding				
Patients	0	7 (16.3)	2 (13.3)	9 (9.1)
Care providers	0	1 (2.3)	0	1 (1.0)
Outcome assessors	23 (56.1)	26 (60.5)	12 (80.0)	61 (61.6)
Analysis				
Full ITT performed[b]	7 (17.1)	12 (27.9)	5 (33.3)	24 (24.2)
Modified ITT performed[c]	1 (2.4)	0	1 (6.7)	2 (2.0)
Per-protocol analysis performed	12 (29.3)	11 (25.6)	5 (33.3)	28 (28.3)
Unclear	21 (51.2)	21 (48.8)	4 (26.7)	46 (46.5)
Sample size justification or study power reported	13 (31.7)	18 (41.9)	5 (33.3)	36 (36.4)

Abbreviation: ITT, intention-to-treat analysis.

[a] All data are expressed as number (%).

[b] All randomized participants were included in the analysis according to their original group assignment.

[c] Analysis excluded participants who never received treatment or who were never evaluated while receiving treatment.

or absence of equivalent placebo effect between control and intervention groups), and low rate of blinding patients and care providers.

Due to the high percentage of individualized interventions (91%), the description and performance of interventions are complex. Therefore, reporting the experience of care providers is important to increase the generalizability of results.

The choice of the control intervention and the blinding of patients could be intimately linked issues in RCTs of orthoses. Ideally, a similar placebo effect should be ascertained in control and intervention groups. This objective is often difficult to achieve because in many cases, a credible placebo orthosis does not exist. Therefore, to minimize bias induced by a deception effect, the patient's preference and the treatment credibility should be routinely recorded, and blinding the patients to the study hypothesis by the use of the Zelen or modified Zelen design is favored. These methodological issues are also site specific: for example, for many upper limb orthoses, no credible placebo orthosis exists, whereas for foot orthoses or insoles, the issue is more to ascertain that the flat or neutral placebo insert, commonly manufactured with the same material as that for the intervention group, has no specific therapeutic effect.

Blinding of assessors is difficult when patients are not blinded and depends on the outcome measures chosen. For most RCTs assessing spine devices (80%), the primary outcome measures were radiographic measurements, which can be analyzed by an external evaluator. Therefore, the issue is not the credible blinding of assessors but, rather, the clinical relevance of choosing radiographic assessment as the primary outcome measure. With functional outcomes, self-reported questionnaires and objective information such as number of days of pain and number of calendar days of sick leave are favored.

22.3 Practical Examples

In this chapter, we focus on specific methodological issues of three RCTs assessing orthoses, to shed light on the CONSORT recommendations for nonpharmacological treatments published in 2008 [43].

22.3.1 Splint for Base-of-Thumb Osteoarthritis

Context: Some guidelines recommend splinting for base-of-thumb osteoarthritis (BTOA), despite lack of evidence of efficacy of this orthosis.

Objective: To assess the efficacy and acceptability of splinting for BTOA [13].

Primary outcome: Change in pain level assessed on a visual analogue scale from baseline to 1 month.

The design was an open-label, parallel group, multicenter, RCT, considering two treatment groups:

- Custom-made neoprene splint
- Usual care

The trial's main limitations are specific to nonpharmacological trials:

- The authors did not blind patients (no credible placebo splint exists), healthcare providers, or outcome assessors to the study intervention and did not blind patients to the study hypotheses. However, the success of such methods has not been demonstrated, and the methods increase the complexity of the organization and cost of trials. The authors also did not record treatment credibility. Consequently, participants under usual care being aware of their treatment assignment may have contributed to their worse outcomes. However, evolution of BTOA was comparable in the two trial arms during the first month; no patients in the control group crossed over to wearing a splint; and patients in each arm had the same number of visits, so these effects were likely marginal. The use of co-interventions was higher in the control group, which could have reduced the apparent treatment effect attributable to the splint alone.
- The trial was conducted in two tertiary care teaching hospitals, and splints were custom-made by trained occupational therapists. Therefore, findings may not be generalizable to other settings.

22.3.2 Effectiveness of Foot Orthoses to Treat Plantar Fasciitis

Context: Foot orthoses are commonly used in the conservative treatment of plantar fasciitis. However, systematic reviews of RCTs on this subject have concluded poor evidence of effectiveness and that further investigation with rigorous RCTs is needed.

Objective: To evaluate the short- and long-term effectiveness of foot orthoses to treat plantar fasciitis [28].

Primary outcome: Change in pain and function at 3 and 12 months.

A three-arm, participant-blinded RCT:

- Sham orthoses
- Prefabricated orthoses
- Customized orthoses

This trial's main limitations are specific to both nonpharmacological treatment and more general treatment (shared between nonpharmacological and pharmacological treatments):

1. Nonpharmacological-specific limitations
   a. The material and techniques used to fabricate or size the prefabricated orthoses were well described, but patients were not given recommendations on how and when to wear the orthoses and treatment compliance was not recorded.
   b. The success of blinding of patients was not recorded nor was treatment credibility.
2. General limitations
   a. Surprisingly, despite efforts to make treatment and placebo orthoses look as similar as possible so the participants were blinded to the device, the assessor was not blinded. Even if the outcome measures were self-reported by patients, this absence of blinding of the assessor could create a bias.
   b. Authors stated that an intention-to-treat analysis was performed, but the number of patients analyzed was not provided in tables to verify this assumption.
   c. A high rate of contamination and broken protocol (23%) was reported at 12 month follow-up.

22.3.3 Lumbar Supports to Prevent Recurrent Low Back Pain among Home Care Workers

Context: In a previous uncontrolled feasibility study, the authors found that home care workers who had frequent episodes of low back pain reported adherence rates of 61%–81% with lumbar supports and a 45% decrease in pain intensity when using these devices.

Objective: To evaluate the effectiveness of adding worker-directed use of lumbar supports to a short course on healthy working methods to reduce low back pain and work absenteeism among home care workers with a history of low back pain [40].

Primary outcome: Number of days of low back pain and sick leave over 12 months.

An RCT with two treatment groups:

- Intervention group: usual care and lumbar support (one of four types)
- Control group: usual care

As in the previous example (Section 22.3.2), the methodological issues were specific to nonpharmacological treatment and common to pharmacological treatment.

1. Specific limitations
   a. No specific information was given to patients on how to wear and position the lumbar supports.
   b. Patients in the control group and care providers were not blinded, which increased the chance of measurement bias, especially because self-reported outcome measures were used. Participants had positive expectations of the lumbar supports, which could have overestimated results.
   c. Patient preference and expectations were recorded but not treatment credibility.
   d. Contamination was marginal: 2 of 177 patients wore a privately bought lumbar support.
   e. Adverse effects were not reported.
   f. Complex interventions are more often assessed with more than one primary outcome than are simple interventions. This trial contained two main outcome measures. Significant effects were found with the outcome number of days with low back pain during 12 months but not number of days on sick leave during 12 months. Therefore, drawing conclusions may be problematic.
2. General limitations
   a. A substantial number of missing data required imputation.
   b. Objective data on sick days due to low back pain were not available, and the reliability of self-reported outcome measures on this topic remains questionable.

22.4 Conclusion

Some methodological pitfalls in reports of RCTs of orthoses are specific to nonpharmacological treatment, and others are common to pharmacological treatment. The most commonly observed pitfalls are a high rate of not reporting allocation concealment, lack of information about care provider skill level and experience, low use of a placebo intervention, low rate of blinding of patients and care providers, low use of intention-to-treat analysis, low reporting of simple size justification (only 36% of recent RCTs), and lack of educating patients on how to wear the orthosis.

Several specific pitfalls are important when interpreting data from such RCTs. Care provider skill level and experience influence the treatment effect. In trials of pharmacological treatment, the effect of the healthcare professional can be considered secondary but is an integral part of the intervention in trials of nonpharmacological treatment. Variation in care provider skills in each arm of a trial can be confounded with the treatment effect [44].

The most important methodological pitfall of RCTs of orthoses may be the absence of blinding. This bias is already known and has been reported for RCTs assessing nonpharmacological treatment [42,45]. Moreover, the lack of blinding could bias treatment effect estimates [46–49]. Several methods to reduce the impact of nonblinding should be used more often in RCTs assessing orthoses; these involve recording treatment credibility and patient preference, blinding the study hypothesis to patients by the use of the Zelen or modified Zelen design, and blinding the outcome assessor to the primary outcome [50,51]. Finally, the rate of contamination could be lowered with a co-intervention.

References

1. Deshaies, L.D. Upper extremity orthoses, in Trombly, C.A., Radomski, M.V. (eds.), *Occupational Therapy for Physical Dysfunction*, 5th edn. Baltimore, MD: Lippincott, Williams and Wilkins, 2002.
2. Hicks, J.E., Leonard, J.A., Nelson, J.R., Fisher, S.V., Esquenazi, A. Prosthetics, orthotics, and assistive devices. 4. Orthotic management of selected disorders. *Arch Phys Med Rehabil*, 70(Suppl. 5), S210–S217, 1989.
3. Yonclas, P.P., Nadler, R.R., Moran, M.E., Kepler, K.L., Napolitano, E. Orthotics and assistive devices in the treatment of upper and lower limb osteoarthritis. *Am J Phys Med Rehabil*, 85, S82–S97, 2006.
4. Walther, M., Werner, A., Stahlschmidt, T., Woelfel, R., Gohlke, F. The subacromial impingement syndrome of the shoulder treated by conventional physiotherapy, self-training, and a shoulder brace: Results of a prospective, randomized study. *J Shoulder Elbow Surg*, 13, 417–423, 2004.
5. Adams, J., Burridge, J., Mullee, M., Hammond, A., Cooper, C. The clinical effectiveness of static resting splints in early rheumatoid arthritis: A randomized controlled trial. *Rheumatology*, 47, 1548–1553, 2008.
6. Struijs, P.A., Korthals-de Bos, I.B., Van Tulder, M.W., Van Dijk, C.N., Bouter, L.M., Assendelft, W.J. Cost effectiveness of brace, physiotherapy or both for treatment of tennis elbow. *Br J Sports Med*, 40, 637–643, 2006.
7. Faes, M., Van Der Akker, B., De Lint, J.A., Kooloos, J.G., Hopman, M.T. Dynamic extensor brace for lateral epicondylitis. *Clin Orthop Pelat Res*, 442, 149–157, 2006.
8. Veehof, M.M., Taal, E., Heijnsdijk-Rouwenhorst, L.M., Van Der Laar, M.A. Efficacy of wrist working splints in patients with rheumatoid arthritis: A randomized controlled study. *Arthritis Rheum*, 59, 1698–1704, 2008.
9. Werner, R.A., Franzblau, A., Gell, N. Randomized controlled trial of nocturnal splinting for active workers with symptoms of carpal tunnel syndrome. *Arch Phys Med Rehabil*, 86, 1–7, 2005.

10. Bulstrode, N.W., Burr, N., Pratt, A.L., Grobbelaar, A.O. Extensor tendon rehabilitation a prospective trial comparing three rehabilitation regimes. *J Hand Surg Br*, 30(2), 175–179, 2005.

11. Sheehan, J.L., Winzeler-Merçay, U., Mudie, M.H. A randomized controlled pilot study to obtain the best estimate of the size of the effect of a thermoplastic resting splint on spasticity in the stroke-affected wrist and fingers. *Clin Rehabil*, 20, 1032–1037, 2006.

12. Silva, P.G., Lombardi, I.J., Breitschwerdt, C., Poli Araujo, P.M., Natour, J. Functional thumb orthosis for type I and II boutonniere deformity on the dominant hand in patients with rheumatoid arthritis: A randomized controlled study. *Clin Rehabil*, 22, 684–689, 2008.

13. Rannou, F., Dimet, J., Boutron, I., et al. Splint for base-of-thumb osteoarthritis, a randomized trial. *Ann Intern Med*, 150, 661–669, 2009.

14. Weiss, S., LaStayo, P., Mills, A., Bramlet, D. Splinting the degenerative basal joint: Custom-made or prefabricated neoprene? *J Hand Ther*, 17, 401–406, 2004.

15. Mowlavi, A., Burns, M., Brown, R.E. Dynamic versus static splinting of simple zone V and zone VI extensor tendon repairs: A prospective, randomized, controlled study. *Plast Reconstr Surg*, 115, 482–487, 2005.

16. Brouwer, R.W., Van Raaij, T.M., Verhaar, J.A., Coene, L.N., Bierma-Zeinstra, S.M. Brace treatment for osteoarthritis of the knee: A prospective randomized multicentre trial. *Osteoarthritis Cartilage*, 14, 777–783, 2006.

17. Toda, Y., Tsukimura, N. A 2-year follow-up of a study to compare the efficacy of lateral wedged insoles with subtalar strapping and in-shoe lateral wedged insoles in patients with varus deformity osteoarthritis of the knee. *Osteoarthritis Cartilage*, 14, 231–237, 2006.

18. Rodrigues, P.T., Ferreira, A.F., Pereira, R.M., Bonfa, E., Borba, E.F., Fuller, R. Effectiveness of medial-wedge insole treatment for valgus knee osteoarthritis. *Arthritis Rheum*, 59, 603–608, 2008.

19. Chuang, S.H., Huang, M.H., Chen, T.W., Weng, M.C., Liu, C.W., Chen, C.H. Effect of knee sleeve on static and dynamic balance in patients with knee osteoarthritis. *Kaohsiung J Med Sci*, 23, 405–411, 2007.

20. Swirtun, L., Jansson, A., Renström, P. The effects of a functional knee brace during early treatment of patients with a nonoperated acute anterior cruciate ligament tear. *Clin J Sport Med*, 15, 299–304, 2005.

21. McDevitt, E., Taylor, D., Miller, M., et al. Functional bracing after anterior cruciate ligament reconstruction. *Am J Sports Med*, 32, 1887–1892, 2004.

22. Collins, N., Crossley, K., Beller, E., Darnell, R., McOoil, T., Vicenzino, B. Foot orthoses and physiotherapy in the treatment of patellofemoral pain syndrome: Randomized clinical trial. *BMJ*, 337, a1735, 2008.

23. Kulig, K., Reichl, S.F., Pomrantz, A.B., et al. Nonsurgical management of posterior tibial tendon dysfunction with orthoses and resistive exercise: A randomized controlled trial. *Phys Ther*, 89, 26–37, 2009.

24. Lamb, S.E., Marsh, J.L., Hutton, J.L., Nakash, R., Cooke, M.W. Mechanical supports for acute, severe ankle sprain: A pragmatic, multicentre, randomised controlled trial. *Lancet*, 373, 575–581, 2009.

25. Boutis, K., Willan, A.R., Babyn, P., Narayanan, U.G., Alman, B., Schuh, S. A randomized, controlled trial of a removable brace versus casting in children with low-risk ankle fractures. *Pediatrics*, 119, e1256–e1263, 2007.

26. Powell, M., Seid, M., Szer, I.S. Efficacy of custom foot orthosis in improving pain and functional status in children with juvenile idiopathic arthritis: A randomized trial. *J Rheumatol*, 32, 943–950, 2005.

27. Petersen, W., Welp, R., Rosenbaum, D. Chronic achilles tendinopathy. A prospective randomized study comparing the therapeutic effect of eccentric training, the AirHeel brace and a combination of them. *Am J Sports Med*, 35, 1659–1667, 2007.

28. Landorf, K.B., Keenan, A.-M., Herbert, R.D. Effectiveness of foot orthoses to treat plantar fasciitis. *Arc Intern Med*, 166, 1305–1310, 2006.

29. Tyson, S.F., Rogerson, L. Assistive walking devices in nonambulant patients undergoing rehabilitation after stroke: The effects on functional mobility, walking impairments, and patient's opinion. *Arch Phys Med Rehabil*, 90, 475–479, 2009.

30. Rome, K., Brown, C.L. Randomized clinical trial into the impact of rigid foot orthoses on balance parameters in excessively pronated feet. *Clin Rehabil*, 18, 624–630, 2004.

31. Burns, J., Landorf, K.B., Ryan, M.M., Crosbie, J., Ouvrier, R.A. Interventions for the prevention and treatment of pes cavus. *Cochrane Database Syst Rev*, (4), CD006154, 2007.

32. Kusumoto, A., Susuki, T., Yoshida, H., Kwon, J. Intervention study to improve quality of life and health problems of community-living elderly women in Japan by shoe fitting and custom-made insoles. *Gerontology*, 53, 348–356, 2007.

33. Campbell, M.J., Carreon, L.Y., Traynelis, V., Anderson, P.A. Use of cervical collar after single-level anterior cervical fusion with plate. *Spine*, 34, 43–48, 2008.

34. Dehner, C., Hartwig, E., Strobel, P., et al. Comparison of the relative benefits of 2 versus 10 days of soft collar cervical immobilization after acute whiplash injury. *Arch Med Rehabil*, 87, 1423–1427, 2006.

35. Kongsted, A., Qerama, E., Kasch, H., et al. Neck collar, "act-as-usual" or active mobilization for whiplash injury? *Spine*, 32, 618–626, 2007.

36. Cakir, B.O., Ercan, I., Cakir, Z.A., Turgut, S. Efficacy of postural restriction in treating benign paroxysmal positional vertigo. *Arch Otolaryngol Head Neck Surg*, 132, 501–505, 2006.

37. Labelle, H., Bellefleur, C., Joncas, J., Aubin, C.-E., Cheriet, F. Preliminary evaluation of a computer-assisted tool for the design and adjustment of braces in idiopathic scoliosis. *Spine*, 32, 835–843, 2007.

38. Wong, M.S., Cheng, J.C.Y., Lam, T.P., et al. The effect of rigid versus flexible spinal orthosis on the clinical efficacy and acceptance of the patients with adolescent idiopathic scoliosis. *Spine*, 33, 1360–1365, 2008.

39. Siebenga, J., Leferink, V.J.M., Segers, M.J.M., et al. Treatment of traumatic thoracolumbar spine fractures: A multicenter prospective randomized study of operative versus nonsurgical treatment. *Spine*, 31, 2881–2890, 2006.

40. Roelofs, P.D., Bierma-Zeinstra, S.M., Van poppel, M.N., et al. Lumbar supports to prevent recurrent low back pain among home care workers: A randomized trial. *Ann Intern Med*, 20, 685–692, 2007.

41. Yee, A.J., Yoo, J.U., Marsolais, E.B., et al. Use of a postoperative lumbar corset after lumbar spinal arthrodesis for degenerative conditions of the spine. A prospective randomized trial. *J Bone Joint Surg Am*, 90, 2062–2068, 2008.

42. Boutron, I., Tubach, F., Giraudeau, B., Ravaud, P. Methodological differences in clinical trials evaluating nonpharmacological and pharmacological treatments of hip and knee osteoarthritis. *JAMA*, 290, 1062–1070, 2003.

43. Boutron, I., Moher, D., Altman, D.G., Schultz, K.F., Ravaud, P. Extending the CONSORT statement to randomized trials of nonpharmacologic treatments: Explanation and elaboration. *Ann Int Med*, 148, 295–309, 2008.
44. Roberts, C. The implications of variation in outcome between health professionals for the design and analysis of randomized controlled trials. *Stat Med*, 18, 2605–2615, 1999.
45. Boutron, I., Tubach, F., Giraudeau, B., Ravaud, P. Blinding was judged more difficult to achieve and maintain in non-pharmacological than pharmacological trials. *J Clin Epidemiol*, 57, 543–550, 2004.
46. Schultz, K.F., Chalmers, I., Hayes, R.J., Altman, D.G. Empirical evidence of bias: Dimensions of methodological quality associated with estimates of treatment effects in controlled trials. *JAMA*, 273, 408–412, 1995.
47. Moher, D., Pharm, B., Jones, A., et al. Does quality of reports of randomised trials affect estimates of intervention efficacy reported in meta-analysis? *Lancet*, 352, 609–613, 1998.
48. Fergusson, D., Glass, K.C., Waring, D., Shapiro, S. Turning a blind eye: The success of blinding reported in a random sample of randomized placebo controlled trials. *BMJ*, 328, 432, 2004.
49. Boutron, I., Estellar, C., Ravaud, P. A review of blinding in randomized controlled trials found results inconsistent and questionable. *J Clin Epidemiol*, 58, 1220–1226, 2005.
50. Boutron, I., Guittet, L., Candice, E., Moher, D., Hrobjartsson, A. Reporting methods of blinding in randomized trials assessing nonpharmacological treatments. *PLoS Med*, 4, 370–380, 2007.
51. Zelen, M. Randomised consent trials. *Lancet*, 340, 375, 1992.

23

Assessing Rehabilitation: Practical Examples

Nadine Elizabeth Foster

Keele University

CONTENTS

23.1 Introduction

Rehabilitation and specifically exercise is a very common nonpharmacological intervention used for the prevention and treatment of a range of health problems, including cardiovascular disease [1], diabetes [2], and musculoskeletal problems [3,4]. Exercise therapy encompasses a broad range of specific interventions and approaches, and can include interventions ranging from specific strengthening exercises, balance and flexibility exercises on land and in water, as well as more general physical activities. Exercise can be advised, prescribed, and delivered by health professionals such as family doctors [5] and physiotherapists [6] or it can be conducted by individuals within their own home or community. Exercise can be delivered in a variety of ways, through individually designed or standardized programs, home exercises, supervised exercises, group or individually tailored programs, and different intensities of exercise with or without the addition of other treatment approaches [7]. Exercise is a good example of a complex, nonpharmacological intervention [8] in that it involves a health professional encouraging an individual to change their behavior over time.

Musculoskeletal pain is a major health problem [9] and the most common types of musculoskeletal pain that impact significantly on functional disability are spinal pain and knee pain [10,11]. Knee pain in older adults is a common disabling problem, managed in the United Kingdom mostly in primary care [12]. Osteoarthritis is the most likely underlying diagnosis and has been shown by radiography to be present in 70% of community dwelling adults aged 50 or more with knee pain [13]. Structural changes before radiography are common in the remainder [14]. In older people, the risk of disability from knee osteoarthritis is as great as the risk of disability from cardiac disease and greater than that due to any other single medical disorder [15]. The aging population and the increase in prevalence of key risk factors for knee osteoarthritis (such as obesity) means that knee pain related to osteoarthritis is a rising problem. The primary prevention of these musculoskeletal conditions has not proved feasible, and thus modern management approaches, including the "core" intervention of exercise [16,17], focus on reducing symptoms, preventing unnecessary disability, and minimizing morbidity. Many of the therapeutic options are nonpharmacological.

Clinical trials and systematic reviews consistently show the benefit of exercise, in a variety of forms, for this patient group [3,4,18,19]. Exercise improves muscle dysfunction and reduces pain and disability without exacerbating joint damage [20]. It can also reduce the risk of other chronic conditions [21]. Recent national and international clinical guidelines [17,22,23] support the overall effectiveness of exercise, placing it as a key component of "core" treatment in primary care. There have been several summaries of systematic reviews of evidence for exercise [3,19,24]. The most recent of these [24] summarized 16 systematic reviews of the role of exercise for musculoskeletal disorders and concluded that exercise therapy is a beneficial component of the management of musculoskeletal conditions for reducing pain and disability.

23.2 Specific Methodological Issues in Assessing This Treatment

The randomized controlled trial (RCT) is the most appropriate tool to obtain unbiased estimates of the effects of interventions and pragmatic trials are broadly defined as RCTs that aim to inform decisions about practice, helping clinicians and policy makers to choose between different care options. A pragmatic trial addresses this type of question "Is this treatment better than the alternative, on average, for a wide range of patients?"

Although there is a plethora of RCTs in the osteoarthritis and exercise literature, many provide examples of poor quality, from simple reporting errors to serious deficiencies and biases in their design, conduct, and analysis. Key features of high-quality RCTs tend to include a relatively simple design, good internal validity (i.e., we can depend upon the results of the trial), good external validity (i.e., we can apply the trial results to the wider clinical population), recruitment of a large sample representative of the patient group often from many centers, randomization of individuals to different treatment approaches, an appropriate control or comparison treatment, concealment of the randomization from both patients and clinicians, a clear focus on a primary outcome, and an intention-to-treat analysis.

However, designing RCTs of nonpharmacological interventions that accurately reflect the characteristics and complexities of clinical care is challenging. Complex interventions try to change the behavior of patients or practitioners, for example, through adhering to healthcare advice or a home exercise program [8]. Thus they depend on a range of factors, including the beliefs and behaviors of the health professional and the beliefs and adherence of the patient. Typically, nonpharmacological therapies are delivered as part of a package of care rather than as a single treatment alone [25], further complicating the design of trials to test treatment effectiveness. The methodological challenges posed by studying "complex" interventions [8] include the need to define the various components of the intervention including their anticipated specific and nonspecific effects, determine the characteristics of patients that may respond to a multimodal intervention, and ensure consistent and high-quality delivery of the treatment program.

23.3 Practical Examples

This chapter draws on two examples of RCTs managed by the Arthritis Research Campaign UK Primary Care Centre at Keele University in the United Kingdom. Both RCTs included a lower limb advice and exercise program led by physiotherapists. First, the Treatment Options for Pain In the Knee (TOPIK) trial [26] involved 325 adults with knee pain in a multicenter RCT that evaluated the effectiveness of two primary care treatment approaches: enhanced pharmacy review and community physiotherapy and compared these with a control intervention. Second, the Acupuncture, Physiotherapy and EXercise (APEX) trial [27] investigated the benefit of adding acupuncture to a package of advice and exercise delivered by physiotherapists, in 37 clinical centers, for pain reduction in 352 older adults with knee pain. Both trials collected data from patients with a 12 month follow-up. Using these recent examples of RCTs investigating the effectiveness of nonpharmacological interventions for patients with clinical knee OA, this chapter provides further detail regarding

- The key challenges in design, conduct, and analysis of high-quality RCTs
- Approaches to addressing both internal validity and external validity by attending to issues of transparency, bias, and clinical applicability
- Ways to improve the quality of RCTs of nonpharmacological interventions

23.3.1 Treatment Options for Pain in the Knee Trial

The context, objective, and primary outcome for the TOPIK trial are summarized in the box as follows.

MANAGEMENT OF KNEE PAIN IN OLDER ADULTS IN PRIMARY CARE

Context: The traditional family doctor-led service to deliver interventions is increasingly unsustainable, and evaluation of alternative models utilizing the skills of other members of the primary healthcare team is needed.

Objective: To evaluate the effectiveness of two primary care treatment approaches for delivering evidence based care to older people with knee pain: enhanced pharmacy review and community physiotherapy.

Primary outcome: A patient-reported evaluation of pain and function (the Western Ontario and McMaster Universities Osteoarthritis Index) at 6 months.

Full details: Hay et al. [26].

23.3.1.1 Trial Planning

23.3.1.1.1 Trial Aim and Design

This trial was designed as a three arm, parallel group, pragmatic RCT to compare three treatment approaches for older adults with knee pain in primary care. At the time of trial design, there were at least two other services that were felt to have the potential to provide systematic, effective care: first, an enhanced pharmacy review service by community pharmacists to optimize drug management of knee pain and reinforce simple self-help messages. Second, a community physiotherapy service that could promote self-management alongside an exercise-based treatment package, in an effort to maximize the benefit of nondrug approaches. We did not wish to compare the efficacy of specific modalities (tablets and exercise) but to evaluate two approaches to delivering care for patients with knee pain (pharmacy and physiotherapy). We therefore conducted a pragmatic RCT to compare the clinical effectiveness, in primary care, of enhanced pharmacy review or community physiotherapy with a control intervention in the treatment of adults aged 55 and over consulting their family doctor with knee pain.

23.3.1.1.2 Control Group Choice

The control group had to be credible to those consulting their family doctor, and could not be a no-treatment control given that this patient group was already seeking care for their symptoms. All patients in the trial had ongoing access to their family doctor (usual primary care) and thus we chose an advice leaflet reinforced by one telephone call by a rheumatology nurse as the control intervention.

23.3.1.1.3 Sample

We recruited participants from 15 general (family) practices. All adults aged 55 years and over who consulted their family doctor with pain, stiffness, or both in one or both knees and who were able to give written, informed consent were invited to participate. We excluded individuals with potentially serious pathology (such as inflammatory arthritis, acute trauma, or malignancy), previous knee joint arthroplasty, those on a waiting list for surgery, and those who had received recent physiotherapy or an intra-articular joint injection to the knee. We purposely did not limit trial participation to only those with radiographic evidence of knee osteoarthritis because we wanted the trial to have a representative sample and to reflect current clinical practice as much as possible, where treatment decisions for people with knee pain seeking health care are made on the basis of presenting symptoms rather than radiographic findings.

23.3.1.1.4 Intervention Protocols

The intervention content was based on recommendations from international guidelines [23,16] and agreed in collaboration with participating clinicians in two workshops before the trial started. The community physiotherapy intervention aimed to encourage patients to engage in an active approach to managing knee pain through education about the safety and importance of exercise, pacing, pain relief, coping strategies, and an individualized exercise program. In total, 19 physiotherapists delivered the intervention, selecting exercises from an agreed list from exercise computer software, including general aerobic and specific muscle strengthening and stretching exercises. The protocol permitted between three to six sessions over 10 weeks and each physiotherapist recorded the treatments in a standardized format, on case report forms, developed for the trial. Thus we could test whether treatments provided were in line with the study protocols; 99 of the 109 participants randomized to physiotherapy attended at least one treatment session, 97 had a home exercise program, and 83 had three or more treatment sessions.

Experienced clinicians delivered the interventions, in accordance with standardized study protocols that were designed to reflect evidence-based practice while retaining sufficient flexibility to ensure that clinicians could develop individualized treatment plans to reflect clinical need [28]. For example, although the physiotherapy intervention protocol permitted up to six treatment sessions with a physiotherapist, the intervention was actually delivered in fewer sessions (a median of four sessions).

23.3.1.1.5 Sample Size Calculation

In order to control for random error in trials, a sufficiently large sample size is needed. The primary outcome was self-reported pain and function at 6 months (using specific subscales of the Western Ontario and McMaster Universities osteoarthritis index or WOMAC) [29]. The psychometric properties of this tool have been extensively studied in knee pain populations in both trials and in postal surveys. We based our sample size on expected changes in WOMAC pain and function scores of 20% between each of the experimental interventions and the control group, assuming that pain and function may improve by 5% in the control group (from previous trials). Full details are provided in Hay et al. [26] but a minimum of 270 participants with post-randomization outcome data at 6 months was sufficient to detect these effects with 80% power at a 5% significance level. We recruited 325 participants to allow for a 20% loss to follow-up at 6 months.

23.3.1.1.6 Statistical Analyses

Analysis was by intention to treat (analysis as per allocated intervention). We calculated estimates of treatment effects (control intervention minus active treatment group) with 95% confidence intervals and used appropriate inferential tests depending on the level of measurement of the data. We used two exploratory sensitivity analyses of the mean WOMAC scores; an analysis of covariance by using multiple linear regression with adjustment for covariates, based on random differences in baseline characteristics and an "on-treatment" or "per-protocol" analysis, restricting the comparison to participants who received their allocated treatment per protocol.

23.3.1.2 Trial Conduct

23.3.1.2.1 Recruitment Methods and External Validity

We aimed to ensure that we successfully recruited the number of patients needed and that we recruited a sample representative of older adults consulting in primary care with knee pain. Thus, we used two methods of recruitment: direct referral from general practitioners (family doctors) and retrospective review of records. We asked the family doctors in the 15 participating practices, during a consultation for knee pain and aided by a prompt appearing on their computer screen when a knee pain–related Read code was entered, to explain the trial to potential participants and give them a study information leaflet. For the retrospective record review, a monthly audit of each practice's computerized records was conducted to identify potential participants not recruited by the direct referral method. A family doctor from each practice then screened lists of potential participants, identified by Read codes, and those that were considered eligible were posted an invitation letter and a study information leaflet. For both methods of recruitment, the study nurse arranged a home visit within 10 working days to gain written consent to randomization and conduct a baseline assessment. Using both of these recruitment methods ensured that all potentially eligible patients were identified and invited to participate.

We assessed external validity by comparing the demographic characteristics of patients obtained through direct referral from family doctors who were not randomized in the trial with those of trial participants. Within trial participants, we compared the recruitment characteristics and treatment allocation across high and low recruiting practices and participants recruited through direct referrals and review of records. These tests of external validity showed that trial participants were similar to nontrial participants and that treatment allocation and baseline characteristics were similar between participants recruited through direct referral and those recruited by record review. Thus, we could be confident that the sample recruited was representative of the wider population of older adults consulting their family doctor with knee pain.

23.3.1.2.2 Randomization and Allocation Concealment

It is important that intervention arms are sufficiently similar at the beginning of an RCT. In order to achieve this, we used a computerized random number generator to produce a predetermined random allocation sequence in blocks of six by family practice. Each participant was allocated a unique study number that corresponded with that on a sealed opaque envelope that contained information about participants' allocated treatment. This was given to consenting participants after their baseline assessment, during a home visit of the study nurse. Through this method, participants were randomly assigned to each of the three treatment approaches. The success of randomization was evident in the comparable intervention groups at the beginning of the TOPIK trial.

It is important, however, also to conceal the randomization outcome and to assess the success of this in RCTs. In the TOPIK trial, our processes ensured that none of the family doctors, physiotherapists, or pharmacists had any influence over which patients were allocated to which intervention.

Concealment of treatment from the study nurse was tested and considered to be effective (for only 15 of 325 participants was treatment allocation revealed).

23.3.1.2.3 *Blinding*

In order to maintain blinding of the study nurse, participants were instructed not to open their envelope in her presence but to wait until after the nurse had completed the home visit and left. Study nurses and researchers who collected, entered, and analyzed data were unaware of treatment allocation. By necessity, participants and the health professionals delivering the treatments were not blind to allocation.

23.3.1.2.4 *Co-Interventions*

In addition to other secondary outcome measures, we recorded participants' use of co-interventions at each follow-up time point. We collected data on self-reported consultations with the family doctor or other health professional for knee pain and on self-reported medication use. These showed that a higher proportion of participants in the control group than in the physiotherapy group reported consulting their family doctor for knee pain during the follow-up. Self-reported use of nonsteriodal anti-inflammatory drugs and simple analgesia was significantly lower in the physiotherapy group than in the control group. Overall, the physiotherapy intervention seemed to produce a shift in consultation behavior away from the traditional family doctor-led model of care.

23.3.1.2.5 *Primary Outcome and Follow-Up*

The primary outcome was WOMAC pain and function at 6 months. Follow-up was at 3, 6, and 12 months after randomization, by postal questionnaire. We followed a standardized process for follow-ups, sending postal reminders to participants and achieved very good follow-up rates (91% for the control group, 95% for the pharmacy group, and 88% for the physiotherapy group at 6 months). We compared those lost to follow-up at 6 months with those who responded and found that they tended to be male, slightly older, and have higher baseline pain and function scores. The results of the TOPIK trial are therefore least likely to apply to this population of consulters.

23.3.2 Acupuncture, Physiotherapy and Exercise Trial

The context, objective, and primary outcome for the APEX trial are summarized in the box as follows. Given that many of the design and process challenges of this trial were similar to those of the previous example, the TOPIK trial, only those issues that were different or for which different decisions were made are detailed.

ACUPUNCTURE AS AN ADJUNCT TO EXERCISE-BASED PHYSIOTHERAPY FOR KNEE PAIN IN OLDER ADULTS IN PRIMARY CARE

Context: Exercise-based physiotherapy is more effective than usual primary care for older adults with knee pain. People with knee pain and osteoarthritis want further nonpharmacological options for pain relief.

Objective: To investigate whether acupuncture is a useful adjunct to exercise-based physiotherapy for knee pain in older adults.

Primary outcome: A patient-reported evaluation of pain (the Western Ontario and McMaster Universities Osteoarthritis Index pain subscale) at 6 months.

Full details: Hay et al. [30] and Foster et al. [27].

23.3.2.1 Trial Planning

23.3.2.1.1 Trial Aim and Design

This trial was designed as a three arm, parallel group, pragmatic RCT to test the additional benefit of acupuncture over and above an exercise-based physiotherapy program. Given that previous research had confirmed the beneficial effects of exercise for this patient group, we designed the trial to ensure that all intervention arms received a good package of advice and exercise, led by physiotherapists. A no-treatment group would have been impossible to justify given that current clinical guidelines and previous trials recommend advice and exercise for this patient group. One of two forms of acupuncture was added to this package, either true acupuncture using penetrating needles and needle manipulation or using sham, nonpenetrating acupuncture needles. The APEX trial compared the clinical effectiveness of adding acupuncture to an advice and exercise package led by physiotherapists in the treatment of 352 adults aged 50 and over who had been referred because of knee pain from their family doctor to physiotherapy services.

23.3.2.1.2 Trial Interventions and Their Impact on Trial Decisions

Each physiotherapist who participated in the trial was trained to deliver any of the three interventions (advice and exercise alone, advice and exercise plus true acupuncture, advice and exercise plus nonpenetrating acupuncture) in pretrial workshops, in which the protocol for each intervention was agreed. Details of the content of the interventions have been provided in detail in a protocol paper [30] and in the results paper [27]. The interventions were delivered within 10 working days of randomization by 67 experienced physiotherapists, trained in acupuncture to at least minimum national standards. Thus, we collaborated with a large group of practicing clinicians who delivered the interventions and we were able to clearly describe our care providers in some detail. We collected brief descriptive data from each of the participating physiotherapists in order to be able to describe their characteristics (e.g., two-thirds had been qualified as physiotherapists for >10 years and over half had been using acupuncture for >3 years).

We audited the delivery of the interventions, again using case report forms. Using a standardized proforma, the physiotherapists recorded the number and duration of treatment sessions that each participant received, plus details of the advice and exercises prescribed, the location and number of acupuncture points (where applicable), and any adverse reactions. The sensation that needling (true or nonpenetrating) evoked from each treatment in the acupuncture groups was also recorded. Using these procedures, we determined that treatments provided were in line with agreed protocols and there were very few treatment protocol violations (three in the advice and exercise group and two in the advice, exercise, and nonpenetrating acupuncture group).

Given the inclusion of sham, nonpenetrating acupuncture needles in one of the intervention groups, potential participants for the APEX trial had to be naïve to acupuncture (i.e., they must never have experienced acupuncture before for their present or any past complaints). In addition, we included careful measurement of treatment credibility at early follow-ups (at 2 weeks by telephone and at 6 weeks by postal questionnaire). This showed that patients receiving either acupuncture intervention were significantly more confident that treatment could help their knee problem than those receiving advice and exercise alone but that all three intervention packages were perceived as highly credible by participants.

23.3.2.1.3 Attention to Nonspecific Effects

Previous data from musculoskeletal pain studies demonstrate relationships between treatment preferences and expectations and patients' clinical outcomes [31,32]. Given this, we included several measures of preferences and expectations, of both the participating patients (through self-report questionnaires) and physiotherapists (through self-report questionnaires and over the telephone when physiotherapists phoned the Research Centre randomization service to find out which intervention the patient was allocated to). We investigated the relationship between (a) patient, (b) therapist, and (c) matched patient–therapist preferences and expectations on clinical outcomes using univariate and multivariate analyses. There was no evidence of a relationship between patients' treatment preferences or expectations and pain reduction. We found weak evidence, from secondary outcomes, that patients' expectations, both general and treatment-specific, are related to clinical outcome from exercise and acupuncture [33]. There was no clear evidence that when patients received the treatment for which both they, and their therapist, held high expectations, that the change in pain was greater than when there was no match [33].

23.3.2.1.4 Blinding

Researchers, who collected, entered, and analyzed data were unaware of treatment allocation. By necessity, the physiotherapists delivering the interventions were not blind to allocation.

23.3.2.1.5 Statistical Analyses

We stated, a priori, our analysis plan for the APEX trial in the published protocol [30] (the main intention-to-treat analysis and a "per-protocol" analysis of the primary outcome).

23.3.2.2 Trial Conduct

23.3.2.2.1 Ethical Issues

Trial participants were not told they may receive a sham intervention nor were they asked to guess which treatment they had received. The information leaflet explained that participants would receive physiotherapy advice and exercise and "may receive acupuncture, using one of two different types of acupuncture needle" without specifying the needles' mode of action (penetrating compared with nonpenetrating) to maximize the effectiveness of participant blinding. In order for the trial to be a robust comparison of all three treatment packages, it was felt important to try to maximize potential participants beliefs in the authenticity of the interventions, a recommendation in a recent Cochrane review for trials with sham interventions [34]. We worked with all of the research ethical approval committees and National Health Service (NHS) Trust Research and Development procedures that approved the trial to ensure that they understood the importance of this aspect of the APEX trial [35].

23.3.2.2.2 Quality of the Interventions

It was important that trial participants received high-quality care, and thus, all three intervention arms included a similar advice and exercise package shown to be effective in previous trials. We know from the APEX trial case report forms that most participants had treatment sessions that included supervised exercise and a review of their home exercise program. We also captured data on self-reported exercise adherence that showed adherence remained over 50% in each group over the 12 month follow-up.

In two of the three intervention arms, participants also received acupuncture (either true or nonpenetrating acupuncture). Treatment could be adjusted according to individual patients' needs within boundaries set and agreed within the study protocols.

All participants had an initial clinical physiotherapy assessment and treatment session of up to 40 min duration. During this session, the physiotherapist identified and recorded potential acupuncture points to be used should the participant be randomized to receive acupuncture (true or nonpenetrating). This was carried out as part of the overall physical examination of the knee problem. The therapist did not draw the participant's attention to the localization of acupuncture points to avoid raising their expectations about the possibility of receiving acupuncture. The advice and exercise package commenced during this initial treatment visit. Randomization occurred after this initial physiotherapy session thus ensuring that the initial physiotherapy assessment and advice and exercise package was performed blind to subsequent treatment allocation.

23.3.2.2.3 Supporting Participating Clinicians

Given that we had 67 physiotherapists delivering treatments within the APEX trial, it was important to engage and support them to participate. The APEX trial team used a variety of strategies including pretrial workshops to agree the design, interventions, and outcome measures at which national experts were invited to give plenary lectures, forming part of the physiotherapists' continuing professional development activities. Regular trial newsletters helped to maintain contact and engagement throughout the trial recruitment phase as did visits to participating clinical centers from the research team. We organized workshops at the end of the trial to present and discuss results and implications for clinical practice and to provide participating therapists with copies of slide presentations that they could give to other colleagues within their own organizations, encouraging local dissemination of trial results.

We supported physiotherapy service participation in the trial in several key ways; from study nurse support to identify potentially eligible patients from physiotherapy referrals, to written service level agreements between our University Research Centre and each participating NHS Trust physiotherapy service that provided a comprehensive written contract. The agreement defined and allocated roles and responsibilities for the trial, the funding arrangements, and approvals in place specified the number of patients that we aimed to recruit at each site and the service support and related costs that would be provided to support each physiotherapy service participate [35]. This facilitated participation in the trial by minimizing the impact on routine clinical services.

23.3.2.2.4 External and Internal Validity

Overall, 1061 potentially eligible participants were identified, of whom 709 were ineligible or did not wish to participate. Those patients who were screened but not randomized were slightly older than those randomized (65 versus 63 years old) but both groups had 61% of women. We could therefore be confident that the trial participants were reasonably representative of all patients being referred from family doctors to physiotherapy services. Treatment allocation and recruitment characteristics were similar between the higher and lower recruiting centers.

The trial follow-up rate was excellent (94% for the advice and exercise group; 93% for the advice, exercise, and true acupuncture group; and 97% for the advice, exercise, and nonpenetrating acupuncture group at 6 months) and thus we had an almost complete dataset for our primary analyses. Nonresponders were more likely to be male, slightly younger, and have lower baseline scores for pain and function.

23.3.2.2.5 *Assessment of Harm*

Adverse events were recorded; there were none in the advice and exercise group or in the advice, exercise, and nonpenetrating exercise group. Five minor adverse events were reported for participants receiving true acupuncture (pain, sleepiness, fainting, nausea, and swelling around the treated knee).

23.3.2.2.6 *Trial Monitoring and Reporting*

In line with good practice [36], the trial was monitored by a Data Monitoring and Ethics Committee on a regular basis (six monthly) and no interim analysis of any outcomes was undertaken. The trial was reported in line with recommended guidelines (www.consort-statement.org/) and full details of the patient information leaflet and intervention protocols were published as additional material on the journal website (see www.bmj.com/cgi/content/full/335/7617/436?view=long&pmid=17699546).

23.4 Conclusion

This chapter has highlighted many key challenges in conducting randomized trials of rehabilitation and exercise interventions and has provided clear examples of ways in which they can be addressed. Key solutions include clear agreement and description of interventions in detailed protocols, agreement and specification of the trial processes and outcomes, careful reporting of the characteristics of participating clinicians, and full reporting of the trial in line with recommended guidance.

Despite the challenges, it is clear that high-quality, randomized trials are possible within the field of rehabilitation and exercise to help inform clinical and healthcare policy decisions about the care of patients. The two example trials in this chapter provide high-quality evidence about nonpharmacological interventions for patients with musculoskeletal conditions. Their combination of high internal and external validity means that their findings are able to inform wider clinical practice and to contribute to future systematic reviews and clinical guidelines. They also provide a strong research platform on which to base future high-quality trials of exercise and rehabilitation approaches for musculoskeletal pain patients.

There are many further questions specifically about exercise interventions that could be addressed in future trials. For example, trials are needed that investigate specific types and doses of exercise, the longer-term benefits of exercise, strategies for increasing adherence with exercise and physical activity in general, and the cost-effectiveness of exercise therapy [37]. However, rather than seek to simply see yet more trials in this field, we need to focus on high-quality trials in particular as only they can ultimately help improve the care of patients. Such future trials may need to consider innovative trial designs and to routinely incorporate health economic analysis. They are likely to require large sample sizes, and thus larger collaborative networks with robust research infrastructure support and experienced research staff, all recent recommendations from related literature [38,39].

Acknowledgment

The author would like to acknowledge colleagues and patients who contributed to the design, conduct, and analysis of the TOPIK and APEX trials, particularly Prof. Elaine Hay.

References

1. Jolliffe, J., Rees, K., Taylor, R.R.S., Thompson, D.R., Oldridge, N., Ebrahim, S. Exercise-based rehabilitation for coronary heart disease. *Cochrane Database Syst Rev*, (1), Art. No.: CD001800, 2001. DOI: 10.1002/14651858.CD001800.
2. Thomas, D., Elliott, E.J., Naughton, G.A. Exercise for type 2 diabetes mellitus. *Cochrane Database Syst Rev*, (3). Art. No.: CD002968, 2006. DOI: 10.1002/14651858. CD002968.pub2.
3. Smidt, N., de Vet, H.C.W., Bouter, L.M., Dekker, J. Effectiveness of exercise therapy: A best evidence summary of systematic reviews. *Aust J Physiother*, 51, 71–85, 2005.
4. Fransen, M., McConnell, S. Exercise for osteoarthritis of the knee. *Cochrane Database Syst Rev*, (4), Art. No.: CD004376, 2008. DOI: 10.1002/14651858. CD004376.pub2.
5. Cottrell, E., Roddy, E., Foster, N.E. The attitudes, beliefs and behaviours of GPs regarding exercise for chronic knee pain: A systematic review. *BMC Fam Pract*, 11(1), 4, 2010.
6. Holden, M.A., Nicholls, E.E., Hay, E.M., Foster, N.E. Physical therapists' use of therapeutic exercise for patients with clinical knee osteoarthritis in the United Kingdom: In line with current recommendations? *Phys Ther*, 88(10), 1109–1121, 2008.

7. Hayden, J.A., Van Tulder, M.W., Tomlinson, G. Systematic review: Strategies for using exercise therapy to improve outcomes in chronic low back pain. *Ann Intern Med*, 142, 776–785, 2005.

8. Craig, C., Dieppe, P., Macintyre, S., Mitchie, S., Nazareth, I., Petticrew, M. Developing and evaluating complex interventions: the new Medical Research council guidance. *BMJ*, 337: a1655, 2008.

9. White, K.P., Harth, M. The occurrence and impact of generalized pain. Bailliere's Best Practice & Research. *Clin Rheumatol*, 13, 379–389, 1999.

10. Elliott, A.M., Smith, B.H., Penny, K.I., Smith, W.C., Chambers, W.A. The epidemiology of chronic pain in the community. *Lancet*, 352, 1248–1252, 1999.

11. Breivik, H., Collett, B., Ventafridda, V., Cohen, R., Gallacher, D. Survey of chronic pain in Europe: Prevalence, impact on daily life, and treatment. *Eur J Pain*, 10(4), 287–333, 2006.

12. Peat, G., McCarney, R., Croft, P. Knee pain and osteoarthritis in older adults: A review of community burden and current use of primary health care. *Ann Rheum Dis*, 60, 91–97, 2001.

13. Duncan, R.C., Hay, E.M., Saklatvala, J., Croft, P.R. Prevalence of radiographic osteoarthritis: It all depends on your point of view. *Rheumatology*, 45, 757–760, 2006.

14. Cibere, J., Trithart, S., Koec, J.A., et al. Pre-radiographic knee osteoarthritis is common is people with knee pain: Results from a population-based study. *Arthritis Rheum*, 52(9 Suppl.), S509 (Abstract), 2005.

15. Guccione, A.A., Felson, D.T., Anderson, J.J., et al. The effects of specific medical conditions on the functional limitations of elders in the Framingham study. *Am J Public Health*, 84, 351–358, 1994.

16. Pencharz, J.N., Grigoriadis, E., Jansz, G.F., Bombardier, C. A critical appraisal of clinical practice guidelines for the treatment of lower-limb osteoarthritis. *Arthritis Res*, 4, 36–43, 2002.

17. NICE. Osteoarthritis: The care and management of adults with osteoarthritis. National Institute of Health and Clinical Excellence, 2008. http://www.nice.org.uk/nicemedia/pdf/CG59NICEguideline.pdf

18. Thomas, K.S., Muir, K.R., Doherty, M., Jones, A.C., O'Reilly, S.C., Bassey, E.J. Home based exercise programme for knee pain and knee osteoarthritis: Randomised controlled trial. *BMJ*, 325, 752–756, 2002.

19. Taylor, N.F., Dodd, K.J., Shields, N., Bruder, A. Therapeutic exercise in physiotherapy practice is beneficial: A summary of systematic reviews 2002–2005. *Aust J Physiother*, 53, 7–16, 2007.

20. Hurley, M.V. Muscle dysfunction and effective rehabilitation of knee osteoarthritis: What we know and what we need to find out. *Arthritis Rheum*, 49, 444–452, 2003.

21. Van Baar, M.E., Assendelft, W.J., Dekker, J., Oostendorp, R.A., Bijlsma, J.W. Effectiveness of exercise therapy in patients with osteoarthritis of the hip or knee. A systematic review of randomized clinical trials. *Arthritis Rheum*, 42, 1361–1369, 1999.

22. Roddy, E., Zhang, W., Doherty, M., et al. Evidence-based recommendations for the role of exercise in the management of osteoarthritis of the hip or knee—The MOVE consensus. *Rheumatology (Oxford)*, 44, 67–73, 2005.

23. Jordan, K.M., Arden, N.K., Doherty, M., et al. EULAR Recommendations 2003: An evidence-based approach to the management of knee osteoarthritis: Report of a Task Force of the Standing Committee for International Clinical Studies including therapeutic trials (ESCISIT). *Ann Rheum Dis*, 62, 1145–1155, 2003.

24. Dziedzic, K., Jordan, J.L., Foster, N.E. Land and water-based exercise therapies for musculoskeletal conditions. *Best Pract Res Clin Rheumatol*, 22(3), 407–418, 2008.

25. Dieppe, P. Short report: Complex interventions. *Musculoskeletal Care*, 2(3), 180–186, 2004.

26. Hay, E.M., Foster, N.E., Thomas, E., et al. Effectiveness of community physiotherapy and enhanced pharmacy review for knee pain in people aged over 55 presenting to primary care: Pragmatic randomised trial. *BMJ*, 333, 995–998, 2006.

27. Foster, N.E., Thomas, E., Barlas, P., et al. Acupuncture as an adjunct to exercise based physiotherapy for osteoarthritis of the knee: Randomised controlled trial. *BMJ*, 335(7617), 436, 2007.

28. Phelan, M., Blenkinsopp, A., Foster, N.E., Thomas, E., Hay, E.M. Pharmacist-led medication review for knee pain in older adults: Content, process and outcomes. *Int J Pharm Pract*, 16, 1–10, 2008.

29. Bellamy, N. *WOMAC Osteoarthritis Index. A User's Guide*. London, Ontario, Canada: London Health Services, McMaster University, 1996.

30. Hay, E., Barlas, P., Foster, N., Hill, J., Thomas, E., Young, J. Is acupuncture a useful adjunct to physiotherapy for older adults with knee pain? The "Acupuncture, Physiotherapy and Exercise" (APEX) Study. *BMC Musculoskelet Disord*, 5, 31, 2004.

31. Linde, K., Witt, C.M., Streng, A., et al. The impact of patient expectations on outcomes in four randomized controlled trials of acupuncture in patients with chronic pain. *Pain*, 128, 264–271, 2007.

32. Preference Collaborative Review Group. Patients' preferences within randomised trials: Systematic review and patient level meta-analysis. *BMJ*, 337, a1864, 2008.

33. Foster, N.E., Thomas, E., Hill, J.C., Hay, E.M. The relationship between patient and practitioner preferences and clinical outcomes in a trial of exercise and acupuncture for knee osteoarthritis. *Eur J Pain*, 14(4), 402–409, 2010.

34. Manheimer, E., Cheng, K., Linde, K., et al. Acupuncture for peripheral joint osteoarthritis. *Cochrane Database Syst Rev*, (1), CD001977, 2010.

35. Hill, J., Foster, N., Hughes, R., Hay, E. Meeting the challenges of research governance. *Rheumatology (Oxford)*, 44(5), 571–572, 2005.

36. MRC (Medical Research Council) Clinical Trials Series. *MRC Guidance for Good Clinical Practice in Clinical Trials*. London, U.K.: MRC, 1998.

37. Jordan, J.L., Holden, M.A., Mason, E.E., Foster, N.E. Interventions to improve adherence to exercise for chronic musculoskeletal pain in adults. *Cochrane Database of Syst Rev*, (1), CD005956, 2010.

38. Relton, C., Torgerson, D.J., O'Cathain, A., Nicholl, J.P. Rethinking pragmatic RCTs: Introducing the 'cohort multiple RCT design'. *BMJ*, 340, c1066, 2010.

39. Foster, N.E., Dziedzic, K.S., van der Windt, D.A.W.N., Fritz, J.M., Hay, E.M. Research priorities for non-pharmacological therapies for common musculoskeletal problems: Nationally and internationally agreed recommendations. *BMC Musculoskelet Disord*, 10, 3, 2009.

24

Assessing Psychotherapy: Practical Examples

Paula P. Schnurr

National Center for Posttraumatic Stress Disorder and
Dartmouth Medical School

CONTENTS

24.1 Introduction

Psychotherapy consists of techniques for treating distress, psychiatric disorder, and behavioral problems, or for enhancing well-being and personal growth, through discussion between a patient and a therapist. The practice is multidisciplinary, and may be performed by a psychologist, psychiatrist, social worker, counselor, psychiatric nurse, or other individuals with advanced mental health training. Treatment typically requires multiple sessions; for example, the US National Comorbidity Survey Replication used various national standards to define minimally adequate treatment as eight sessions of at least 30 min each [1].

24.2 Specific Methodological Issues in Assessing This Treatment

The fact that psychotherapy is inherently dynamic creates what is perhaps its most distinctive characteristic: the interaction between patient and therapist is the treatment, whereas for other types of interventions, such as medications and surgery, the interaction is the context in which the treatment is delivered. Yet, despite this fact, methodological issues in psychotherapy research are similar to those encountered in trials of other nonpharmacological interventions [2], such as the difficulty of blinding and the difficulty of placebo control. Like these interventions, psychotherapy is complex and the interventionist's skill can significantly influence how the intervention is delivered. Quality control thus must be built into the study process. Issues of particular importance in psychotherapy research involve the choice of a comparison condition; equating the comparison condition with an active treatment; assigning therapists to conditions; manualization; training, supervision, and monitoring; and group treatment [2].

24.2.1 Choice of a Comparison Condition

Four types of comparison groups are used in psychotherapy trials [2,3] (see Table 24.1).

In a *wait-list design*, patients are assigned to receive no treatment for the duration of the active treatment. This design controls for most threats to internal validity if randomization is used for assignment, but is useful only at the initial stages of research because a difference between the treated group and the waitlist group cannot be attributed specifically to the particular treatment. Stronger inferences are possible in a study that uses a *nonspecific design*, in which comparison patients receive a treatment that includes the nonspecific elements of good psychotherapy or care as usual. Since this design controls for the general benefits of psychotherapy, any difference between the active treatment and the comparison group may be attributed to the treatment's specific benefits. *Component control designs* involve variation of the elements of a treatment to determine its active ingredients. In a *dismantling design*, groups that contain only some or one of a treatment's elements are compared with the complete treatment to

TABLE 24.1

Suggested Guidelines for Choosing a Comparison Condition

Comparison Condition	Question	Existing Evidence	Example
Wait list	Does Treatment A have benefit?	No/few controlled studies	Exposure therapy vs. wait list
Nonspecific comparison treatment or treatment as usual	Is the effect of Treatment A greater than the effect of simply going to therapy or getting usual care?	Wait list–controlled studies	Exposure therapy vs. present-centered therapy
Component control treatments (dismantling, additive, and parametric designs)	Why does Treatment A work? What are the active ingredients? Does adding Treatment A to Treatment B have greater benefit than either one alone?	Nonspecific comparison studies and a rationale for isolating active ingredients or combining effective treatments	Exposure therapy vs. cognitive restructuring vs. exposure plus cognitive restructuring
Other active treatment	Is Treatment A better than Treatment C?	Evidence that at least one of the treatments is effective (e.g., a "standard" treatment)	Exposure therapy vs. antidepressant medication

Source: Adapted from Schnurr, P.P., *J. Trauma Stress*, 20, 779, 2007.

determine the relative effectiveness of the elements. This kind of design is useful typically after a treatment's benefit has been established. In an *additive design*, the combined benefits of two or more established treatments are compared with the benefits of each separate treatment. In a parametric design, the amount of the active element is systematically manipulated to determine the effects of dose. The final type of design in psychotherapy research involves comparison between active treatments, including medications and other strategies. This kind of design permits varying amounts of control depending on how treatments can be equated on nonspecific elements.

24.2.2 Equating a Comparison Condition with an Active Treatment

It is important to equate a comparison condition with an active treatment on structural elements that could provide an alternative explanation for the treatment's effects. These elements include (but are not limited to) the number and length of sessions, format (individual or group), type of therapist, homework, and credibility. A meta-analysis of clinical trials of psychotherapy [4] found that the effect size was significantly larger in studies with an inequivalent structure ($d = 0.47$) than in studies with an equivalent structure ($d = 0.15$). Baskin et al. [4] recommended equating a comparison condition with active treatment on the number and length of sessions, format (individual vs. group), and therapist skill and training; an exception would be intentional manipulation, such as a study comparing group and individual formats. Another exception concerns their recommendation that patients in a comparison condition should be allowed to discuss the problem for which they are being treated, for example, adult survivors of childhood sexual abuse should be allowed to talk about their abuse. Although this sounds like a reasonable recommendation, a comparison group that provides a credible rationale for focusing on other issues, such as a patient's current problems caused by a trauma (e.g., [5,6]), could have equivalent credibility. Schnurr [2] (pp. 785–786) recommended that treatment integrity be preserved and suggested a standard of "reasonableness" for equating treatments—that "a reasonable person would accept differences between treatments as true differences."

24.2.3 Assigning Therapists to Conditions

Since differences between therapists in skill and competence can affect patient outcome, outcomes of patients treated by a given therapist therefore may be correlated. In a meta-analysis of therapist effects in psychotherapy research, Crits-Cristoph et al. [7] reported that therapist effects accounted for 8.6% of the variance outcome (range = 0.0%–48.7%). Use of a therapy manual and greater therapist experience were associated with smaller therapist effects.

Therapist effects should be accounted for in sample size projection and analysis (Chapter 5 of Section 1.2; [8–10]). The other issue to consider is how therapists are assigned to the treatment they deliver. There are advantages and disadvantages of different possible approaches [2,11,12]. One approach is to have each therapist administer all treatments. This approach is typically used in small, single-site trials in which there are few therapists. Having the same therapists deliver all treatments may seem to provide adequate control for therapist effects, but therapists may not deliver all treatments with equivalent skill and enthusiasm and may have difficulty keeping the treatment protocols distinct from one another. Careful ongoing supervision is necessary to detect and correct any of these problems as they arise. Another approach is to have different therapists deliver different treatments. This approach is necessary if experts are required because it is not feasible to create psychotherapy experts during the time frame of a typical research project. If experts are not required, randomization can be used to assign therapists in larger trials with a larger number of therapists. This provides the same methodological benefits as randomizing subjects, but therapists (like subjects) must be willing to accept randomization. Also, if different therapists are used to deliver each treatment, a patient cannot be randomized unless there is an available therapist in each condition. As in the case when therapists deliver all treatments, careful training and supervision are important to ensure that therapists follow study treatment protocols.

24.2.4 Manualization

Manuals are used to ensure consistent treatment delivery, study replication, and dissemination in clinical practice. They also are necessary for training, supervision, and fidelity monitoring. Manuals should contain the theoretical background for a treatment, rationale to offer patients, and procedures to be used, including detailed guidelines for each session and specific prompts and suggestions for delivering the protocol and addressing protocol deviations [3]. Manuals for different treatments should be as equated in detail and credibility, although they may vary in length, for example, if one treatment has different content in each session and another does not.

24.2.5 Training, Supervision, and Monitoring

Training, supervision, and monitoring are important elements of quality control in psychotherapy research. Training may involve in-person instruction, virtual instruction (by telephone or teleconference), and reading. The amount and format of training should be comparable across conditions but may differ as needed, for example, less training may be required to train therapists to use existing skills according to a manual and more may be needed to teach new skills. Therapists should treat one or more practice cases while receiving supervision to ensure they attain the desired standard of proficiency before treating study patients. The supervision may be based on videotape, audiotape, or therapist report and vary in both frequency and format (individual vs. group; phone or in-person). Supervision is also used during a study to help therapists maintain proficiency and manual adherence. Fidelity monitoring by an expert who is independent of therapist training and supervision should be used to assess therapist skill and adherence across conditions. Ideally, fidelity should be rated based on a sample of session videotapes, although audiotapes are acceptable. A fidelity scale should include elements unique to a particular treatment, proscribed in that treatment, and necessary but not unique. Since treatments vary in their unique and proscribed elements, a fidelity scale also should contain items that permit comparisons across treatments (e.g., global ratings of adherence and skill) or that can be converted to aggregate percentages.

24.2.6 Group Treatment

Psychotherapy may be delivered in group and individual formats. The group format creates clustering effects like those resulting from therapists that need to be addressed in sample size projection and analysis (see Chapter 5 of Section 1.2; [2,11,13]). Until recently, most studies of group psychotherapy have failed to take group clustering into account, and as a result, have probably overestimated the benefits of group therapy. A recent study [13] reanalyzed findings from 33 trials of group therapies on a list of empirically supported treatments produced by the American Psychological Association. After the degrees of freedom were corrected to treat groups, rather than subjects, as the unit of analysis, only 50.8%–68.2% of the previously significant findings were significant, depending on assumptions about the tests. With the additional assumption that the within-group intraclass correlation was 0.05, only 35.2%–43.3% of the tests were significant.

Issues regarding randomization in group-based interventions are discussed in Chapter 1.3.1 of Section 1.3. Adherence needs to be defined differently in group interventions because participants who complete the treatment protocol may miss sessions. Erratic participation can affect group process, and therefore, the integrity of the treatment, so investigators need to define what allowable absences are. Depending on the strategy used for randomization, logistic issues also arise when therapy is conducted in a group format. If a cohort of participants is to be assembled before they can be randomized to treatment groups (e.g., [5]), participants must wait until the entire cohort has been enrolled before treatment begins. In this case, investigators can minimize dropout by intensifying recruitment efforts. An additional remedy is to provide brief support, such as phone calls, to keep participants engaged while they wait for the groups to start. Another logistic issue that occurs with group therapy is that all members must be able to attend group at the same time, so if randomization occurs prior to the meeting time being established, some randomized patients may not be able to receive treatment. This problem can be addressed by making agreement with possible times an inclusion criterion.

24.3 Practical Examples

Posttraumatic stress disorder (PTSD) is an anxiety disorder that occurs secondary to exposure to a life-threatening event such as assault, rape, combat, accident, or disaster [14]. The lifetime prevalence in the United States is ~6% [15] and is elevated in populations with a high likelihood of exposure, such as military personnel, emergency workers, and civilians exposed to war. PTSD is a significant concern for individuals and healthcare systems engaged in the treatment of military veterans, among whom the disorder is prevalent. Cognitive-behavioral therapy (CBT) and selective serotonin reuptake inhibitors are recommended treatments [16,17].

In the following, we illustrate the methodological challenges of conducting psychotherapy research, using two studies of CBT for treating PTSD in male and female military veterans [5,6]. We studied CBT because it has the largest effects according to meta-analysis [18,19]. The studies were designed as practical trials [20], with clinically relevant comparison groups, diverse patient populations, a range of representative practice settings, and measures that broadly captured a range of outcomes. Further details about the designs have been published elsewhere [11,12].

24.3.1 VA Cooperative Study #420: Group Psychotherapy for Posttraumatic Stress Disorder

CSP #420: TRAUMA-FOCUSED GROUP THERAPY FOR PTSD IN MALE VIETNAM VETERANS [5]

Context: Group therapy was the most widely practiced treatment for posttraumatic stress disorder in the U.S. Department of Veterans Affairs healthcare system. However, results of clinical trials showed that the most effective approach was individual exposure therapy, in which a patient repeatedly recounts a traumatic event to emotionally process the event.

Objective: To evaluate the effectiveness of group therapy that included individual exposure treatment.

Primary outcome: PTSD severity according to a clinician-rated structured interview.

In this study, 360 male Vietnam veterans with PTSD at 10 VA Medical Centers were randomly assigned to participate in one of two group treatments, delivered according to a manualized protocol: CBT (trauma-focused group therapy) or a comparison treatment (present-centered group therapy). The groups had six members each and were conducted weekly for 30 sessions over 7 months and then monthly for 5 months, for a total of 35 sessions; sessions lasted 90 min except for 12 sessions of the CGT, which lasted 120 min. Outcomes were assessed before treatment and then at 7 and 12 months, and then, at 18 months (2/3 of participants) and 24 months (1/3 of participants). There were no differences between conditions on any outcome. Although PTSD severity and other measures improved from baseline, average improvement was modest in both treatments. Dropout was higher in trauma-focused group therapy.

24.3.2 VA Cooperative Study #494: Individual Psychotherapy for Posttraumatic Stress Disorder

---

CSP #494: COGNITIVE-BEHAVIORAL THERAPY FOR PTSD IN FEMALE VETERANS AND ACTIVE DUTY PERSONNEL [6]

Context: No study had evaluated treatment for PTSD among women who have served in the military, even though PTSD prevalence is elevated in these women. Cognitive-behavioral treatments for PTSD had the strongest evidence base but were not widely used.

Objective: To evaluate the effectiveness of prolonged exposure, a type of CBT, for the treatment of PTSD in female veterans and active duty personnel.

Primary outcome: PTSD severity according to a clinician-rated structured interview.

In this study, 284 female veterans and active duty personnel with PTSD at 12 sites were randomly assigned to participate in either CBT (prolonged exposure; [21]) or a comparison treatment (present-centered therapy). The sites included nine VA Medical Centers, two VA community counseling centers, and one military hospital. Each treatment consisted of 10 90 min sessions, delivered according to a manualized protocol. Outcomes were assessed before and after treatment, and then 3 and 6 months later. Relative to women who received present-centered therapy, women who received prolonged exposure had greater reduction of PTSD symptoms and were more likely to no longer meet diagnostic criteria and achieve total remission. Dropout was higher in prolonged exposure.

24.3.3 Specific Issues and Solutions in Designing CSP #420 and CSP #494

Choice of a comparison condition: The comparison treatments in CSP #420 and CSP #494 were designed to control for the nonspecific benefits of treatment so that the effect of trauma-focused group therapy and prolonged exposure could be attributed, respectively, to the unique elements of these treatments. The present-centered treatments did not focus on trauma and did not include exposure, cognitive restructuring, or other components of the active treatments. Instead, the present-centered treatments focused on patients' current problems, which are often the basis for seeking treatment, and evidence showed were often the focus of treatment in VA settings (e.g., [22]). Anecdotal evidence from both studies indicated that therapists and patients found the present-centered approach to be credible and clinically acceptable. Also, in CSP #494, which measured patients' expectations before treatment and satisfaction afterward, there were no differences between treatments in expectations or satisfaction.

Equating a comparison condition with the active treatment: In CSP #420 and CSP #494, the present-centered treatment was equated with the trauma-focused treatment in the format and number of sessions, session length (with the exception noted in the following), use of a manual by therapists, provision of psychoeducation at the beginning of treatment, and inclusion of home-work. Prolonged exposure includes a strong emphasis on the rationale for focusing on trauma, so present-centered therapy included a similarly strong rationale for focusing on the present. This illustrates how treatments can be equated while preserving the integrity of each. As another example, the present-centered treatments in CSP #420 and CSP #494 included homework, although the type differed from the homework in the active treatments and was consistent with the present-centered content. One further example illus-trates how structural elements may need to differ in order to preserve treat-ment integrity. Therapy groups typically last 60 min. Ninety minutes were needed for trauma-focused group therapy, and 120 min were needed for the 12 of the 30 weekly sessions (because of the session content). We felt that 90 min sessions of present-centered group therapy would be feasible because such groups sometimes run longer, but that stretching sessions to 120 min would be impractical and likely would negatively affect the group process. Therefore, the all present-centered sessions were 90 min, which meant that over the course of 1 year and 35 total sessions patients who received present-centered therapy had 52.5 h of treatment and patients who received trauma-focused treatment had 58.5 h of treatment. If the trauma-focused group had a superior outcome, we felt it would be implausible to attribute the outcome to the additional 6 h.

Assigning therapists to conditions: Both studies randomized therapists to the treatment they delivered. We chose this strategy because there were a large number of therapists (four or more at each of the 10 sites in CSP #420 and 12 sites in CSP #494) and therapist were not required to be experts in cognitive behavioral techniques; the latter decision was based on our desire to maximize the generalizability of study findings to VA therapists. All therapists in CSP #420 were satisfied with their treatment assignment. Initially, some therapists in CSP #494 who were assigned to present-centered therapy raised concerns about delivering what they considered to be a less effective therapy, but we resolved these concerns for all but one thera-pist (one of the few who was an expert in CBT and who routinely used CBT in practice). As mentioned earlier, if different therapists deliver each treatment, a patient cannot be randomized unless a therapist is available in each condition. In CSP #494, enrolled patients sometimes had to wait for therapists to have openings in their caseload, and one patient was not treated because these problems could not be resolved. We could have avoided, or at least reduced, such complications if we had hired therapists specifically for the study rather than using VA therapists who conducted the study treatment as part of their routine clinical work.

Manualization: All treatments in CSP #420 and CSP #494 were delivered according to detailed manuals. Due to the more detailed structure of the trauma-focused group therapy and prolonged exposure sessions, the manuals for these treatments were longer than the manuals for their respective comparison treatments. This is an example of how treatment integrity should be preserved when trying to equate conditions. All manuals had comparable detail, but adding unnecessary material to the comparison manuals just to increase their length could have hurt, rather than helped, the credibility of these manuals.

Training, supervision, and monitoring: All treatment sessions in both studies were videotaped. In CSP #420, therapists participated in a 2 day in-person training session for the treatment to which they were assigned and those assigned to the trauma-focused condition additionally conducted a pilot group in order to practice the new skills they had learned. Pilot groups were not used for the present-centered condition because the treatment utilized the therapists' existing skills. During the pilot and during the study, therapists received weekly telephone supervision with their co-therapist (groups were led by two therapists). A similar procedure was used in CSP #494, although the training for prolonged exposure lasted 5 days vs. 2 in present-centered therapy, and therapists in both conditions treated 1–2 pilot cases. Weekly individual supervision was provided to therapists throughout the pilot and study phases; if a therapist was judged to have attained a high level of proficiency, frequency was reduced to every other week. In both studies, monthly group conference calls for each treatment were used to support the training and supervision and to discuss issues related to the study protocol.

In CSP #420, a senior clinician who was independent of training and supervision rated three sessions from each of the 60 groups (10%). He rated 32 specific elements for adherence on a 5-point scale ranging from "not enough" to "too much," and rated the same elements for therapist competence on a 5-point scale ranging from "poor" to "highly competent." Adherence and competence were high, but were slightly better in trauma-focused treatment than in present-centered treatment. However, adherence and competence were unrelated to patient outcomes. In CSP #494, a senior clinician who was independent of training and supervision rated 11.7% of the videotapes. The ratings included a 5-point scale ranging from "poor" to "excellent" for therapists' competence and adherence to essential manual elements that were unique to that approach and not unique to that approach. Proscribed elements were rated present/absent and were converted to a percentage for each tape. Adherence and competence were high and did not differ between treatments.

Group treatment: In CSP #420, each of the 10 sites ran three cohorts (pairs) of six-person groups. Twelve patients were recruited for a cohort and then randomized to receive either trauma-focused or present-centered treatment; although the treatment was delivered in groups, the randomization was done individually using permuted blocks of four in three blocks of PTSD severity scores. This strategy created significant pressure on study personnel to enroll patients as quickly as possible. Sites were staffed by a full-time assistant to focus on recruitment and scheduling and another full-time position to focus screening and assessment. Prior to the beginning of a cohort, staff worked with referral sources to develop a list of potential participants. Sources were asked details about potential participants' inclusion and exclusion diagnoses to efficiently rule out ineligible participants. Our strategy also created pressure for patients because they had to give up most forms of other therapy (except self-help and medications) while waiting for study treatment. We began case management as soon as a participant was enrolled to minimize dropout and provide interim care. We think this was effective because only 3.7% of eligible participants dropped out before randomization even though some had to wait up to 3 months to receive treatment. Due to the delay, we readministered the primary outcome measure (PTSD symptom severity) before treatment began for anyone whose initial assessment had occurred >30 days prior.

24.4 Conclusion

In psychotherapy research, it is important to choose a comparison group that is appropriate given the state of knowledge about a treatment and the question being asked. Treatment comparison groups typically contain legitimate therapeutic elements and are not analogous to a pill placebo. Waitlist controls are useful at the earliest stages of research but then a comparison treatment that controls for nonspecific therapeutic benefit is needed, sometimes to be followed by research to isolate mechanisms or by comparisons with other treatments. The comparison group should be equivalent to the active treatment on factors that plausibly could explain any difference from the active treatment in outcome, such as amount of treatment, format, and credibility. Another key issue is the way therapists are assigned to the treatment they deliver. In small studies with few therapists, it is preferable to have therapists deliver both/all treatments. In larger studies, it is preferable to randomize therapists to deliver a single treatment. Therapists also may be assigned to a single treatment if expertise in a particular technique is required. Whichever option is chosen, careful training and supervision are needed to build quality control into the study process. Detailed manuals are essential to ensure that treatment is delivered optimally and reliably to all participants. However, an independent fidelity monitor should assess the success of such quality control efforts.

Lastly, since psychotherapy is sometimes conducted in groups, there may be logistic issues to address in addition to those related to sample size estimation and analysis.

An important article in the field of psychotherapy outcome research reminds us that, "the fundamental goal of any between-group experimental design and its associated methodology is to hold all factors consistent other than the one variable about which cause-and-effect conclusions are to be drawn" [3, p. 249]. The complexity of psychotherapy interventions makes this goal challenging but not impossible to attain.

References

1. Wang, P.S., Lane, M., Olfson, M., et al. Twelve-month use of mental health services in the United States: Results from the National Comorbidity Survey Replication. *Arch Gen Psychiatry*, 62, 629–640, 2005.
2. Schnurr, P.P. The rocks and hard places in psychotherapy outcome research. *J Trauma Stress*, 20, 779–792, 2007.
3. Borkovec, T.D. Between-group therapy outcome research: Design and methodology. *NIDA Res Monogr*, 137, 249–289, 1993.
4. Baskin, T.W., Tierney, S.C., Minami, T., Wampold, B.E. Establishing specificity in psychotherapy: A meta-analysis of structural equivalence of placebo controls. *J Cons Clin Psychol*, 71, 973–979, 2003.
5. Schnurr, P.P., Friedman, M.J., Foy, D.W., et al. A randomized trial of trauma focus group therapy for posttraumatic stress disorder: Results from a Department of Veterans Affairs Cooperative Study. *Arch Gen Psychiatry*, 60, 481–489, 2003.
6. Schnurr, P.P., Friedman, M.J., Engel, C.C., et al. Cognitive-behavioral therapy for posttraumatic stress disorder in women: A randomized controlled trial. *JAMA*, 297, 820–830, 2007.
7. Crits-Cristoph, P., Baranckie, K., Kurcias, J.S., et al. Meta-analysis of therapist effects in psychotherapy outcome studies. *Psychother Res*, 1, 81–91, 1991.
8. Elkin, I., Falconnier, L., Martinovich, Z., Mahoney, C. Therapist effects in the National Institute of Mental Health Treatment of Depression Collaborative Research Program. *Psychother Res*, 16, 144–160, 2006.
9. Kim, D.M., Wampold, B.E., Bolt, D.M. Therapist effects in psychotherapy: A random-effects modeling of the National Institute of Mental Health Treatment of Depression Collaborative Research Program data. *Psychother Res*, 16, 161–172, 2006.
10. Lutz, W., Leon, C.C., Martinovich, Z., et al. Therapist effects in outpatient psychotherapy: A three-level growth-curve approach. *J Cons Clin Psychol*, 54, 32–39, 2007.
11. Schnurr, P.P., Friedman, M.J., Lavori, P.W., Hsieh, F.Y. Design of Department of Veterans Affairs Cooperative Study No. 420: Group treatment of posttraumatic stress disorder. *Cont Clin Trials*, 22, 74–88, 2001.
12. Schnurr, P.P., Friedman, M.J., Engel, C.C., et al. Issues in the design of multisite clinical trials of psychotherapy: CSP #494 as an example. *Cont Clin Trials*, 26, 626–636, 2005.

13. Baldwin, S.A., Murray, D.M., Shadish, W.R. Empirically supported treatments or type I errors? Problems with the analysis of data from group-administered treatments. *J Cons Clin Psychol*, 73, 924–935, 2005.
14. American Psychiatric Association. *Diagnostic and Statistical Manual of Mental Disorders*, 4th edn. Washington, DC: Author, 1994.
15. Kessler, R.C., Chiu, W.T., Demler, O., et al. Prevalence, severity, and comorbidity of 12-month DSM-IV disorders in the National Comorbidity Survey Replication. *Arch Gen Psychiatry*, 62, 617–627, 2005.
16. American Psychiatric Association. Practice guideline for the treatment of patients with acute stress disorder and posttraumatic stress disorder. *Am J Psychiatry*, 161(Supp.), 3–31, 2004.
17. VA/DoD Clinical Practice Guideline Working Group. *Management of Posttraumatic Stress* (Office of Quality and Performance Publication 10Q-CPG/PTSD-04). Washington, DC: Departments of Veterans Affairs and Defense, 2003.
18. Bradley, R.G., Greene, J., Russ, E., et al. A multidimensional meta-analysis of psychotherapy for PTSD. *Am J Psychiatry*, 162, 214–227, 2005.
19. Van Etten, M.L., Taylor, S. Comparative efficacy of treatments for posttraumatic stress disorder: A meta-analysis. *Clin Psychol Rev*, 5, 126–144, 1998.
20. Tunis, S.R., Stryer, D.B., Clancy, C.M. Practical clinical trials. Increasing the value of clinical research for decision making in clinical and health policy. *JAMA*, 290, 1624–1632, 2003.
21. Foa, E.B., Rothbaum, B.O. *Treating the Trauma of Rape: Cognitive-Behavioral Therapy for PTSD*. New York: Guilford, 1998.
22. Rosen, C.S., Chow, H.C., Finney, J.F., et al. VA practice patterns and practice guidelines for treating posttraumatic stress disorder. *J Trauma Stress*, 17, 213–222, 2004.

25

Assessing Psychosocial Interventions for Mental Health: Practical Examples

Graham Thornicroft
King's College London

Michele Tansella
University of Verona

CONTENTS

25.1 Introduction

The main reason for using the randomized controlled trial (RCT) design is that the public health is likely to suffer as a result of avoiding such high-quality evidence [1]. The origin of the RCT is generally attributed to Sir Austin Bradford Hill, and Bull has provided an extensive account of their historical development [2]. The first well-documented RCT of medical treatment was organized by the Medical Research Council (MRC) in the United Kingdom [3]. Nevertheless, similar methodologies, called "experiments," had been used outside medicine by psychologists before this, and have therefore a much longer tradition than the clinical trial [4]. Following the earlier work of Sir Ronal Fisher in 1926 on agricultural research, the contribution of Sir Bradford Hill (the prime motivator behind the MRC trials before 1950) is considered pre-eminent in making systematic the methodology for conducting RCTs [5]. For a full history of the clinical trial, see the James Lind Library at www.jameslindlibrary.org.

RCTs have well-recognized advantages and disadvantages [6–10]. In particular, the key advantage is the minimization of both known and unknown confounders by the random allocation of individuals or groups of individuals [11,12]. Since the seminal paper by Schwartz and Lellouch, it has been common to distinguish between efficacy trials (which tend to be explanatory) and effectiveness trials (sometimes otherwise called large simple, pragmatic, practical, or management trials) [9,13–15]. This categorical distinction has its uses, although for some purposes we may rather see efficacy and effectiveness trials as falling along a continuum. Efficacy trials, which usually precede effectiveness studies, refer to those conducted under more ideal, experimental conditions, while effectiveness trials are RCTs carried out in more routine clinical conditions [11,16–18]. Nevertheless, some important questions, for example, the impact of clinical guidelines, may only be researchable in real-world settings, and will therefore bypass the efficacy study stage [19]. Cochrane has defined effectiveness, at the patient level, as assessing whether an intervention does more good than harm when provided under usual circumstances of healthcare practice [11]. In relation to psychosocial mental health interventions within primary care, Wells has defined effectiveness trials as those which "duplicate as closely as possible the conditions in the target practice venues to which study results will be applied" [20–22].

25.2 RCTs and Mental Disorders

RCTs have been applied in relation to research on mental illnesses over the last 50 years [4]. It is likely that the first controlled clinical trial was published in 1955 [9,23]. It reported a double-blind study of reserpine versus placebo, conducted at the Maudsley Hospital in London and involving 54 patients with symptoms of anxiety and depression. The first large-scale RCT appeared some years later, in 1965. This study was a multicenter MRC study on the treatment of depression and compared the effects of electro-convulsive therapy, a tricyclic antidepressant, a monoamine oxidase inhibitor antidepressant and placebo in hospitalized psychiatric patients [24]. Although mental health trials started only a few years after the ground-breaking MRC study of streptomycin in 1948, more recently other fields of medicine have made much greater use of RCT methods. In comparison with oncology for example, until the last decade, with 21,408 currently active trials registered for the treatment of cancer and 3021 for mental disorders in April 2009 [25].

There has been increasing interest in the importance of effectiveness trials in psychosocial interventions in psychiatry [26–29]. While the technical quality of efficacy trials is usually relatively high because such trials are tightly controlled by medicinal product regulatory authorities, the same standards of research conduct are less often applied to effectiveness trials. In fact, there are few published criteria relevant to the improvement of the quality of effectiveness RCTs in psychosocial trials psychiatry [30], and in Table 25.1 we propose seven such criteria for the assessment of psychosocial interventions which are discussed in more detail below.

25.3 Specific Methodological Issues in Assessing This Treatment

The recognition and treatment of the majority of people with mental illnesses, especially depression and anxiety-related disorders, remains a task that falls mostly to primary care. Von Korff and Goldberg reviewed 12 different RCTs of enhanced care for major depression in primary care settings, which include psychosocial interventions. They found that those directed solely toward training and supporting General Practitioners (GPs), or family physicians have not been shown to be effective. They argued that interventions should focus on low-cost case management, coupled with flexible and accessible working relationships between the case manager, the primary care doctor, and the mental health specialist. In other words, the whole process of care needs to be enhanced and reorganized to include the following key elements: active follow-up by the case manager, monitoring treatment adherence and patient outcomes, adjustment of treatment plan if patients do not improve, and referral to a specialist when necessary [31]. The range of services required for comprehensive mental health care can be described under the following five headings [32].

TABLE 25.1

Criteria to Assess RCTs of Psychosocial Interventions

1. *Study question*

   Who defines the aim of the study?

   What process is used to identify the question addressed?

   Is the study question expressed in an answerable way (as a clear hypothesis)?

   Prior evidence of intervention effect size

   Is the answer to this question really unknown?

   Why is this question important now?

   Is there initial evidence from efficacy trials or effectiveness studies (observational or trials)?

   What is the public health importance of the policy or practice question addressed?

   What is the clinical necessity of the question?

   Sample size and statistical power for primary/secondary aims and related hypotheses

2. *Reference population*

   What is the reference group (or subgroup) to which the trial results should be generalized?

   What are their sociodemographic, and clinical characteristics?

   What are the ethnic and cultural characteristics of the target group?

   What is resource level in this population?

   What is the nature and standard and coverage of health and social care?

   At what time point is population identified?

3. *Patient sample*

   What are their sociodemographic, and clinical characteristics?

   What are the inclusion criteria?

   Not invited to participate rate

   Nonparticipation rate

   Patient preferences

   What are the exclusion criteria?

   How far does the sample reflect the target population?

   What level of heterogeneity is there?

   Selection of incident or prevalent cases (true incidence/prevalence or treated incidence/prevalence)

   What are the rates of adherence and nonadherence to treatment as recommended?

4. *Study settings*

   Characteristics and representativeness of professional staff

   Levels of resources available

   Research-oriented culture

   Staff morale and sustainability of intervention

   Incentives for research collaboration

   Opportunities for data linkage

   Center/professional nonparticipation

5. *Study intervention*

   Is intervention acceptable?

   Total time needed to deliver intervention

   Frequency of interventions

TABLE 25.1 (continued)

Criteria to Assess RCTs of Psychosocial Interventions

Simplicity/complexity of the intervention

Single/multicomponent intervention

Is intervention manualized?

Do usual professional staff deliver the intervention during the study?

Can treatment process be measured (fidelity)?

Degree of fit/feasibility for current practice

Exit strategy, who pays after the end of study

6. *Control condition*

Treatment as usual or specific control

Acceptability to patients of control condition

Cost and feasibility of control condition

Variation between control condition within and between sites (fidelity)

Are the key characteristics of the control condition well-described?

7. *Bias*

Does contamination take place?

Degree of blinding

Choice of primary and secondary outcomes

Perspectives prioritized in outcome choice

Time(s) at which outcomes measured

Total length of follow-up and late effects

Sources of outcome data

Respondent burden

Consent rate

Recruitment rate

Attrition/dropout and follow-up rates

25.3.1 Outpatient/Ambulatory Clinics

There is surprisingly little evidence on all of these key characteristics of outpatient care [33], but there is a strong clinical consensus in many countries that they are a relatively efficient way to organize the provision of assessment and treatment providing that the clinic sites are accessible to local populations. Nevertheless, these clinics are simply methods of arranging clinical contact between staff and patients, and so the key issue is the *content* of the clinical interventions, namely to deliver treatments that are known to be evidence-based [34].

25.3.2 Community Mental Health Teams

These are the basic building block for community mental health services, and these are discussed in more detail in Section 25.4. A series of studies and systematic reviews, comparing community mental health teams (CMHTs) with a variety of local usual services, suggests that there are clear benefits to the introduction of generic community-based multidisciplinary teams: they can improve engagement with services, increase user satisfaction, increase met needs, and improve adherence to treatment, although they do improve symptoms or social function [35–40].

25.3.3 Acute Inpatient Care

There is relatively weak evidence base on many aspects of inpatient care, and most studies are simply descriptive accounts [41]. There are therefore few systematic reviews in this field, one of which found that there were no differences in outcomes between routine admissions and planned short hospital stays [42].

25.3.4 Long-Term Community-Based Residential Care

The evidence deinstitutionalization is largely based upon nonrandomized (and sometimes uncontrolled) pre–post comparisons carried out in relation to hospital closure [43,44]. The TAPS study in London [45], for example, completed a 5 year follow-up on over 95% of 670 discharged long-stay nondemented patients and found:

- Two-thirds of the patients were still living in their new residence.
- There was no increase in the death rate or the suicide rate.
- Very few patients became homeless, and none was lost to follow up from a staffed home.
- Over one-third were briefly readmitted, and at follow-up 10% of the sample were in hospital.
- Patients' quality of life was greatly improved by the move to the community.
- There was little difference between total hospital and community costs, and overall community care is more cost-effective than long-stay hospital care.

25.3.5 Occupation and Day Care

There is little scientific research of these traditional forms of day care and a review of over 300 papers, for example, found no relevant RCTs. Nonrandomized studies have given conflicting results and for areas with medium levels of resources it is reasonable at this stage to make pragmatic decisions about the provision of rehabilitation and day-care services if the more differentiated and evidence-based options discussed below are not affordable [46,47].

25.4 Practical Examples

25.4.1 Community Mental Health Teams

A series of studies and systematic reviews, comparing CMHTs with a variety of local usual services, suggests that there are clear benefits to the introduction of generic community-based multidisciplinary teams: they can improve engagement with services, increase user satisfaction, increase met needs, and improve adherence to treatment, although they do improve symptoms or social function [35,37,48].

25.4.2 Assertive Community Treatment Teams

These provide a form of specialized mobile outreach treatment for people with more disabling mental disorders, and have been clearly characterized [49]. There is now strong evidence that assertive community treatment (ACT) can produce the following advantages in high level of resource areas: (i) reduce admissions to hospital and the use of acute beds, (ii) improve accommodation status and occupation, and (iii) increase service user satisfaction. ACT has not been shown to produce improvements in mental state or social behavior. ACT can reduce the cost of inpatient services, but does not change the overall costs of care [50–54].

25.4.3 Early Intervention Teams

There has been considerable interest in recent years in the prompt identification and treatment of first or early episode cases of psychosis. Much of this research has focused upon the time between first clear onset of symptoms and the beginning of treatment, referred to as the "duration of untreated psychosis" (DUP), whereas other studies have placed more emphasis upon providing family interventions when a young person's psychosis is first identified [55,56]. There is now emerging evidence that longer DUP is a predictor of worse outcome for psychosis; in other words, if patients wait a long time after developing a psychotic condition before they receive treatment, then they may take longer to recover and have a less favorable long-term prognosis. Few controlled trials have been published of such interventions, and so a Cochrane systematic review [57] has concluded as follows:

> We identified insufficient trials to draw any definitive conclusions. The substantial international interest in early intervention offers an opportunity to make major positive changes in psychiatric practice, but making the most of this opportunity requires a concerted international programme of research to address key unanswered questions [58–64].

25.4.4 Alternatives to Acute Inpatient Care

In recent years, three main alternatives to acute inpatient care have been developed: acute day hospitals, crisis houses, and home treatment/crisis resolution teams. *Acute day hospitals* are facilities that offer programs of day treatment for those with acute and severe psychiatric problems, as an alternative to admission to inpatient units. A systematic review of nine RCTs has established that acute day hospital care is suitable about 30% of people who would otherwise be admitted to hospital, and offers advantages in terms of faster improvement and lower cost. It is reasonable to conclude that acute day hospital care is an effective option when demand for inpatient beds is high [47].

25.4.5 Home Treatment/Crisis Resolution Teams

Home treatment/crisis resolution teams are mobile CMHTs offering assessment for patients in psychiatric crises and providing intensive treatment and care at home [65,66]. A Cochrane systematic review [67] found that most of the research evidence is from the United States or the United Kingdom, and concluded that home treatment teams reduce days spent in hospital, especially if the teams make regular home visits and have responsibility for both health and social care [68,69].

25.5 Specific Issues and Possible Solutions in Planning a Trial of Nonpharmacological Treatment

25.5.1 Study Question

In planning an effectiveness trial, the first task is to ask a clear and important question, relevant to clinical practice (e.g., symptom severity) and/or to public health (e.g., burden of disability) [70]. The background to defining the study question will usually be clinical uncertainty about whether two or more interventions produce different or similar outcomes, for example, in comparing pharmacological products [71]. Indeed, some ethicists have argued that all trials should be based upon some degree of equipoise (genuine uncertainty about which is the best treatment). In this case, the likely outcome is that one intervention will be shown to be better than the other.

By comparison, when a trial is designed to show no difference in outcomes between two or more interventions, then this type of study is called an equivalence trial [72]. Equivalence trials are harder to conduct and need to be much bigger than both efficacy and effectiveness trials to demonstrate equivalence, rather than failing to show a difference between where there is one "type 2" error.

The choice of the primary hypothesis for a psychosocial trial is one of the most central decisions. This will lead to the selection of the primary outcomes measures, and to the necessary sample size, but will itself follow an a priori decision on the single most important question to investigate [13,73]. The selection of primary and secondary outcomes can be seen along a spectrum from simple, easily collectable, and often dichotomous variables on one hand (such as medication change, relapsed/not relapsed, or admitted/nonadmitted) to more complex outcomes (such as symptoms, quality of life, carer impact or disability) on the other. The former (larger, simpler, and more realistic trials) will allow smaller treatment effect sizes to be detected [74]. The latter will be more costly, time-consuming, and may yield a lower follow-up rate, but will produce a far greater richness of data to understand the interrelationships between key variables, and so the possible mechanisms of action of the intervention. In each trial, a trade-off therefore needs to be struck between simplicity and complexity in outcome selection.

Simple questions, such as time to hospital readmission, or bed-day use, are more often selected as primary outcomes in effectiveness than in efficacy trials. This is both because these are important for service provision, and because such administrative outcomes allow high completion rates at follow up, especially when patient interviews are not possible. In addition, it is more common to include costs among the outcome measures of effectiveness studies so that the cost-effectiveness can be measured, for example, in assessment of cognitive behavioral treatments [75,76]. Further, it is regrettable that many trials in psychiatry have been too small to provide sufficient statistical power to answer the questions they have addressed [77].

25.5.2 Reference Population

In contrast to efficacy RCTs, psychosocial effectiveness trials by definition aim to establish the external validity of their results in terms of the relevance of the findings to a wider reference population of similarly affected individuals. Interestingly, effectiveness trials rarely specify or even discuss this wider target group, and so do not allow judgments to be made about how far the patient sample does in fact represent the intended reference group [78]. For a satisfactory understanding of the wider reference patient population, we would need, for example, to report the number of screened, randomized, and excluded patients, with reasons for exclusion, and to know (i) key clinical and sociodemographic characteristics, (ii) features of the relevant healthcare-related services available to the reference population, (iii) service coverage to population subgroups, and (iv) the wider context, such as the overall level of resources in those sites. Resource level can be described, according to the WHO, as high, medium, or low [70]. These three levels are relevant to individual countries, as well as to different population groups, recognizing that there are disadvantaged areas or groups within all countries, even those that have the best resources and services.

25.5.3 Patient Sample

Having decided upon the wider reference patient population, psychosocial effectiveness trials then need to recruit patients who reflect that target group, using relatively wide inclusion and few exclusion criteria, for example, in assessing psychological interventions [79–81]. This will usually mean far higher heterogeneity between patients in effectiveness than in efficacy studies, and this may be determined by clinically meaningful eligibility criteria, for example, by explicitly including cases with comorbid disorders. High rates of participation are also necessary; otherwise, the trial will move along the continuum toward the status of an efficacy study [78]. In addition, patient motivation may vary. Typically, in psychosocial effectiveness trials is it important that the eligibility criteria be as simple as possible, so that these should be as generalizable as the intervention. Study groups other than patients may be also important for some research questions, such as the identification of effective interventions to support family members of people with schizophrenia [82].

25.5.4 Study Settings

Another important aspect of psychosocial effectiveness trials is that their results should be relevant not only to other patients, but also to other staff and healthcare settings. To establish if this is the case, it will be necessary to assess key staff characteristics, including training, clinical experience, available resources, and how far the context is research-oriented. It may also be important to measure staff morale and burnout, if the particular intervention is sustainable over time, for example, whether it places high demands upon clinicians. One implication of this is that effectiveness trials will often need to include multiple, real-world study sites, such as case management trials in primary care for the combined treatment of depression [83].

25.5.5 Study Intervention

The choice of the intervention is of central importance in effectiveness studies [13]. Typically, they evaluate psychosocial interventions of proven efficacy [6]. A critical question is as follows: "Can the experimental intervention, if effective, be realistically generalized?" For example, is the intervention ethically and practically acceptable to patients and to clinicians? Further, is the intervention sufficiently manualized (especially for more complex treatment packages) to be taught and practiced in a similar way by routine clinical staff? In other words, can treatment fidelity be assured, both in the short term and in the long term? Using routine staff to deliver the intervention within effectiveness trials may be informative in these respects. Indeed it can be argued that the degree of fidelity of practitioners in using, for example, a manualized intervention or a treatment guideline is in itself a key variable in effectiveness trials. It may be easier to be clear about the adherence to protocol of pharmacological interventions than about the characteristics of psychosocial treatments.

If effective, is the new intervention affordable in ordinary practice, in high, medium, and low resources countries and regions? Who will pay for the continuation of a successful project after the initial research funding ends? A related question better considered at the start than at the end of a study is whether there will be a market for the new technology, and who will own its intellectual property? This may not be a straightforward issue, as in some cases mental health trials evaluate complex interventions that combine multiple components (each with their own intellectual property implications), for example, for depression [22,31], self-harm [40], psychotic disorders [40,84], or within a specific setting, such as a day hospital [85]. Such everyday treatment settings may also require more flexible treatment regimens, for example, in pharmacological trials, by explicitly allowing treatments and doses to vary, according to the patients' clinical condition and consistent with accepted clinical practice [86].

25.5.6 Control Condition

Trial protocols and reports commonly do not pay sufficient attention to the control condition. Interestingly, the results of psychosocial RCTs are often discussed in terms of the impact of a new treatment or intervention, while in fact any differences between the experimental and the control conditions are as much due to the latter as the former.

The choice of the control condition is often difficult to make. For example, in a trial of supported work placement, is it preferable to choose usual local rehabilitation services or a manualized and specific control interventions as the comparison [87]? A "treatment as usual" control condition may also need to be viewed cautiously, as it may not be fully representative of the range of settings in the intended reference population [50,52]. Further, the content of the "treatment as usual" should be consistent with the range of practices current in routine clinical sites relevant to the larger reference population. One must also ask, "Is the control acceptable to patients, carers, and staff?"

25.5.7 Bias

Bias can be defined as any factor or process that tends to deviate the results or conclusions of a trial systematically away from the truth, leading to an under- or overestimate of the effect of an intervention [6,88,89]. In fact, the twin key strengths of the RCT design are random allocation (to reduce selection bias) and blinding (to reduce information bias) [17,90]. In a study with random allocation, the differences between treatment groups behave like the differences between random samples from a single population. It means that each patient has a similar chance of being given each treatment, but the treatment to be given cannot be predicted. Bias is an ever-present threat at every stage of a psychosocial trial [6], and some types of bias are more specific to effectiveness than to efficacy trials. Blinding, for example, may be compromised in an effectiveness trial. Blinding (not knowing patient's status in a trial) can apply to treating staff, patients, researchers, or to any combination of these, and may therefore be termed nil, single, double, or triple blinding. However, in a comparison of home or hospital treatment, for example, it is likely that blinding of staff and patients is not feasible, and that even single blinding of researchers is difficult or impossible. For this reason, blinding may be less complete in complex and in psychosocial interventions than in pharmacological trials.

Every effort needs to be made to preserve at least the single blindness of researchers to patient status, for example, by basing them away from the clinical site. However, even blindness of the research staff may be difficult or impossible in some types of effectiveness trial, for example, those comparing either antidepressant or psychological treatment for mild depression disorders in primary care [91], or treatments of bipolar disorder in routine psychiatric practice [26]. Even if full blinding is not possible in an effectiveness trial, nevertheless it is vital that concealment of allocation must be maintained.

A further sources of bias in effectiveness trials is a high attrition rate (between baseline and follow-up), which may relate to the nature of the assessments (invasive or noninvasive tests), the study burden (number and duration of the measures), or to the acceptability of the experimental and control treatments. Specifically, bias may occur as a result of attrition when there are different attrition rates between treatment and control groups, or when there are different reasons for attrition between these two groups. Bias may also be increased by contamination between conditions. This means that the distinction between the two treatment or intervention packages may be diminished or even lost by blurring what staff deliver in practice. Where this risk is substantial, then a cluster randomization design needs to be considered, although in general terms, for ethical, methodological, and reporting reasons, including an increased risk of spuriously "significant" findings [92], individual level randomization is preferable wherever possible [93,94].

25.6 Conclusions

If "efficacy is the potential effectiveness of a treatment" [95], then a clear direction of travel is established in moving from efficacy to effectiveness studies of psychosocial interventions, and then to the implementation of new interventions which do work [96]. In other words, after core clinical effects have been proven in efficacy trials, effectiveness trials occur at a later stage when attention moves to questions regarding generalizability, costs, and broader effects. Further, it will be necessary in future to conduct more effectiveness trials (including large simple trials) [26], which have sufficient statistical power to provide precise answers to assist clinicians in making treatment decisions [73]. Further, for most healthcare decision makers clinical effectiveness is a necessary but insufficient analysis. Very often cost-effectiveness is the key dimension in deciding whether an intervention provides value for money and should be widely implemented. The assessment of direct, indirect, and hidden costs is therefore often a vital complement to the assessment of the outcomes of interventions, to allow their cost-effectiveness to be measured [97,98].

In future effectiveness trials of psychosocial interventions in mental health, both for studies of individual treatments and for service evaluations, are likely to improve if investigators carefully consider, and come to well-justified decisions about, the questions presented under the seven headings of this chapter. Well-conducted effectiveness RCTs of psychosocial interventions can maintain external validity by virtue of good design, for instance by adopting some of the approaches we suggest in this chapter [99,100].

Paradoxically, RCTs are both relatively simple in concept and relatively complex in practice. The design of psychosocial effectiveness trials necessarily involves a series of trade-offs, or compromises, between competing attractions, for example, in choosing between few or many outcome measures, or between a "treatment as usual" or an active control condition. With great insight Bradford Hill wrote that "the essence of a successful controlled clinical trial lies in its minutiae—in a painstaking and sometimes very dull, attention to every detail" [5]. The need for such attention to detail in clinical trials is no less applicable today.

Acknowledgment

This chapter was written with support for Graham Thornicroft by a National Institute for Health Research (NIHR) Applied Programme grant awarded to the South London and Maudsley NHS Foundation Trust, and in relation to the NIHR Specialist Mental Health Biomedical Research Centre at the Institute of Psychiatry, King's College London and the South London and Maudsley NHS Foundation Trust.

Declaration of Interests

No financial support was received for this work and there are no financial conflicts of interests of the authors in the writing of this chapter.

References

1. Edwards, S., Lilford, R.J., Braunholz, D., Jackson, J., Hewison, J., Thornton, J. Ethical issues in the design and conduct of randomised controlled trials. *Health Technol Assess*, 2(15), i–vi, 1–132, 1998.

2. Bull, J.P. The historical development of clinical therapeutic trials. *J Chronic Dis*, 10, 218–248, 1959.

3. Medical Research Council. Streptomycin treatment of pulmonary tubercolosis. *BMJ*, ii, 769–782, 1948.

4. Tansella, M. The scientific evaluation of mental health treatments: An historical perspective. *Evid Based Mental Health*, 5(1), 4–5, 2002.

5. Bradford Hill, A. Medical ethics and controlled trials. *BMJ*, 1043–1049, 1963.

6. Jadad, A.R. *Randomised Controlled Trials*. London, U.K.: BMJ Books, 1998.

7. Black, N. Why we need observational studies to evaluate the effectiveness of health care. *BMJ*, 312(7040), 1215–1218, 1996.

8. Slade, M., Priebe, S. Are randomised controlled trials the only gold that glitters? *Br J Psychiatry*, 179, 286–287, 2001.

9. Everitt, B., Wessely, S. *Clinical Trials in Psychiatry*. Oxford, U.K.: Oxford University Press, 2003.

10. Thomas, J., Harden, A., Oakley, A., et al. Integrating qualitative research with trials in systematic reviews. *BMJ*, 328(7446), 1010–1012, 2004.

11. Cochrane, A. *Effectiveness and Efficiency: Random Reflections on Health Services.* London, U.K.: Nuffield Provincial Hospitals Trust, 1972.

12. Kleijnen, L., Goetzsche, P., Kunz, R., Oxman, A., Chalmers, I. So what's so special about randomisation? In Maynard, A., Chalmers, I. (eds.), *Non Random Reflections on Health Services Research*. London, U.K.: BMJ Publishing, pp. 93–106, 1997.

13. Schwartz, D., Lellouch, J. Explanatory and pragmatic attitudes in therapeutic trials. *J Chronic Dis*, 20(20), 637–648, 1967.

14. Peto, R., Collins, R., Gray, R. Large scale randomised evidence. *Ann NY Acad Sci*, 703, 314–340, 1993.

15. Oliver, S. Exploring lay perspectives on questions of effectiveness, in Maynard, A., Chalmers, I. (eds.), *Non Random Reflections on Health Services Research*. London, U.K.: BMJ Publications, 1997, pp. 272–291.

16. Haynes, B. Can it work? Does it work? Is it worth it? The testing of healthcare interventions is evolving. *BMJ*, 319(7211), 652–653, 1999.

17. Lilienfeld, A. The Fielding H. Garrison Lecture: Ceteris paribus: The evolution of the clinical trial. *Bull Hist Med*, 56, 1–18, 1982.

18. Pocock, S. *Clinical Trials: A Practical Approach*. London, U.K.: Wiley, 1983.

19. Andrews, G. Randomised controlled trials in psychiatry: Important but poorly accepted. *BMJ*, 319(7209), 562–564, 1999.

20. Proudfoot, J., Goldberg, D., Mann, A., Everitt, B., Marks, I., Gray, J.A. Computerized, interactive, multimedia cognitive-behavioural program for anxiety and depression in general practice. *Psychol Med*, 33(2), 217–227, 2003.

21. Wells, K.B. Treatment research at the crossroads: The scientific interface of clinical trials and effectiveness research. *Am J Psychiatry*, 156(1), 5–10, 1999.

22. Wells, K.B., Sherbourne, C., Schoenbaum, M., et al. Impact of disseminating quality improvement programs for depression in managed primary care: A randomized controlled trial. *JAMA*, 283(2), 212–220, 2000.

23. Davies, D.L., Shepherd, M. Reserpine in the treatment of anxious and depressed patients. *Lancet*, ii, 117–120, 1955.
24. Medical Research Council. Report by its Clinical Committee: Clinical trial of the treatment of depressive illness. *BMJ*, i, 881–886, 1965.
25. National Institutes of Health Clinical Trials Register. nihclinicaltrials.gov. Washington, DC: National Institutes of Health, 2005.
26. Geddes, J. Large simple trials in psychiatry: Providing reliable answers to important clinical questions. *Epidemiol Psichiatr Soc*, 14, 122–126, 2005.
27. Lagomasino, I.T., Dwight-Johnson, M., Simpson, G.M. The need for effectiveness trials to inform evidence-based psychiatric practice. *Psychiatr Serv*, 56, 649–651, 2005.
28. March, J.S., Silva, S.G., Compton, S., Shapiro, M., Califf, R., Krishnan, R. The case for practical clinical trials in psychiatry. *Am J Psychiatry*, 162, 836–846, 2005.
29. Stroup, S. Practical clinical trials for schizophrenia. *Epidemiol Psichiatr Soc*, 14, 132–136, 2005.
30. Boutron, I., Moher, D., Altman, D.G., Schulz, K.F., Ravaud, P. Extending the CONSORT statement to randomized trials of nonpharmacologic treatment: Explanation and elaboration. *Ann Intern Med*, 148(4), 295–309, 2008.
31. Von Korff, M., Goldberg, D. Improving outcomes in depression. The whole process of care needs to be enhanced. *BMJ*, 323, 948–949, 2001.
32. Thornicroft, G., Tansella, M. Components of a modern mental health service: A pragmatic balance of community and hospital care: Overview of systematic evidence. *Br J Psychiatry*, 185, 283–290, 2004.
33. Becker, T. Out-patient psychiatric services, in Thornicroft, G., Szmukler, G. (eds.), *Textbook of Community Psychiatry*. Oxford, U.K.: Oxford University Press, 2001, pp. 277–282.
34. Roth, A., Fonagy, P. *What Works for Whom? A Critical Review of Psychotherapy Research*. New York: Guildford Press, 1996.
35. Burns, T. Generic versus specialist mental health teams, in Thornicroft, G., Szmukler, G. (eds.), *Textbook of Community Psychiatry*. Oxford, U.K.: Oxford University Press, 2001, pp. 231–241.
36. Simmonds, S., Coid, J., Joseph, P., Marriott, S., Tyrer, P. Community mental health team management in severe mental illness: A systematic review. *Br J Psychiatry*, 178, 497–502, 2001.
37. Thornicroft, G., Wykes, T., Holloway, F., Johnson, S., Szmukler, G. From efficacy to effectiveness in community mental health services. PRiSM Psychosis Study. 10. *Br J Psychiatry*, 173, 423–427, 1998.
38. Tyrer, P., Morgan, J., Van Horn, E., et al. A randomised controlled study of close monitoring of vulnerable psychiatric patients. *Lancet*, 345(8952), 756–759, 1995.
39. Tyrer, P., Evans, K., Gandhi, N., Lamont, A., Harrison-Read, P., Johnson, T. Randomised controlled trial of two models of care for discharged psychiatric patients. *BMJ*, 316(7125), 106–109, 1998.
40. Tyrer, P., Thompson, S., Schmidt, U., et al. Randomized controlled trial of brief cognitive behaviour therapy versus treatment as usual in recurrent deliberate self-harm: The POPMACT study. *Psychol Med*, 33(6), 969–976, 2003.
41. Szmukler, G., Holloway, F. In-patient treatment, in Thornicroft, G., Szmukler, G. (eds.), *Textbook of Community Psychiatry*. Oxford, U.K.: Oxford University Press, 2001, pp. 321–337.

42. Johnstone, P., Zolese, G. Systematic review of the effectiveness of planned short hospital stays for mental health care. *BMJ*, 318(7195), 1387–1390, 1999.

43. Shepherd, G., Murray, A. Residential care, in Thornicroft, G., Szmukler, G. (eds.), *Textbook of Community Psychiatry*. Oxford, U.K.: Oxford University Press, 2001, pp. 309–320.

44. Thornicroft, G., Bebbington, P. Deinstitutionalisation—From hospital closure to service development. *Br J Psychiatry*, 155, 739–753, 1989.

45. Leff, J. *Care in the Community. Illusion or Reality?* London, U.K.: Wiley, 1997.

46. Catty, J., Burns, T., Comas, A. *Day Centres for Severe Mental Illness (Cochrane Review)*. The Cochrane Library, Issue 1. Oxford, U.K.: Update Software, 2003.

47. Marshall, M., Crowther, R., Almaraz-Serrano, A., et al. Systematic reviews of the effectiveness of day care for people with severe mental disorders: (1) acute day hospital versus admission; (2) vocational rehabilitation; (3) day hospital versus outpatient care. *Health Technol Assess*, 5(21), 1–75, 2001.

48. Tyrer, S., Coid, J., Simmonds, S., Joseph, P., Marriott, S. *Community Mental Health Teams (CMHTs) for People with Severe Mental Illnesses and Disordered Personality (Cochrane Review)*. Oxford, U.K.: Update Software, 2003.

49. Scott, J., Lehman, A. Case management and assertive community treatment, in Thornicroft, G., Szmukler, G. (eds.), *Textbook of Community Psychiatry*. Oxford, U.K.: Oxford University Press, 2001, pp. 253–264.

50. Burns, T., Creed, F., Fahy, T., Thompson, S., Tyrer, P., White, I. Intensive versus standard case management for severe psychotic illness: A randomised trial. UK 700 Group. *Lancet*, 353(9171), 2185–2189, 1999.

51. Burns, T., Fioritti, A., Holloway, F., Malm, U., Rossler, W. Case management and assertive community treatment in Europe. *Psychiatr Serv*, 52(5), 631–636, 2001.

52. Fiander, M., Burns, T., McHugo, G.J., Drake, R.E. Assertive community treatment across the Atlantic: Comparison of model fidelity in the UK and USA. *Br J Psychiatry*, 182, 248–254, 2003.

53. Marshall, M., Lockwood, A. *Assertive Community Treatment for People with Severe Mental Disorders (Cochrane Review)*. The Cochrane Library, Issue 1. Oxford: Update Software, 2003.

54. Phillips, S.D., Burns, B.J., Edgar, E.R., et al. Moving assertive community treatment into standard practice. *Psychiatr Serv*, 52(6), 771–779, 2001.

55. Addington, J., Coldham, E.L., Jones, B., Ko, T., Addington, D. The first episode of psychosis: The experience of relatives. *Acta Psychiatr Scand*, 108(4), 285–289, 2003.

56. Raune, D., Kuipers, E., Bebbington, P.E. Expressed emotion at first-episode psychosis: Investigating a carer appraisal model. *Br J Psychiatry*, 184, 321–326, 2004.

57. Marshall, M., Lockwood, A. Early Intervention for psychosis. *Cochrane Database Syst Rev*, (2), CD004718, 2004.

58. Friis, S., Larsen, T.K., Melle, I., et al. Methodological pitfalls in early detection studies—The NAPE Lecture 2002. Nordic Association for Psychiatric Epidemiology. *Acta Psychiatr Scand*, 107(1), 3–9, 2003.

59. Harrigan, S.M., McGorry, P.D., Krstev, H. Does treatment delay in first-episode psychosis really matter? *Psychol Med*, 33(1), 97–110, 2003.

60. Larsen, T.K., Friis, S., Haahr, U., et al. Early detection and intervention in first-episode schizophrenia: A critical review. *Acta Psychiatr Scand*, 103(5), 323–334, 2001.

61. Marshall, M., Rathbone, J. Early intervention for psychosis. *Cochrane Database Syst Rev*, (4), CD004718, 2006.

62. McGorry, P.D., Yung, A.R., Phillips, L.J., et al. Randomized controlled trial of interventions designed to reduce the risk of progression to first-episode psychosis in a clinical sample with subthreshold symptoms. *Arch Gen Psychiatry*, 59(10), 921–928, 2002.
63. McGorry, P.D., Killackey, E.J. Early intervention in psychosis: A new evidence based paradigm. *Epidemiol Psichiatr Soc*, 11(4), 237–247, 2002.
64. Warner, R., McGorry, P.D. Early intervention in schizophrenia: Points of agreement. *Epidemiol Psichiatr Soc*, 11(4), 256–257, 2002.
65. Johnson, S., Nolan, F., Hoult, J., et al. Outcomes of crises before and after introduction of a crisis resolution team. *Br J Psychiatry*, 187, 68–75, 2005.
66. Johnson, S., Nolan, F., Pilling, S., et al. Randomised controlled trial of acute mental health care by a crisis resolution team: The north Islington crisis study. *BMJ*, 331(7517), 599, 2005.
67. Catty, J., Burns, T., Knapp, M., et al. Home treatment for mental health problems: A systematic review. *Psychol Med*, 32(3), 383–401, 2002.
68. Glover, G., Arts, G., Babu, K.S. Crisis resolution/home treatment teams and psychiatric admission rates in England. *Br J Psychiatry*, 189(5), 441–445, 2006.
69. Joy, C.B., Adams, C.E., Rice, K. Crisis intervention for people with severe mental illnesses. *Cochrane Database Syst Rev*, (4), CD001087, 2006.
70. World Health Organisation. *World Health Report 2001. Mental Health: New Understanding, New Hope.* Geneva: World Health Organization, 2001.
71. Barbui, C., Hotopf, M. Forty years of antidepressant drug trials. *Acta Psychiatr Scand*, 104(2), 92–95, 2001.
72. Djulbegovic, B., Clark, M. Scientific and ethical issues in equivalence trials. *JAMA*, 285(9), 1206–1208, 2001.
73. Essock, S.M., Drake, R.E., Frank, R.G., McGuire, T.G. Randomized controlled trials in evidence-based mental health care: Getting the right answer to the right question. *Schizophr Bull*, 29(1), 115–123, 2003.
74. Walwyn, R., Wessely, S. RCTs in psychiatry: Challenges and the future. *Epidemiol Psichiatr Soc*, 14, 127–131, 2005.
75. Kuipers, E., Fowler, D., Garety, P., et al. London-east Anglia randomised controlled trial of cognitive-behavioural therapy for psychosis. III: Follow-up and economic evaluation at 18 months. *Br J Psychiatry*, 173, 61–68, 1998.
76. Chisholm, D., Sanderson, K., Ayuso-Mateos, J.L., Saxena, S. Reducing the global burden of depression: Population-level analysis of intervention cost-effectiveness in 14 world regions. *Br J Psychiatry*, 184(5), 393–403, 2004.
77. Gilbody, S.M., House, A.O., Sheldon, T.A. Outcomes research in mental health. Systematic review. *Br J Psychiatry*, 181, 8–16, 2002.
78. Bauer, M.S., Williford, W.O., Dawson, E.E., et al. Principles of effectiveness trials and their implementation in VA Cooperative Study #430: 'Reducing the efficacy-effectiveness gap in bipolar disorder.' *J Affect Disord*, 67(1–3), 61–78, 2001.
79. Shapiro, S.H., Weijer, C., Freedman, B. Reporting the study populations of clinical trials. Clear transmission or static on the line? *J Clin Epidemiol*, 53(10), 973–979, 2000.
80. Barrowclough, C., Haddock, G., Tarrier, N., et al. Randomized controlled trial of motivational interviewing, cognitive behavior therapy, and family intervention for patients with comorbid schizophrenia and substance use disorders. *Am J Psychiatry*, 158(10), 1706–1713, 2001.

81. Haddock, G., Barrowclough, C., Tarrier, N., et al. Cognitive-behavioural therapy and motivational intervention for schizophrenia and substance misuse. 18-month outcomes of a randomised controlled trial. *Br J Psychiatry*, 183, 418–426, 2003.

82. Szmukler, G., Kuipers, E., Joyce, J., et al. An exploratory randomised controlled trial of a support programme for carers of patients with a psychosis. *Soc Psychiatry Psychiatr Epidemiol*, 38(8), 411–418, 2003.

83. Wells, K., Sherbourne, C., Schoenbaum, M., et al. Five-year impact of quality improvement for depression: Results of a group-level randomized controlled trial. *Arch Gen Psychiatry*, 61(4), 378–386, 2004.

84. Walsh, E., Gilvarry, C., Samele, C., et al. Reducing violence in severe mental illness: Randomised controlled trial of intensive case management compared with standard care. *BMJ*, 323(7321), 1093–1096, 2001.

85. Creed, F., Mbaya, P., Lancashire, S., Tomenson, B., Williams, B., Holme, S. Cost effectiveness of day and inpatient psychiatric treatment: Results of a randomised controlled trial. *BMJ*, 314(7091), 1381–1385, 1997.

86. Hotopf, M., Churchill, R., Lewis, G. Pragmatic randomised controlled trials in psychiatry. *Br J Psychiatry*, 175, 217–223, 1999.

87. Drake, R.E., McHugo, G.J., Bebout, R.R., et al. A randomized clinical trial of supported employment for inner-city patients with severe mental disorders. *Arch Gen Psychiatry*, 56(7), 627–633, 1999.

88. Fletcher, R.H. Evaluation of interventions. *J Clin Epidemiol*, 55(12), 1183–1190, 2002.

89. Sackett, D.L. Bias in analytic research. *J Chronic Dis*, 32, 51–63, 1979.

90. Maynard, A., Chalmers, I. *Non Random Reflections on Health Services Research*. London, U.K.: BMJ Publishing, 1997.

91. Peveler, R., Kendrick, T. Selective serotonin reuptake inhibitors: THREAD trial may show way forward. *BMJ*, 330, 420–421, 2005.

92. Medical Research Council. *Cluster Randomised Trials: Methodological and Ethical Considerations*. London, U.K.: Medical Research Council, 2002.

93. Gilbody, S., Whitty, P. Improving the delivery and organisation of mental health services: Beyond the conventional randomised controlled trial. *Br J Psychiatry*, 180, 13–18, 2002.

94. Schoenbaum, M., Unutzer, J., Sherbourne, C., et al. Cost-effectiveness of practice-initiated quality improvement for depression: Results of a randomized controlled trial. *JAMA*, 286(11), 1325–1330, 2001.

95. Andrews, G. Efficacy, effectiveness and efficiency in mental health service delivery. *Aust N Z J Psychiatry*, 33(3), 316–322, 1999.

96. Campbell, M., Fitzpatrick, R., Haines, A., et al. Framework for design and evaluation of complex interventions to improve health. *BMJ*, 321(7262), 694–696, 2000.

97. Chisholm, D. Keeping pace with assessing cost-effectiveness: Economic efficiency and priority-setting in mental health. *Aust N Z J Psychiatry*, 39(8), 645–647, 2005.

98. Wang, P.S., Simon, G., Kessler, R.C. The economic burden of depression and the cost-effectiveness of treatment. *Int J Methods Psychiatr Res*, 12(1), 22–33, 2003.

99. Wells, K., Miranda, J., Bruce, M.L., Alegria, M., Wallerstein, N. Bridging community intervention and mental health services research. *Am J Psychiatry*, 161(6), 955–963, 2004.

100. Gilbody, S., Wahlbeck, K., Adams, C. Randomized controlled trials in schizophrenia: A critical perspective on the literature. *Acta Psychiatr Scand*, 105(4), 243–251, 2002.

26

Designing and Evaluating Interventions to Change Health-Related Behavior Patterns*

Charles Abraham

University of Exeter

CONTENTS

26.1 Introduction

Health and longevity are determined by many factors including genetics, nutrition, family environment, work environments, the distribution of wealth within countries, and the prevalence of stressful working environments (cf. [1,2]). Within this context, individual behavior patterns have an important impact on individual and public health. The Alameda County study followed 7000 people over 10 years and showed that sleep, exercise, drinking alcohol, and eating habits predicted mortality [3]. Similarly, Khaw et al. [4] found that controlling for age, gender, body mass index, and socioeconomic status, a group of 20,000 people, those who smoked and consumed more than moderate quantities of alcohol, were not physically active, and did not eat five portions of fruits and vegetables a day, were more than four times more likely to have died than those not engaging in these four behavior patterns-over an 11 year observation period.

Individual behavior patterns affect national sick leave and accident costs as well as health services demand. In the United Kingdom, the Wanless [5,6] reports on resources required to provide high-quality health services

---

* The preparation of this paper was supported by the National Institute for Health Research (NHR), UK. However, the views expressed are those of the author(s) and not necessarily those of the MRC, the NIHR or the UK Department of Health.

concluded that despite continued service investment, public health would not improve and could deteriorate without greater public engagement with health-promoting behaviors. Behavioral interventions can improve public health and simultaneously cut health service spending [7]. Thus, on economic grounds alone national health services need to invest in interventions to induce health-promoting behavior patterns and discourage health-damaging behaviors. Such investments may target individuals, organizations, and industries and may focus on information provision, motivation enhancement including economic rewards, skills training, and legislation.

Considerable resource has been devoted to designing behavior change interventions and evidence from a recent meta-analysis of >1000 evaluations of health behavior change interventions demonstrates that theory- and evidence-based behavior change interventions can be effective [8]. The expertise to design, evaluate, and faithfully implement behavior change interventions exists but, unfortunately, multiple barriers impede progress toward establishment of evidence-based health promotion. In this chapter, I will consider three key threats to the development and implementation of interventions to change individual behavior patterns, namely,

1. Intervention design may not optimize correspondence between the behavior change techniques included in an intervention and the needs of the targeted population.

2. Interventions may not be evaluated or evaluated using methods that lack internal and external validity casting doubt on the generalizability of findings and curtailing implementation and sustainable of intervention delivery in practice.

3. Interventions are often poorly described so that they cannot be replicated by other researchers and may be inaccurately implemented in practice.

26.2 Designing Behavior Change Interventions

Careful design of behavior change interventions, including evidence-based selection of the optimal combination of behavior change techniques for particular populations, is prerequisite to effectiveness (see, e.g., [9]). Intervention mapping [10,11] provides a useful guide to systematic behavior change intervention design.

The first stage is *needs assessment*, that is, establishing who needs to change what. For example, having recognized that a particular population needs to reduce calorie intake, what level of intake reduction is required and which behavior change intervention will be most effective in targeting individual consumers or food manufactures? In either case, what particular changes are needed to change eating behavior?

The second stage involves *setting change objectives*, that is, specifying the behavior changes the intervention is designed to induce. It also involves identifying potentially modifiable determinants of those behaviors. For example, an intervention designed to promote condom use amongst sexually active teenagers may first focus on changing norms in relation to carrying condoms when attending social events. In this case, increased condom carrying, the primary change objective, may be achieved most effectively by targeting young people's beliefs about condom carrying, including beliefs about whether other young people carry condoms and approve of condom carrying (see [12] for further details on applying models of modifiable determinants of behavior).

Completion of the second planning stage should also clarify key outcome measures for intervention evaluation. So, in the earlier example, reliable and valid measures of condom carrying and measures of condom carrying beliefs will be required. It will be hypothesized that the latter will mediate the impact of the intervention on the former, that is, the effect of the intervention on condom carrying will be accounted for by the impact of the intervention on beliefs. Note that mediation analysis [13] of this kind allows testing of psychological models of determinants during intervention evaluation. For example, if target beliefs are successfully changed but the behavior is not, this implies that the wrong modifiable determinants were selected, suggesting that stage 2 should be re-visited and an alternative determinants model adopted. Unfortunately, behavior change intervention evaluations too infrequently include such model testing analyses [14].

The third stage involves *identifying change processes and effective change techniques*, that is, linking the determinants identified in stage 2 to change mechanisms likely to shift modifiable determinants in the desired direction. Having understood the crucial change processes, corresponding behavior change techniques (BCTs) must be selected (see Ref. [15] for further details). This is a critical aspect of intervention design. It entails development of a good model of the processes that regulate the target behavior. So, for example, if adolescent normative beliefs are targeted, providing information about teenagers' views and behaviors based on reliable research can be effective when adolescents underestimate normative support for a behavior. Another approach would be to create role models and use opinion leaders to influence norms (see, e.g., [16]).

Stage 3 involves reviewing the relevant literature. A wide range of BCTs are available. Abraham and Michie [17] identified 26 distinct BCTs used in interventions described in 195 published works. This was developed further by Abraham et al. [18] who present a list of 40 distinct BCTs used in previous interventions, grouped according to the change target that each BCT is likely to impact on; change targets that may be identified during intervention planning (see Ref. [15]).

Three meta-analytic studies provide evidence on the potential effectiveness of particular BCTs. Michie et al. [19] reviewed of interventions designed to promote physical activity or healthy eating. Drawing upon the BCTs defined by Abraham and Michie [17], and Carver and Scheier's [20] control theory, Michie et al. found that prompting self-monitoring combined with other BCTs designed to promote goal setting and enhance self-regulatory skills, specifically, (1) prompting intention formation, (2) prompting specific planning, (3) prompting reviews of behavioral goals, and (4) providing feedback on performance–were associated with greater effectiveness. The average Cohen's *d* for interventions including these specified BCTs was 0.42 compared with 0.26 for interventions without self-monitoring and one or more of the other four BCTs. Similarly, Webb and Sheeran [21] found that interventions including information provision, goal setting, modeling, and skill training yielded small to medium effects on behavior change (with *d*'s around 0.3) while interventions including use of contingent rewards and provision of social support were more effective (with *d*'s between 0.5 and 0.6).

In a meta-analysis of HIV prevention, Albarracín et al. [22] found that the most effective interventions provided (1) information, (2) arguments to promote positive attitudes toward condom use, (3) behavioral skills relevant to condom use, and (4) self-regulatory or skills training. In addition, provision of condoms and HIV counseling and HIV testing enhanced intervention effectiveness. In contrast, inclusion of threat or fear appeals did not enhance effectiveness. Albarracín et al. also found some approaches to be effective with one target group but not another. Arguments targeting normative beliefs were found to enhance intervention effectiveness when the target audience was under 21 years of age but to reduce effectiveness amongst older recipients. Thus age moderated the relationship between inclusion of normative arguments and intervention effectiveness [13].

Finally, in systematic reviews examining the effectiveness of school-based health promotion interventions targeting sexual, substance abuse, and nutrition behaviors, Peters et al. [23] found strong evidence that effective school-based interventions shared five characteristics, namely,

1. Use of theory in intervention design
2. Targeting social influences, especially social norms
3. Targeting cognitive and behavioral skills
4. Training those delivering the intervention
5. Including multiple components

These findings simultaneously emphasize the importance of planning and, in particular, a good theoretical model of modifiable determinants, change processes by which determinants can be modified and target-relevant BCTs capable of effecting these changes (see e.g., [15]).

The fourth stage of intervention mapping involves the translation of the theoretical model of determinants, change processes, and corresponding change techniques into a *practical plan*. This will involve constructing and reviewing materials and methods (e.g., videos, leaflets, lesson plans, etc.) to implement the selected change techniques in the target setting in a manner that will engage the target audience. It is critical to pretest materials by running small-scale experiments and assess whether exposure to the intervention materials has the intended effect on representative participants prior to investing in the final practical design.

The fifth stage is *adoption and implementation*. This involves identifying those who will adopt and use the intervention (e.g., teachers in schools or worksite managers), negotiating its use with them and producing the intervention in a manner that facilitates accurate implementation (e.g., by providing training courses). It is crucial that those responsible for adoption of the intervention (head teachers, managers, mothers, teenagers, etc.) are persuaded. Consequently, opinion-leading representatives of this group should be involved from stage 3 onward. Even effective interventions have no impact if they are not adopted.

The sixth and last stage of the intervention mapping process is *evaluation*, which we will consider in more detail below.

26.3 Evaluating Behavior Change Interventions

Without evaluation, we do not know whether a behavior change intervention was effective and whether or not it should be used again. If it is worth designing an intervention, it is worth evaluating it. Intervention evaluation should, at a minimum, report (1) how effective the intervention was (e.g., using Cohen's d [34] or other effect sizes statistics), (2) how any observed effectiveness was achieved (e.g., using mediation analysis), and (3) for whom the intervention was effective (i.e., the target group should be clearly specified and moderation analysis conducted, if warranted) [24].

If an intervention study has poor internal validity, then we cannot have confidence in the effectiveness of the intervention; so rigorous, defensible evaluation is foundational to future adoption and implementation. A useful evaluation will have appropriate power, use appropriate control groups, control for confounds (by randomization, multilevel modeling, and other methods including controlling for pre-intervention measures), use appropriate outcome measures (including clinically relevant measures and measures used in mediation analyses), and apply appropriate analyses including intention-to-treat analysis (see [12] for further details).

When an intervention is not effective, it is important to know why. Mediation analyses may show that the determinant model designed at IM stage 2 was inaccurate implying that re-design is required. Interventions can, however, also fail because they are not delivered as designed. For example, if critical BCTs are not implemented then the intervention may fail to change important determinants and, consequently, the target behavior. To clarify why an intervention has failed, a process evaluation is required. This involves examining whether delivery in practice matched delivery design. This may involve interviewing those delivering and receiving the intervention, using surveys to assess their experience of the delivery and observing intervention groups to check for implementation fidelity.

Quantitative evaluation is critical to question 1 (listed above) but qualitative evaluation can be crucial to pre-intervention assessment of the target population, to process evaluation and to answering question 2, that is, identifying what features of the intervention were critical to effectiveness (see, e.g., [25]).

Interventions are adopted by users when they are perceived to be useful, easy to implement, sustainable, and affordable in the setting in which they have been tested (see IM stage 5). Yet even rigorously evaluated, effective interventions may fail on these criteria and so have little real-world impact. This is highlighted by research using the Reach, Effectiveness, Adoption, Implementation and Maintenance (RE-AIM) framework (e.g., [26,27]).

Reach refers to how many of the target population were involved in an evaluation and how representative they were. For example, if an intervention was evaluated using economically advantaged participants, then questions would arise as to whether it would also be effective those who are less well off. A clear answer to question 3 given earlier is required to assess the potential "reach" of an intervention.

Effectiveness refers to the range of effects an intervention might have. For example, even if it changed the target behavior, did it enhance overall quality of life and were there any unintended consequences (e.g., did participants find it onerous or upsetting?)?

Adoption refers to whether the users (e.g., doctors, teachers, nurses, or managers) are persuaded of the utility of the intervention. This is likely to depend on how easy it is implement, whether they or their clients like it, and whether it is compatible with their other primary goals, including minimizing cost.

Implementation refers to the ease and feasibility of faithful delivery. If an intervention is complex, expensive, or requires specialist training or teams of people to deliver it then it is less likely to be faithfully implemented in real-world settings.

Maintenance refers to the longer-term sustainability of the intervention in real-world settings. For example, if a community does not have the resources to deliver an intervention then, no matter how efficacious, it will be dropped or changed over time. If the complexity of an intervention leads to it being changed or adapted, then critical BCTs may be omitted so compromising effectiveness.

Unless these RE-AIM considerations are prioritized in IM stages 2–4, then potentially effective behavior change interventions may never be widely implemented or may be used only for short periods, thereby, undermining their impact. RE-AIM emphasizes that along with internal validity, evaluations must have good external validity if they are to ensure that behavior change interventions are not only effective but are readily implemented and sustained where they are needed.

26.4 Describing and Sharing Behavior Change Interventions

Some progress has been made in standardizing the description of intervention evaluations through acceptance by journal editors of the CONSORT [28,29] and TREND statements [30], which specify the details that should be included in published reports. Davidson et al. [31] have helpfully augmented these guidelines by proposing that evaluations of behavioral interventions should also report (1) the content or elements of the intervention, (2) characteristics of the those delivering the intervention, (3) characteristics of the recipients, (4) the setting (e.g., worksite), (5) the mode of delivery (e.g., face to face), (6) the intensity (e.g., contact time), (7) the duration (e.g., number sessions over a given period), and (8) adherence to delivery protocols. Continued standardization of reporting could greatly accelerate the science of behavior change.

Unfortunately, however, standardized reporting of behavior change intervention is not currently required by journal editors so descriptions of behavior change interventions in peer-reviewed journals are highly variable in length and structure and in the language used to describe intervention components and behavior change techniques. Clarity concerning the "content or elements" [31] of behavior change interventions is especially problematic because, although CONSORT guidelines specify that evaluators should report "precise details of interventions [as] actually administered" [27, p. 1192], there is no standardized vocabulary with which to describe behavior change techniques employed by designers of behavior change interventions (see [16]). As Davidson et al. [31] rightly note "Often reports fail to describe the actual behavioral intervention techniques used; instead they provide details regarding treatment format (e.g., the number of sessions, type of treatment)" (p. 165).

Consequently, intervention descriptions included in journal articles do not usually allow other researchers or practitioners to accurate replicate the evaluated intervention. Adoption of standardized descriptions of BCTs such as those provided by Abraham and Michie [17] and Abraham [15] would greatly facilitate communication between behavior change researchers and between researchers and practitioners.

Although standardization of the structure of intervention descriptions and the language used to describe intervention contents would be a major step forward, pressure on journal space means that many interventions cannot be adequately described in journal articles. This was illustrated by a small pilot study comparing BCTs described in published evaluations with BCTs described in detailed protocols or manuals for the same interventions. Abraham and Michie [17] found that only two-thirds of the BCTs identified in manuals were also included in the published evaluations. Thus replication and implementation fidelity of effective interventions by other researchers or practitioners depends critically on availability of detailed manuals or protocols describing exactly how interventions were designed and implemented and including relevant instructions and materials.

Surprisingly and worryingly, detailed manuals or protocols are not always available. While conducting systematic reviews of the content of behavior change interventions (in order to relate intervention content to effectiveness), I receive many e-mails from researchers informing me that, even for interventions published over the last 5 years, no manual or protocol is available either because a manual was never written or because it has been lost over time. Consequently many published, effective evaluations can no longer be faithfully replicated or implemented. This means that substantial proportions of the science of behavior change are being lost each year. Imagine, by comparison, a chemist who had published a paper reporting an experiment in which a new and useful compound was created but who later declared that (s)he had lost the laboratory notes (or not kept adequate notes) and so was unable to describe how the compound was made. This would not be acceptable in chemistry and should not be acceptable in behavioral science. Important ethical issues also arise in relation to the use of public funds to develop and evaluate interventions that cannot later be replicated or implemented for public benefit.

This problem can be easily solved. Journal editors should insist on simultaneous submission of detailed manuals with papers reporting behavior change intervention evaluations. Indeed if, as appears to be the case, descriptions of interventions in manuscripts do not adequately describe interventions, then reviewers surely need to be able to refer to such manuals to accurately judge the quality of an intervention evaluation report. Some journals have already adopted this policy and, hopefully, more will follow. A good example of such an editorial policy is provided by the editor of *Addiction* [32].

Not only do we need to provide standardized descriptions of behavior change interventions that the specify behavior change techniques employed, but also we need to provide equivalent descriptions of interventions received by active control groups, including those receiving routine care (cf. [31]). Without such descriptions, reviews of well-conducted randomized controlled trails that compare innovative behavior change interventions against routine care may reach misleading conclusions. For example, if two interventions are tested in comparison with routine care and one generates a Cohen's d of 0.0 while the other generates a d of 0.5, most reviewers would assume that the former was ineffective while the latter showed a small medium effect size that could deliver an important improvement on current care at a population level. This interpretation depends on an assumption that is rarely true across trials conducted in different locations, namely, that routine care is the same across the two trials. If, for example, the routine care against which the former intervention was tested was considerably more effective than that against which the latter was tested then it is possible that the former intervention is more effective than the latter. This means that reviewers (including meta-analysts) and commissioners of services may wrongly interpret the research literature. The potential for such misinterpretation was illustrated by de Bruin et al. [33] who examined the intervention content of routine care in trials of interventions designed to enhance adherence to highly active antiretroviral therapy. This is an important focus for behavior change because ~50% of patients do not sustain the high levels of medication adherence required for optimal viral suppression. de Bruin and colleagues showed that the content of routine care varied considerably and, predictably, that the inclusion of BCTs in routine care predicted the effectiveness of routine care. de Bruin et al. concluded that rather than testing further interventions, clinicians should strive to implement existing best practice. This research demonstrates that it is only meaningful to compare the results of trials of behavior change interventions tested against active control groups when we also have detailed descriptions of those control groups. Again, this is something that journal editors need to insist on, in order to improve the science of behavior change intervention.

Concern regarding the scientific and practice implications of poor reporting of behavior change interventions in journals and in supplementary manuals and protocols led a group of researchers and editors to issue a consensus statement after meetings at the 2008 and 2009 annual conferences of the European Health Psychology Society. A summary of the consensus statement issued by the Workgroup for Intervention Development and Evaluation Research (WIDER) is provided in Table 26.1. The full statement and a list of WIDER members can be downloaded from the "WIDER recommendations" link at http://interventiondesign.co.uk/?page_id=9.

TABLE 26.1

Summary of WIDER Recommendations

---

1. *Detailed description of interventions in published papers*
   Instructions to authors should specify that behavior change intervention evaluations describe (1) characteristics of those delivering the intervention, (2) characteristics of the recipients, (3) the setting (e.g., worksite, time, and place of intervention), (4) the mode of delivery (e.g., face-to-face), (5) the intensity (e.g., contact time), (6) the duration (e.g., number of sessions and their spacing over a given period), (7) adherence/fidelity to delivery protocols, and (8) a detailed description of the intervention content provided for each study group.

2. *Clarification of assumed change process and design principles*
   Instructions to authors should specify that behavior change intervention evaluations describe (1) the intervention development, (2) the change techniques used in the intervention, and (3) the causal processes targeted by these change techniques in as much detail as possible, unless these details are easily available elsewhere (e.g., in a prior publication).

3. *Access to intervention manuals/protocols*
   Before publishing a behavior change intervention evaluations report, editors ask authors to submit a protocol or manual describing the intervention or, alternatively, specify where such a manual can be easily and reliably accessed by readers.

4. *Detailed description of active control conditions*
   Instructions to authors should specify that behavior change intervention evaluations describe the content of active control groups in as much detail as is possible (e.g., the techniques used) in a similar manner to the description of the content of the intervention itself.

26.5 Conclusion

Behavior change is urgently needed in relation to health-related behavior patterns as well as in other areas such as sustainable energy use. There is clear evidence that behavior change interventions can be effective. However, effectiveness depends on careful design based on a detailed knowledge of (1) the determinants of behavior change and how these are distributed within the target population, (2) the processes by which behavior changes, and (3) behavior change techniques capable of shaping those processes. Further work on the second two areas and especially the third is required to establish a science of behavior change. Change techniques included in interventions and the active control groups to which they are compared are not currently described in a standardized format, which can be understood by all researchers. Moreover, descriptions in journals are often incomplete and detailed manuals and protocols are often not readily available impeding accurate replication and adoption of effective interventions. Use of standardized descriptions of behavior change techniques [15,16] by researchers and intervention designers as well as insistence on adherence to CONSORT guidelines (e.g., [28]) and WIDER recommendations would do much to accelerate understanding and enhance effectiveness in the science of behavioral change.

References

1. Alder, N.E., Boyce, T., Chesney, M.A., et al. Socio-economic status and health: The challenge of the gradient. *Am Psychol*, 49, 15–24, 1994.
2. Wilkinson, R.G. Health inequalities: Relative or absolute material standards? *British Medical Journal*, 314, 591–595, 1997.
3. Belloc, N.B., Breslow, L. Relationship of physical health status and health practices. *Prev Med*, 9, 469–421, 1972.
4. Khaw, K.T., Wareham, N., Bingham, S., Welch, A., Luben. R., Day, N. Combined impact of health behaviours and mortality in men and women: The EPIC-Norfolk prospective population study. *PLoS Med*, 5(1), e12, 2008. Open access—DOI:10.1371. http://medicine.plosjournals.org/perlserv/?request=get-document&doi=10.1371/journal.pmed.0050012
5. Wanless, D. *Securing Our Future Health: Taking a Long-Term View*. London, U.K.: HMSO, 2002.
6. Wanless, D. *Securing Good Health for the Whole Population*. London, U.K.: HMSO, 2004.
7. Friedman, R., Sobel, D., Myers, P., Caudill, M., Benson, H. Behavioral medicine, clinical health psychology and cost offset. *Health Psychol*, 14, 509–518, 1995.
8. Johnson, B.T., Scott-Sheldon, L.A.J., Carey, M.P. Meta-synthesis of health behaviour change meta-analyses. *American Journal of Public Health*, 100 (11): 2193–2198, 2010.

9. Mullen, P.D., Green, L.W., Persinger, G. Clinical trials of patient education for chronic conditions: A comprehensive meta analysis. *Prev Med*, 14, 75–81, 1985.

10. Bartholomew, L.K., Parcel, G.S., Kok, G., Gottlieb, N.H. *Planning Health Promotion Programs. An Intervention Mapping Approach*. San Francisco, CA: Jossey-Bass, 2006.

11. Kok, H., Schaalama, H., Ruiter, R.A.C. & van Empelen, P. Intervention mapping: A protocol for applying health psychology theory to prevention programmes. *Journal of Health Psychology*, 9, 85–98, 2004.

12. Abraham, C., Conner, M., Jones, F., O'Conner, D. *Health Psychology*. London, U.K.: Hodder, 2008.

13. Baron, R.M., Kenny, D.A. The moderator–mediator variable distinction in social psychological research: Conceptual, strategic and statistical considerations. *J Pers Soc Psychol*, 51, 1173–1182, 1986.

14. Rothman, A.J. Is there nothing more practical than a good theory? Why, innovations and advances in health behavior change will arise if interventions are used to test and refine theory. *Int J Behav Nutr Phys Act*, 1, 11, 2004 (http://www.ijbnpa.org/content/1/1/11).

15. Abraham, C. Mapping change mechanisms and behaviour change techniques: A systematic approach to promoting behaviour change through text. In C. Abraham, and M. Kools (Eds.). *Writing Health Communication*: An Evidence-Based Guide for Professionals. SAGE Publications Ltd., London, 2011.

16. Schaalma, H., Kok, G. A school HIV-prevention program in the Netherlands, in Bartholomew, L.K., Parcel, G.S., Kok, G., Gottlieb, N.H. (eds.), *Planning Health Promotion Programs. An Intervention Mapping Approach*. San Francisco, CA: Jossey-Bass, 2006, pp. 511–544.

17. Abraham, C., Michie, S. A taxonomy of behavior change techniques used in interventions. *Health Psychol*, 27, 379–387, 2008.

18. Abraham, C., Kok, G., Schaalma, H., Luszczynska, A. Health promotion, in Martin, P.R., Cheung, F., Kyrios, M., et al. (eds.), *The International Association of Applied Psychology Handbook of Applied Psychology*. Oxford, U.K.: Wiley-Blackwell, 2011.

19. Michie, S., Abraham, C., Whittington, C., McAteer, J., Gupta, S. Identifying effective techniques in interventions: A meta-analysis and meta-regression. *Health Psychol*, 28, 690–701, 2009.

20. Carver, C.S., Scheier, M.F. Control theory: A useful conceptual framework for personality-social, clinical and health psychology. *Psychol Bull*, 92, 111–135, 1998.

21. Webb, T.L., Sheeran, P. Does changing behavioral intentions engender behavior change? A meta-analysis of the experimental evidence. *Psychol Bull*, 132, 249–268, 2006.

22. Albarracín, D., Gillete, J.C., Earl, A.N., Glasman, L.R., Durantini, M.R. A test of major assumptions about behavior change: A comprehensive look at the effects of passive and active HIV-prevention interventions since the beginning of the epidemic. *Psychol Bull*, 131, 856–897, 2005.

23. Peters, L.H.W., Kok, G., Ten Dam, G.T.M., Buijs, G.J., Paulussen, T.G.W.M. Effective elements of school health promotion across behavioral domains: A systematic review of reviews. *BMC Public Health*, 9, 182, 2009. DOI:10.1186/1471-2458-9-182.

24. Michie, S., Abraham, C. Interventions to change health behaviours: Evidence-based or evidence inspired? *Psychol Health*, 19, 29–49, 2004.

25. Abraham, C., Gardner, B. What psychological and behaviour changes are initiated by "expert patient" training and what training techniques are most helpful? *Psychol Health*, 10, 1153–1165, 2009.

26. Glasgow, R.E., Bull, S.S., Gillette, C., Klesges, L.M., Dzewaltowski, D.M. Behavior change intervention research in healthcare settings: A review of recent reports with emphasis on external validity. *Am J Prev Med*, 23, 62–69, 2002.

27. Green, L.W., Glasgow, R.E. Evaluating the relevance, generalization, and applicability of research: Issues in translation methodology and external validity. *Eval Health Prof*, 29, 126–153, 2006.

28. Moher, D., Schultz, K.F., Altman, D.G.; the CONSORT Group. The CONSORT statement: Revised recommendations for improving the quality of reports of parallel-group randomized trials. *Lancet*, 357, 1191–1194, 2001.

29. Boutron, I., Moher, D., Altman, D.G., Schulz, K.F., Ravaud, P. for the CONSORT Group. Extending the CONSORT statement to randomized trials of nonpharmacological treatment: Explanation and elaboration. *Ann Intern Med*, 148, 295–309, 2008.

30. Des Jarlais, D.C., Lyles, C., Crepaz, N. Improving the reporting quality of nonrandomized evaluations of behavioral and public health interventions: The TREND statement. *Am J Public Health*, 94, 361–366, 2004.

31. Davidson, K.W., Goldstein, M., Kaplan, R.M., et al. Evidence-based behavioral medicine: What is it and how do we achieve it? *Ann Behav Med*, 26, 161–171, 2003.

32. West, R. Providing full manuals and intervention descriptions: Addiction policy. *Addiction*, 103, 1411, 2008.

33. de Bruin, M., Viechtbauer, W., Schaalma, H.P., Kok, H., Abraham, C., Hospers, H.J. Standard care impact on effects of highly active antiretroviral therapy adherence interventions: A meta-analysis of randomized controlled trials. *Ann Intern Med*, 170, 240–250, 2010.

34. Cohen, J. A power primer. *Psychol Bull*, 112, 155–159, 1992.

27

Assessing Hypnotherapy: Practical Examples

Isabelle Marc

Laval University

CONTENTS

27.1 Introduction

In 2007, almost 4 out of 10 adults had used complementary and alternative medicine in the past 12 months. Among them, mind–body interventions increased in popularity for a large variety of medical conditions [1].

Administered or self-practiced, mind–body interventions, including meditation, hypnosis, self-imagery, yoga, and deep breathing, have entered the realm of clinical interest and research. Sharing many physiological characteristics, such interventions are indeed complex [2].

The hypnotic state is characterized by heightened concentration, suggestibility, and relaxation [3]. Some individuals achieve this state more readily than others. It may generally be facilitated by a trained therapist, but traditional procedures for induction are not a prerequisite. Hypnotherapy induces mental self-relaxation associated with active, focused attention that fosters the acceptance of appropriate suggestions directed toward changes in pain, anxiety, or other experiences [4–6]. For example, when the goal is to attain an analgesic effect, appropriate suggestions may focus on modulating the perception of pain components (i.e., intensity and unpleasantness). Accepting the experiential content of a suggestion facilitates incorporation of the feelings suggested to achieve positive changes. Moreover, hypnosis allows the subject to distance himself/herself from his/her symptoms, to use his/her internal resources while maintaining control over himself/herself. Interest in hypnosis for health care is supported by the fact that recent clinical studies indicate its application in medicine as a safe technique with real efficacy in pain and anxiety. Being relatively inexpensive, it may reduce health costs [7–9].

Hypnosis has been particularly well-studied in the context of pain. In addition to the nonspecific effects of hypnotic suggestions on the individual's expectancies toward pain relief, experimental research provides evidence that hypnotherapy has specific physiological outcomes. Brain activity, in response to experimental pain, has been demonstrated by positron emission tomography to be modulated by specific hypnotic suggestions, including suggestions for decreasing pain [10]. Supported by physiological data, several randomized clinical trials (RCTs) have provided evidence of benefits from short- or long-term hypnoanalgesia interventions for various health conditions and populations [11].

RCTs have been conducted to evaluate hypnotherapy in various medical areas, such as oncology, cardiology, gastroenterology, pediatrics, obstetrics, and gynecology or during surgical or medical procedures. Moderate evidence exists for the relief of symptoms in the treatment of irritable bowel syndrome [12,13]. Hypnotherapy reduces the pain and anxiety of patients undergoing burn-wound debridement, bone marrow aspiration, or childbirth [14–16]. Nausea, chronic pain, and other symptoms related to cancer may also be controlled with hypnosis [9]. The benefits include a decrease in the perception of pain intensity and unpleasantness in association with a reduced need for medication [8,17–19]. As part of an integrative approach, mind–body interventions may have positive effects on psychological functions (perceived stress, anxiety, emotions) and quality of life, and may particularly enhance the patients' capacity for self-knowledge as well as his/her ability to cope with chronic conditions [20,21]. Future research may explore the role of mind–body therapy to promote patient resilience, self-care, and healthy life habits. More data are needed on hypnotherapy benefits in health care and patient–care provider relationships.

27.2 Specific Methodological Issues in Assessing This Treatment

Despite increasing interest in the therapeutic or preventive use of medical hypnotherapy, epidemiological studies in the field are still insufficient—to review and update information for the public and recommendations for physicians. The evidence-based approach to designing hypnotherapy trials is appropriate. At least 30 clinical trials on the topic have been published since 2000, and we recently evaluated their methodology quality [22] with the Checklist for Nonpharmacological Trials [23]. The recommendations of the CONSORT statement for nonpharmacological treatments when designing an RCT help with study design quality [24]. Nevertheless, specifics need to be underlined to improve the overall design and reports of such studies. Major challenges (e.g., randomization and blinding) that investigators undergo when planning a hypnotherapy RCT are linked to the nature of complex integrative interventions.

We will illustrate and discuss such challenges and specificities according to two recent RCTs in the field. One of them is based on our own experience [19], and the other is a model in the field reported by such studies [8].

27.3 Practical Examples

27.3.1 Hypnotic Analgesia Intervention

HYPNOTIC ANALGESIA INTERVENTION DURING FIRST-TRIMESTER PREGNANCY TERMINATION [19]

Context: Elective pregnancy terminations are among the most common outpatient surgical procedures performed on women. The effective management of abortion-related pain and anxiety requires a combination of pharmacological and nonpharmacological approaches.

Objective: To evaluate the utility of a hypnoanalgesia intervention in pain and anxiety management during surgical pregnancy termination in comparison with standard care.

Primary outcome: The participant's need for analgesic and anxiolytic medication (intravenous, conscious sedation) and self-reported assessment of pain and anxiety during the procedure.

27.3.1.1 Study Design

This was a randomized trial. As reported, randomization was successful in minimizing the bias related to baseline differences between groups. An RCT design was appropriate for evaluating the effectiveness of a therapeutic intervention on specific outcomes with limited risk for patient with the hypnotic intervention. But results on the therapeutic effects needed to be balanced with the other multidimensional benefits of hypnotherapy. Randomization may affect important data related to these benefits (*challenge*).

For example, in our study, women's satisfaction with care was assessed as a secondary outcome [33]. Many argue that randomization decreases satisfaction and acceptance [25]. Thus, to evaluate such an outcome, an RCT may not be the best study design. Moreover, especially in hypnotherapy RCTs, it may be true if items evaluating women's views about care are specifically related to hypnotic suggestions. In our study [33], it seems unlikely that women's views directly reflected analgesic suggestions as there was no difference in pain-related items between groups.

27.3.1.2 Intervention Standardization

A standardized versus individualized hypnotic intervention may decrease the potential effect of hypnotherapy (*challenge*). Nevertheless, for reproducibility, the hypnotherapy intervention was standardized with a written protocol [26]. Intervention standardization included the training of personnel and care providers, so a preliminary study was conducted before the main trial [27]. We also carefully discussed all co-interventions before the trial. For example, the standardization of initial fentanyl and midazolam doses to be administered on patient request was discussed with the medical staff. The challenge was to standardize co-interventions in such a way as to maintain or improve usual standard care (*challenge*). For example, in the two groups, the physicians who performed the surgery were allowed to communicate with the patient during the procedure. Consequently, the physician was not blind to the participant group. We thought that blinding the physicians who performed the surgery (e.g., by installing a screen between them and the patients) would badly influence the physician/patient relationship in the standard care group during the procedure. This would disfavor patients in the standard care group with usual care, which would be less supportive than usual. Nevertheless, the risk of bias for the nonblinding of physicians was minimized by the fact that as a primary outcome, medication was requested by the patients, not by the physicians.

27.3.1.3 Choice of the Comparator and Lack of Blinding

As reported, hypnotic intervention was compared in this study to supportive standard care by a nurse accompanying the patient during the procedure. There was no formal attention intervention group that controlled for the extra attention given to the patients. In addition, blinding of the participants and hypnotherapist was not feasible (*challenge*). Thus, we cannot exclude that a bias in self-assessment in the hypnosis group was introduced by patients wanting to please the practitioner (i.e., social demands). Nevertheless, it was reasonable to think that patients who requested less medication because they wanted to please the practitioner would experience a higher level of pain or anxiety. So, to increase the strength of evidence, the primary outcome was a combination of three conditions: reduced anesthesia without increased pain and anxiety. The three conditions had to be achieved to draw a conclusion on the success of the intervention.

27.3.1.4 Blinding of the Assessor

Outcomes have to be evaluated by blind and trained assessors. Nevertheless, in various clinical settings, this is not always feasible. In our study [19], medication need was assessed by chart review (as an objective outcome), and we measured pain and anxiety on self-reported scales during the procedure. To ensure validity of the measure, pain and anxiety in the operating room need to be assessed immediately following the painful procedure. Consequently, the assessor was not blind to the participant group. To minimize bias, the assessor was trained to present the scales with the same empathy in both groups. Furthermore, the results on pain and anxiety measures during the procedure were correlated with a retrospective self-assessment of these outcomes in the recovery room.

27.3.1.5 Recruitment

We anticipated no major problem in conducting the main trial. A pilot study was helpful in determining the recruitment rate and standardizing the way of introducing the study to eligible patients [27]. Study information had to be balanced between the need to convince patients to participate in a randomized trial and to eliminate misconceptions about hypnosis. In this RCT, the participation rate was around 55%, but a significant percentage of women refused to participate because they were skeptical about the technique. On the other hand, we voluntarily did not insist on the potential benefits of hypnosis to avoid creating excessive expectations after the patients had been randomly assigned to the groups (*challenge*). Anyway, we assessed, in both groups at baseline, the participants' expectations of pain and anxiety levels as well as expectations of pain and anxiety relief by hypnosis and/or medication during the procedure. Some authors have recommended the assessment of expectations of symptom relief before randomization as well as their changes along the study as the nonspecific effects of hypnotherapy [28].

27.3.2 Adjunctive Nonpharmacological Analgesia for Invasive Medical Procedures

---

**ADJUNCTIVE NONPHARMACOLOGICAL
ANALGESIA FOR INVASIVE MEDICAL
PROCEDURES: A RANDOMIZED TRIAL [8]**

Context: Patients undergoing percutaneous vascular and renal procedures experience distress with pain and anxiety.

Objective: To compare standard care, structured attention, and self-hypnotic relaxation to diminish perceived pain and anxiety, to reduce the amount of conscious sedation, and to make the procedure safer.

Primary outcome: Pain, anxiety, and drug use during the procedure; hemodynamic stability; procedure time.

27.3.2.1 Intervention

Both studies illustrated the heterogeneity of the hypnotherapy intervention components (*challenge*). They underlined the need for authors to report detailed interventions in protocols. As mentioned in the Methods section of this paper, self-hypnosis in Lang et al.'s [8] study included progressive muscle relaxation, self-imagery, deep breathing but not direct suggestions for decreasing pain and anxiety (in contrast to our study). Heterogeneity affects external validity and is a limitation of study data synthesis and interpretation. Moreover, hypnosis is a mental state that is not completely on demand. It is a mental process that requires the patient's commitment in the process, and discrepancies between administered and received interventions are possible. Hence, it is difficult to assess participants' adherence to treatment based on their sole participation in the session. Finally, hypnotherapy comprises words and, as underlined by Lang et al. [8], standardization of the verbatim is of prime importance to avoid contamination between groups (as an attention group).

27.3.2.2 Care Provider's Expertise

In Lang et al.'s [8] study, four providers were trained to deliver all interventions in the two groups (hypnosis and structured attention). The providers were initially inexperienced, but their training was closely supervised and measured. Interestingly, in Lang et al.'s [8] study, adherence to the intervention protocol was ensured by analyzing the videotaped sessions. More generally, we believe that investigators should avoid the same provider delivering more than one intervention as the provider's beliefs are important for therapy success and may introduce a bias (*challenge*).

27.4 Conclusion

A major limitation of hypnotherapy trials is the difficulty with blinding. Hypnotherapy cannot be administered in a blinded fashion. The lack of ethical and credible control interventions has been underlined by several authors [28,29]. Consequently, the lack of blinding does not allow specific and nonspecific effects of hypnotherapy to be differentiated. The effects of hypnotherapy may be related to (a) a specific action of hypnosis, as indicated by physiological responses to suggestions [10] or (b) a nonspecific action, associated with any treatment, tied to patient expectations or motivation [30]. In addition, the complexity of the hypnotic intervention and interaction between hypnotherapy components add to the challenge of interpreting specific and nonspecific effects.

Furthermore, systematic reviews to evaluate the effectiveness of hypnotherapy are limited by the heterogeneity of such interventions. Some authors have emphasized the need to report details of research interventions [28,31]. A standardized protocol will increase external validity and reproducibility and will limit confusion in the literature to explain inconsistencies in results between research studies [32]. Conversely, assessing a standardized intervention in an RCT needs to be consistent with the use of an individualized approach in clinical practice.

Such holistic interventions are patient-centered, focusing on well-being and symptoms as well as on disease management. Pain, stress, coping, and quality of life are the principal outcomes targeted in research evaluating hypnotherapy. Most of the time, hypnotherapy is considered to be beneficial when it is combined with other conventional treatments to decrease symptom perception or to motivate behavior. The research approach needs to be comprehensive, taking into account the specific and nonspecific actions of hypnotherapy that interact together to produce the whole effect addressing multiple aspects of the disease.

Mind–body interventions, such as hypnotherapy, refer, by definition, to a paradigm of health care that is comprehensive, multifactorial, and patient-centered. Research studies in this area focus on the therapeutic effect of the treatment modality in balance with the burden and risk to patients. Besides, research has to report the broader effectiveness of the whole intervention with effectiveness defined as an objective physiological parameter, well-being, or cost. Adding qualitative measures is relevant, and assessing economic data is important, especially when the interventions request additional resources such as a therapist.

The ultimate goal of research is to guide clinical practice with the highest level of evidence. Mind–body interventions have a low or little risk. The highest level of evidence is not always available. Efficacy must be evaluated by RCTs, but goals of the intervention, accessibility, costs, safety, quality of the intervention, and care providers, and the patient–physician relationship have to be assessed concomitantly to define the level of evidence that could be acceptable for incorporation into practice.

Besides mechanistic studies, the most effective research strategies need to be evaluated to understand the impact of patient-centered medicine like mind–body interventions where the symptoms reported by patients may be as important as a biological marker. High-quality randomized trials in medicine are required to evaluate the role of hypnotherapy in the ability to cope with pain and anxiety in chronic conditions, to motivate healthy behaviors, to reduce stress in patients and practitioners, and to improve quality of care and empathy.

References

1. Barnes, P.M., Bloom, B., Nahin, R.L. Complementary and alternative medicine use among adults and children: United States, 2007. CDC National Center for Health Statistics, Report #12, 2008.
2. Grant, J.A., Rainville, P. Hypnosis and meditation: Similar experiential changes and shared brain mechanisms. *Med Hypotheses*, 65(3), 625–626, 2005.
3. Mott, T., Jr. The role of hypnosis in psychotherapy. *Am J Clin Hypn*, 24(4), 241–248, 1982.
4. Kirsch, I. APA definition and description of hypnosis. Defining hypnosis for the public. *Contemporary Hypnosis*, 11(3), 142–143, 1994.
5. Price, D.D., Barrell, J.J. Mechanisms of analgesia produced by hypnosis and placebo suggestions. *Prog Brain Res*, 122, 255–271, 2000.
6. NIH Technology Assessment Panel on Integration of Behavioral and Relaxation Approaches into the Treatment of Chronic Pain and Insomnia. Integration of behavioral and relaxation approaches into the treatment of chronic pain and insomnia. *JAMA*, 276(4), 313–318, 1996.
7. Lang, E.V., Rosen, M.P. Cost analysis of adjunct hypnosis with sedation during outpatient interventional radiologic procedures. *Radiology*, 222(2), 375–382, 2002.
8. Lang, E.V., Benotsch, E.G., Fick, L.J., et al. Adjunctive non-pharmacological analgesia for invasive medical procedures: A randomised trial. *Lancet*, 355(9214), 1486–1490, 2000.
9. Montgomery, G.H., Bovbjerg, D.H., Schnur, J.B., et al. A randomized clinical trial of a brief hypnosis intervention to control side effects in breast surgery patients. *J Natl Cancer Inst*, 99(17), 1304–1312, 2007.
10. Rainville, P., Duncan, G.H., Price, D.D., Carrier, B., Bushnell, M.C. Pain affect encoded in human anterior cingulate but not somatosensory cortex. *Science*, 277(5328), 968–971, 1997.
11. Ernst, E., Pittler, M.H., Wider, B., Boddy, K. Mind–body therapies: Are the trial data getting stronger? *Altern Ther Health Med*, 13(5), 62–64, 2007.
12. Vlieger, A.M., Menko-Frankenhuis, C., Wolfkamp, S.C., Tromp, E., Benninga, M.A. Hypnotherapy for children with functional abdominal pain or irritable bowel syndrome: A randomized controlled trial. *Gastroenterology*, 133(5), 1430–1436, 2007.

13. Calvert, E.L., Houghton, L.A., Cooper, P., Morris, J., Whorwell, P.J. Long-term improvement in functional dyspepsia using hypnotherapy. *Gastroenterology*, 123(6), 1778–1785, 2002.

14. Frenay, M.C., Faymonville, M.E., Devlieger, S., Albert, A., Vanderkelen, A. Psychological approaches during dressing changes of burned patients: A prospective randomised study comparing hypnosis against stress reducing strategy. *Burns*, 27(8), 793–799, 2001.

15. Liossi, C., White, P., Hatira, P. Randomized clinical trial of local anesthetic versus a combination of local anesthetic with self-hypnosis in the management of pediatric procedure-related pain. *Health Psychol*, 25(3), 307–315, 2006.

16. Mehl-Madrona, L.E. Hypnosis to facilitate uncomplicated birth. *Am J Clin Hypn*, 46(4), 299–312, 2004.

17. Cyna, A.M., McAuliffe, G.L., Andrew, M.I. Hypnosis for pain relief in labour and childbirth: A systematic review. *Br J Anaesth*, 93(4), 505–511, 2004.

18. Butler, L.D., Symons, B.K., Henderson, S.L., Shortliffe, L.D., Spiegel, D. Hypnosis reduces distress and duration of an invasive medical procedure for children. *Pediatrics*, 115(1), e77–e85, 2005.

19. Marc, I., Rainville, J., Masse, B., et al. Hypnotic analgesia intervention during first-trimester pregnancy termination: An open randomized trial. *Am J Obstet Gynecol*, 199(5), e1–e5, 2008.

20. Montgomery, G.H., DuHamel, K.N., Redd, W.H. A meta-analysis of hypnotically induced analgesia: How effective is hypnosis? *Int J Clin Exp Hypn*, 48(2), 138–153, 2000.

21. Patterson, D.R., Jensen, M.P. Hypnosis and clinical pain. *Psychol Bull*, 129(4), 495–521, 2003.

22. Marc, I., Pellend-Marcotte, M.-C., Ernst, E. Do standards for the design and reporting of nonpharmacological trials facilitate hypnotherapy studies? *Int J Clin Exp Hypn* 59(1), 64–81, 2010.

23. Boutron, I., Moher, D., Tugwell, P., et al. A checklist to evaluate a report of a nonpharmacological trial (CLEAR NPT) was developed using consensus. *J Clin Epidemiol*, 58(12), 1233–1240, 2005.

24. Boutron, I., Moher, D., Altman, D.G., Schulz, K.F., Ravaud, P. Extending the CONSORT statement to randomized trials of nonpharmacologic treatment: Explanation and elaboration. *Ann Intern Med*, 148(4), 295–309, 2008.

25. King, M., Nazareth, I., Lampe, F., et al. Conceptual framework and systematic review of the effects of participants' and professionals' preferences in randomised controlled trials. *Health Technol Assess*, 9(35), 1–186, iii–iv, 2005.

26. Marc, I., Rainville, P., Dodin, S. Hypnotic induction and therapeutic suggestions in first-trimester pregnancy termination. *Int J Clin Exp Hypn*, 56(2), 214–228, 2008.

27. Marc, I., Rainville, P., Verreault, R., Vaillancourt, L., Masse, B., Dodin, S. The use of hypnosis to improve pain management during voluntary interruption of pregnancy: An open randomized preliminary study. *Contraception*, 75(1), 52–58, 2007.

28. Jensen, M.P., Patterson, D.R. Control conditions in hypnotic-analgesia clinical trials: Challenges and recommendations. *Int J Clin Exp Hypn*, 53(2), 170–197, 2005.

29. Koyama, T., McHaffie, J.G., Laurienti, P.J., Coghill, R.C. The subjective experience of pain: Where expectations become reality. *Proc Natl Acad Sci USA*, 102(36), 12950–12955, 2005.

30. Price, D.D. Do hypnotic analgesic interventions contain placebo effects? *Pain*, 124(3), 238–239, 2006.
31. Patterson, D.R. Treating pain with hypnosis. *Curr Directions Psychol Sci*, 13(6), 252–255, 2004.
32. Kessler, R., Dane, J.R. Psychological and hypnotic preparation for anesthesia and surgery: An individual differences perspective. *Int J Clin Exp Hypn*, 44(3), 189–207, 1996.
33. Marc, I., Rainville, P., Verreault, R., et al. Women's views regarding hypnosis for the control of surgical pain in the context of a randomized clinical trial. *J Womens Health*, 18(9), 1441–1447, 2009.

Index

Printed and bound by CPI Group (UK) Ltd, Croydon, CR0 4YY

24/10/2024

01778278-0018